Springer Series in Advanced Manufacturing

The **Springer Series in Advanced Manufacturing** includes advanced textbooks, research monographs, edited works and conference proceedings covering all major subjects in the field of advanced manufacturing.

The following is a non-exclusive list of subjects relevant to the series:

1. Manufacturing processes and operations (material processing; assembly; test and inspection; packaging and shipping).
2. Manufacturing product and process design (product design; product data management; product development; manufacturing system planning).
3. Enterprise management (product life cycle management; production planning and control; quality management).

Emphasis will be placed on novel material of topical interest (for example, books on smart manufacturing and sustainable manufacturing) as well as new treatments of more traditional areas.

As advanced manufacturing usually involves extensive use of information and communication technology (ICT), books dealing with advanced ICT tools for advanced manufacturing are also of interest to the Series.

Springer and Professor Pham welcome book ideas from authors. Potential authors who wish to submit a book proposal should contact Anthony Doyle, Executive Editor, Springer, e-mail: anthony.doyle@springer.com.

INDEXED IN SCOPUS

Keng Hsu • Li Yang • Francisco Medina
Jorge Mireles • Mambballykalathil Menon
Michelle Meyer • Steve Fournier
Donald Godfrey

Additive Manufacturing of Metals

Technology, Materials, Design, Production, and Supply Chain

Second Edition

Springer

Keng Hsu
School of Manufacturing Systems
and Network
Arizona State University
Mesa, AZ, USA

Francisco Medina
Dept. of Mechanical Engineering
The University of Texas at El Paso
El Paso, TX, USA

Mambally kalathil Menon
Honeywell Aerospace
Phoenix, AZ, USA

Steve Fournier
Additive Manufacturing
General Atomics Aeronautical Systems inc.
(GA-ASI)
California, CA, USA

Li Yang
University of Louisville
Louisville, KY, USA

Jorge Mireles
The University of Texas at El Paso
El Paso, TX, USA

Michelle Meyer
MatterProviders
Boulder, CO, USA

Donald Godfrey
Consulting
D&B Partners Consulting LLC
Phoenix, AZ, USA

ISSN 1860-5168 ISSN 2196-1735 (electronic)
Springer Series in Advanced Manufacturing
ISBN 978-3-032-04960-5 ISBN 978-3-032-04961-2 (eBook)
https://doi.org/10.1007/978-3-032-04961-2

This Springer imprint is published by the registered company Springer Nature Switzerland AG
The registered company address is: Gewerbestrasse 11, 6330 Cham, Switzerland

Competing Interests The authors have no competing interests to declare that are relevant to the content of this manuscript.

Contents

Chapter 1
Introduction to Additive Manufacturing

1.1 Brief History of AM Development

The idea of producing a 3-dimensional object layer by layer came about long before the development of ideas around additive manufacturing. The first concept patented can perhaps be traced back to Peacock for his patented laminated horseshoes in 1902. Half a century later in 1952, Kojima demonstrated the benefits of layer manufacturing processes. A number of additional patents and demonstrations took place in the time period of 1960s to 1980s that further solidified the idea of producing a 3-dimensional object using a layer-wide approach and in the meantime set the stage for introduction and development of processes based on this principle to produce physical prototypes.

Emerged as the rapid prototyping system in 1987, the SLA-1 (SLA stands for stereolithography apparatus) from 3D Systems (Fig. 1.1) marks the first-ever commercialized system in the world. This process is based on a laser-induced photopolymerization process patented by 3D Systems' founder, Chuck Hall, wherein a UV laser beam is rastered on a vat of photopolymer resin. 3D prototypes are formed by curing the monomer resin layer by layer while in between each layer the build platform submerges deeper into the resin vat.

As 3D Systems' rapid prototyping machine kept evolving, other players in system and materials development in the field gradually surfaced. In 1988, in collaboration with 3D Systems, Ciba-Geigy introduced the first generation of acrylate resins, which marked the genesis of a large part of currently available photopolymer resins in the market. DuPont and Loctite also entered the field in system development and resin business. Meanwhile in Japan, NTT Data CMET and Sony/D-MEC commercialized the Solid Object Ultraviolet Plotter (SOUP) and Solid Creation System (SCS), respectively. These systems were also based on the same photopolymerization principle. In the same time period, the first epoxy-based photo-curable resins were introduced by Asahi Denka Kogyo. Epoxy-based resins remain, to this

K. Hsu et al., *Additive Manufacturing of Metals*, Springer Series in Advanced Manufacturing, https://doi.org/10.1007/978-3-032-04961-2_1

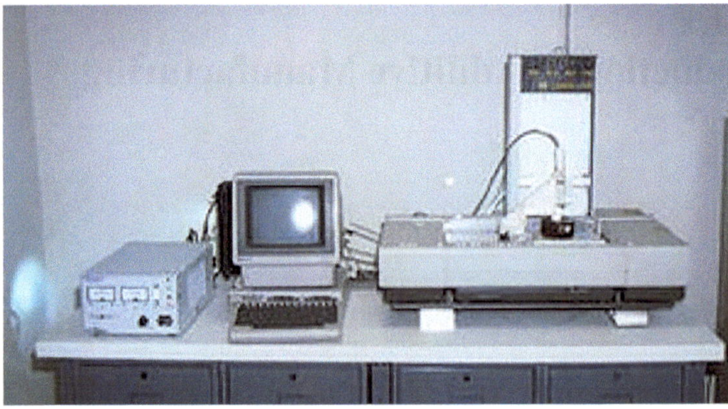

Fig. 1.1 The SLA-1 system by 3D systems launched in 1987

date, another large part of available materials for photopolymerization-based 3D printing methods. In the same time frame, Electro Optical System (EOS) and Quadrax in the European community introduced the first stereolithography-based system, while Imperial Chemical Industries introduced the first photopolymer in the visible wavelength range.

Years 1991 and 1993 marked an important milestone of how the current landscape of additive manufacturing takes shape. Five technologies were commercialized the same year: fused deposition model (FDM) from Stratasys, solid ground curing (SGC) from Cubital, and laminated object manufacturing LOM) from Helisys in 1991; soon after that, selective laser sintering (SLS) from DTM and direct shell production casting (DSPC) from Soligen were introduced. The FDM technology represents a large part of the current landscape of additive manufacturing, while the LOM process now has a small market share. The SGC process, though, did not see large commercial success; its operating principle became the forefather of the projection-based SLA system from a number of OEMs and the continuous liquid interface production (CLIP) technology from Carbon 3D. Both the SLS and the DSPC processes currently occupy a large section of the AM technology market. While the SLS technology has remained largely similar to its initial invention, the DSPC technology has evolved into the systems on which companies such as ExOne and voxeljet base their production machines.

The past two decades marked an accelerated period of AM development. While key existing technologies continued to evolve, new technologies such as the PolyJet material printing, laser-engineered net shaping, aerosol jetting, ultrasonic consolidation (a.k.a. ultrasonic additive manufacturing), selective laser melting of metals, and very recently continuous liquid interface production technology were demonstrated, developed, and commercialized. Also during the same time frame, existing materials were improved, and new materials were demonstrated and commercialized covering polymers, metals, ceramics, composites, foods, and biological materials for a wide range of applications. 3D printed articles are no longer just prototypes.

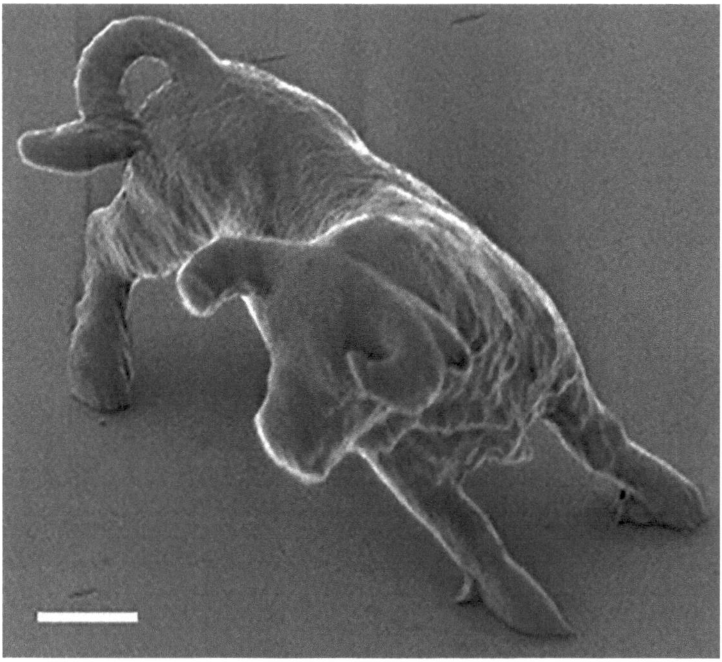

Fig. 1.2 "Micro-bull" produced by the two photon-excitation SLA process by a team of researchers in Japan in 2001. The bull measures 10 μm long. S Kawata et al. 2001 Nature 412,697

They now can be fully functional end-user parts, assemblies, and even complete systems on scales as small as a "micro-bull" in Fig. 1.2 that is smaller than the diameter of a human hair to functional passenger cars (Fig. 1.3), to dwell-able modular homes built by a large gantry system printing concrete, and to the idea of establishing space colonies aided by 3D printing (Fig. 1.4).

A good way of predicting our future direction is to look back and see where we have been. Over the past 30 years, it has been evident that the idea of building a 3-dimensional object layer by layer not only is feasible, but has also proven to be potentially something that can shift the entire manufacturing paradigm. In early 2016, every week, some form of innovation was being introduced, be it a new technology, a new product, a new material, or a new application. It is not difficult to believe that the next decade will see true breakthrough in the "additive" approach of manufacturing goods in that the phrase "design for manufacturing" is no longer needed, but "manufacturing for design" becomes a reality. Total transformation of the manufacturing industry is also in the foreseeable future where the production of a batch of 1,000,000 parts is no longer the sole responsibility of a mass production facility in one or a handful of locations, but hundreds of thousands of "micro factories" in geographically distant locations co-producing in parallel. It is also not absurd to believe that emergency rooms at hospitals can 1 day be "deployable" lifesaving mobile stations wherein biological materials are synthesized and 3D printed onto wounds for real-time recovery of injuries.

Fig. 1.3 Artist's depiction of the idea of "Lunar Colonies" utilizing 3D printing technologies. Photo source: Contour Crafting

Fig. 1.4 3D printed functional passenger vehicle by Local Motors. Photo source: speedhunters.com

1.2 Distinctions and Benefits of Additive Manufacturing

Additive manufacturing processes are fundamentally different from "traditional" manufacturing processes such as cutting, forming, and casting processes. The main difference resides in the fact that in traditional manufacturing processes, shaping of materials takes place across the entire physical domain of the desired part, whereas in additive manufacturing processes, the shaping of material primarily takes place in the formation of the elements (such as voxels, filaments, and layers) which as a whole make up the desired part. The chain of steps in the shaping of elements is implemented in computer-automated environments wherein fabrication of physical 3-dimensional objects from computer-aided design models is accomplished using metallic, polymeric, ceramic, composite, and biological materials.

The distinct process nature of additive manufacturing processes gives rise to a host of advantages over traditional processes. From an application perspective, AM offers high degrees of customization and personalization with little impact on manufacturing complexity and cost as the tooling and associated cost component do not exist for AM processes. In the pilot run and low-volume production environment, material waste, time, and costs associated with materials and inventory are significantly reduced. In addition, geometrically complex, compositionally heterogeneous, and individualized components can be fabricated (for some technologies), while doing so with traditional manufacturing processes can be cost prohibitive. Shown in Fig. 1.5 is an example of an ordered lattice structure not possible by traditional manufacturing processes. The unique characteristics of AM processes foster innovation as it offers short turnaround time for prototyping and drastically lowers the threshold of production of small-volume end-user products. Not bounded by constraints imposed by traditional manufacturing processes, additive manufacturing processes may bring about a paradigm shift in the global manufacturing industry as a whole as it enables concepts such as 3D faxing, cloud-based manufacturing, and on-demand end-user location manufacturing.

1.3 Additive Manufacturing Technologies

1.3.1 Material Extrusion

1.3.1.1 Process Overview

The core process in material extrusion-based AM technologies is the use of an effective 1D strip (commonly called a road) of material to fill in a 2D space to form one layer. Repeating this process layer by layer, one on top of another, allows the formation of a fully defined 3D object. Some of the commercialized technologies based on this process are the fused deposition model (originally patented by Stratasys Inc.) and fused filament fabrication. Shown in Fig. 1.6 are examples of

Fig. 1.5 3D lattice structure of truncated octahedron unit cells made by the powder-bed melting technology

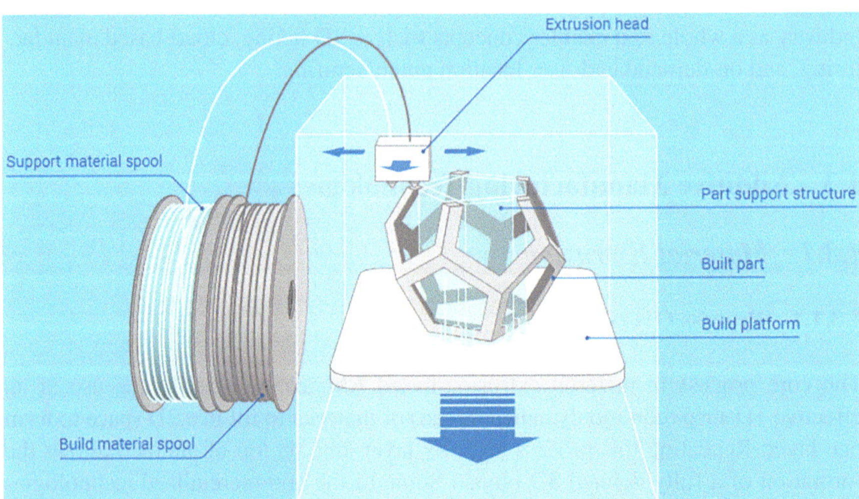

Fig. 1.6 Fused deposition model process

such systems. In these processes, a thermoplastic polymer filament is fed through a heated nozzle in which the polymer is heated to above its glass transition or melting point to allow shaping of filament into a "road." These roads fill a layer which, combined with all other layers, forms the 3D object. One key process characteristic is that the properties of the end product are often anisotropic and highly dependent on the adhesion between each road and all its adjacent roads in the intra- and interlayer directions. The reason is that the properties on the interfaces rely on the thermally activated polymer diffusion process or reptation. This technology is currently the most widespread and has the lowest cost associated with the process and the post-fabrication processes.

As the principle of the extrusion-based process relies on the shaping of a raw material into an extrudate with which an area on a surface is filled to construct one layer of a 3D object, materials with somewhat reversible property changes allow formation and positioning of the polymer roads. In addition to thermoplastics, thermosets, elastomers, polymer matrix composites, highly viscous liquids, slurry, concrete, and biological media have been shown to be 3D printable using this approach.

The extrusion-based technology is currently the most accessible and flexible in terms of cost and scale of implementation. On the one hand, do-it-yourself kits can be sourced easily at very low cost on popular online resellers for hobbyist and educational market users. On the other hand, machines capable of meter-scale engineering structural parts in high-fidelity materials such as ULTEM and PEEK are being sold. Continual advancement of this technology is seen in directions such as material availability, process refinement, and improvement.

1.3.1.2 Process Development

While over the past two decades the FDM technology has seen breakthroughs in many areas of development, there are still technological gaps that need to be bridged to bring the FDM technology to the next level of adoption as a manufacturing tool. An example is the low part strength in the build direction of FDM parts. Though the material and process capabilities of this technology have evolved over the years and are now at a point where end-user products can be directly produced, a main property anisotropy issue is still present in FDM parts with optimized build process parameters: the part strength in the direction normal to the build layers is only 10–65% of that in the directions along the filaments with low predictability. This issue places significant design constraints in the growing number of unique engineering applications of FDM-fabricated parts where dynamic loads or multi-direction static loads are present.

So far, the FDM process has seen research efforts in mainly five areas: part quality improvement, process improvement, new material development, material properties, and applications. Among them, three key areas are relevant to the work presented here: (1) material properties, (2) process improvement, and (3) part quality improvement. In material properties, significant work has been focused in the areas of material testing and use of design of experiments to optimize known process parameters for given part properties. While in process improvement, work has been done to improve support generation process and to establish numerical simulations of the FDM process, research and development work in FDM part quality improvement has seen progress in accuracy, surface finish, build orientation of parts, and, in extension, repeatability. Here, we will provide a review of works most relevant to our proposed effort.

In the context of FDM part mechanical behavior, a significant amount of work has been put in by various groups that focused on the investigation of effects of FDM process parameters including extruder temperature, raster angle, layer thickness, air gaps in between layers and FDM roads, as well as road widths on the tensile and flexural strengths, elastic behaviors, and residual stresses [1–19]. The conclusions from these studies all point to the same direction: each parameter in the FDM process has different effects on different properties of a part, and an optimal set of parameters for one property can result in worsening of other properties. An example is that when a minimal dimensional deviation and surface roughness are desired, lower extruder temperatures or active cooling should be used. However, lowering the extruder temperature or the use of active cooling reduces the overall part strength due to less inter-filament and interlayer bond strength [5].

With the technology evolving for the past few decades mainly under Stratasys, the optimization of process parameters for "best possible" combinations of part dimensional accuracy, surface roughness, and strength is mainly determined in the factory. From an end user's perspective, there is not much that needs to/can be done to improve part qualities. With the original patent expiring in 2009, many low-cost FDM-based 3D printing solutions surfaced and have become an important part of the additive manufacturing revolution taking place in the design community. Irrespective of the level of the FDM machine, for part strength isotropy, the "as-built" tensile strengths of parts in the inter-filament/layer directions fall in the range of 10–65% of that in the direction along the filaments [20]. For the part strength along the directions on a slice/layer, though it also depends on the inter-filament bond strength, it can be remedied by alternating raster angles of adjacent layers (tool path planning) such that in any given direction along the layers, the filament-direction strength can contribute to the overall strength of that layer. Here, a "strength isotropy factor" is used to describe the ratio of the tensile strength of FDM parts in the normal-to-layer direction to the strength in the directions along the filament. By this definition, the strength isotropy values would range from 0 to 1 with 0 being the case where there is no strength in the normal-to-layer direction, and 1 being the case where the strengths are the same in both along-the-layer and across-layer directions. Current FDM parts have a strength isotropy factor ranging from 0.1 to 0.65 with 0.65 being the case with optimized process parameters and a heated build envelope. Almost all work existing in the literature studying part strength properties takes the viewpoint of process parameters and build orientation and their optimization, but only a handful of studies examine the physics of the interlayer bonding process taking place during FDM and its relation to process inputs.

In a handful of studies, the effect of various process parameters on the bond formation between a "hot" polymer filament and a "cold" existing polymer surface has been investigated. The findings of these studies all indicated that the critical factors that determine the extent of the bond strength between a filament and its adjacent ones lie in the temperatures of the nozzle and the build environment, as well as the heat transfer processes in the vicinity of the bond site [21, 22]. Of the two temperatures, the build environment temperature has a more significant effect on the bond strength. While it suggests that one could simply increase the build envelope temperature to increase the inter-filament bond strength, the ramification of doing so is that the part dimensional and structural accuracy and tolerances go way down as the build envelope temperature increases beyond certain points. One team devised a way of heating the entire part surface with hot air during a build process [23]. Though the effect of using hot air was inconclusive due mainly to the approach, the observations very much were in agreement with earlier studies that a critical interface temperature needs to be reached and maintained for a given amount of time for the bond formation between the filament and the existing surface to go through its three stages of formation: wetting, diffusion, and randomization, much like the reptation model introduced [24] and later adopted [25].

In 2016, Hsu et al. demonstrated an in-process laser local pre-deposition heating method, wherein a near-IR laser supplies thermal energy to a focused spot located on the surface of an existing layer in front of the leading side of the extrusion nozzle as it travels [26]. The principle of this process is shown in Fig. 1.7. As the polymer extrudate comes in contact with the laser-heated region of the surface of the existing layer, the wetting, diffusion, and randomization stages needed to form a strong intermolecular-penetrated bond take place to a larger extent as compared to deposition processes without local preheating. In the results reported here, a 50% increase in interlayer bond strength has been observed. Unlike the current build envelope heating method where the highest temperature used is around half of most

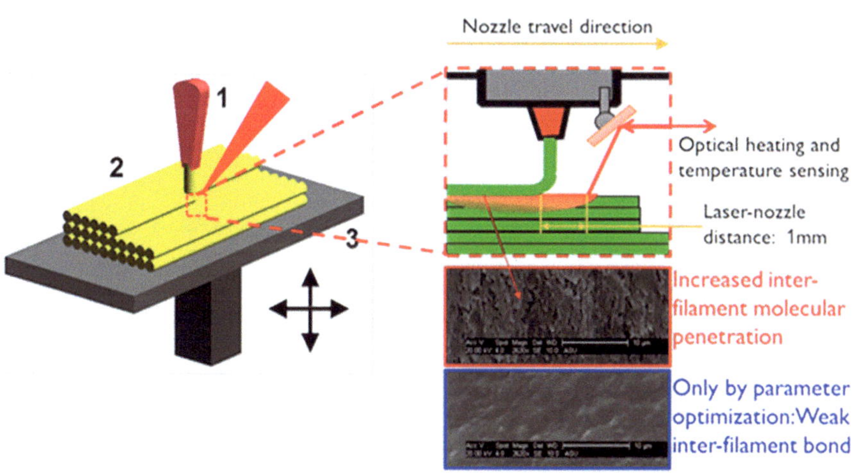

Fig. 1.7 Concept of in-process laser-localized pre-deposition heating. This process is demonstrated to be effective in increasing the interlayer bond strength of FDM parts by more than 50%

polymer's Tg to prevent dimensional and geometrical issues, the laser-based local preheating demonstrated in this report is capable of heating extremely locally at only the actual bond site—to above its Tg without a negative impact on the part dimension and geometry.

Results were obtained for bending loads required to fracture samples built with laser pre-deposition heating (as shown in Fig. 1.8a). These samples were built at speeds of 1–10 mm/s at the laser intensity of 0.75 W. The flexural strengths of these samples were then calculated from the obtained fracture loads and the geometry of these samples. The flexural strengths of samples built at different nozzle speeds (identical to laser scanning speed) at the same laser heating intensity are plotted in

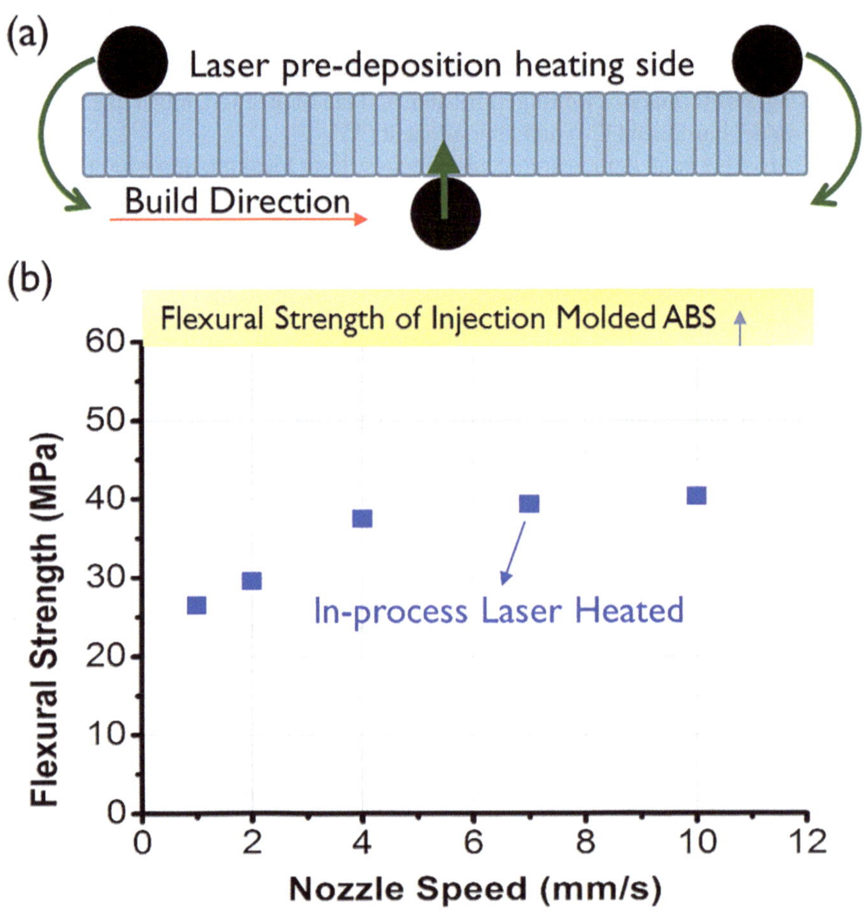

Fig. 1.8 (**a**) Shows a schematic of a laser pre-deposition heating process with arrows indicating the build direction. (**b**) is an X-Y chart titled "Flexural Strength of Injection Molded ABS," displaying data points for flexural strength in MPa against nozzle speed in mm/s. The chart highlights "In-process Laser Heated" data, showing a trend of flexural strength around 30 to 40 MPa as nozzle speed increases from 0 to 10 mm/s

Fig. 1.8b. Our results indicate that, across a range of low laser scanning speeds, the flexural strength of samples built with our laser pre-deposition local heating method increases as the print speed increases; it levels out at above 4 mm/s. This is attributed to the increase in evaporation of material at the high-intensity regions of the laser-illuminated spot as print speed decreases. During FDM, this material evaporation creates a trench into the surface where the incoming extrudate makes contact. If the material flow in the incoming extrudate is not enough to fill the trench, a defect is formed which can later serve as a stress concentrator in the bending test. The defect also reduces the actual cross-sectional surface area to bear the load. In our FDM platform, the laser scanning speed is coupled with the nozzle speed. Therefore, a decrease in nozzle speed results in increased optical energy input into the material surface, causing more material to be evaporated and creating larger defects.

Shown in Fig. 1.9 are the scanning electron micrographs of the fracture surface of the two types of samples (with laser local pre-deposition heated vs. without) used in the bending tests. Fracture surfaces are primarily along the surfaces on which cracks propagate during bending tests and are primarily at the interlayer interfaces.

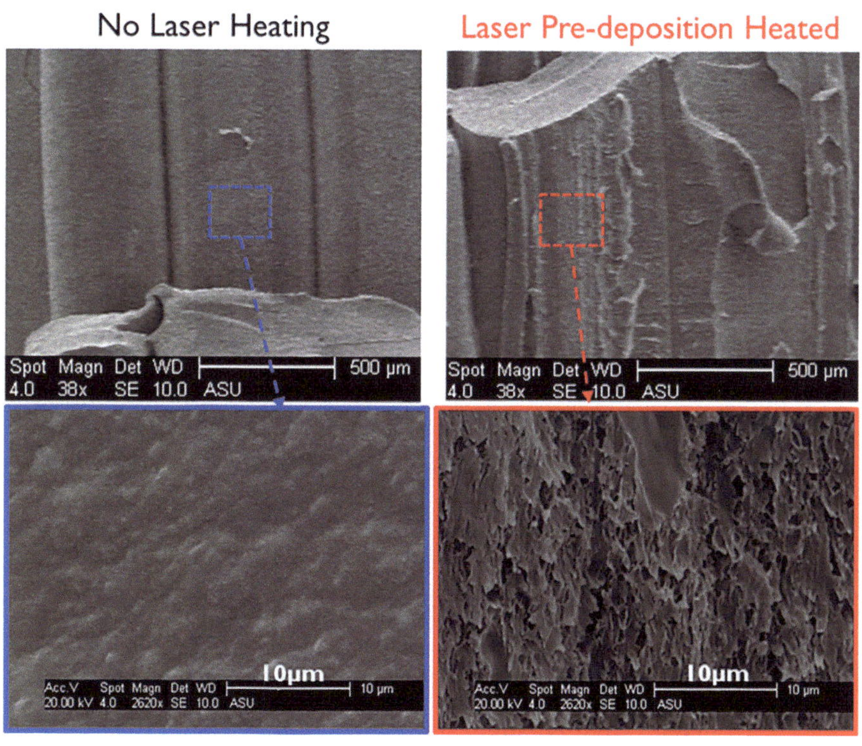

Fig. 1.9 Electron microscopy images comparing material surfaces. The left panel shows a surface with no laser heating, featuring a relatively smooth texture. The right panel displays a surface subjected to laser pre-deposition heating, revealing a rougher texture. Both panels include magnified insets highlighting the surface details, with scale bars indicating 500 micrometers for the main images and 10 micrometers for the insets

The results shown here indicate that with laser pre-deposition heating, the fracture surfaces show rougher morphology than those of control samples. We attribute this to the plastic deformation the material adjacent to the interface goes through before the interlayer interface separates. This indicates that interlayer interfaces in samples built with the proposed in-process addition show a fracture behavior that is similar to that of the ABS material itself. This change in fracture behavior suggests increased amounts of interpenetrated diffusion across these interlayer interfaces. This increase in diffusion across the interface allows the interlayer bond to strengthen as the same interface disappears as a result of interpenetrated diffusion. The fracture surfaces on samples built without the proposed pre-deposition heating method have noticeably smoother morphology. This indicates low crack propagation resistance along these interlayer interfaces and the much smaller degree to which polymer chain interpenetrated diffusion takes place on the interlayer interfaces in the control samples.

Shown in Fig. 1.10a are the load-deflection relations of samples obtained from the three-point bending tests. A number of differences in the behaviors between the laser-heated FDM samples and the controls are observed. First, the control samples exhibit a "brittle fracture" behavior where at the end of the linear relation between bending load and deflection, a sharp drop in load is observed that marks the fracture of the samples. On the other hand, the samples with our in-process local pre-deposition heating fail in a ductile behavior where nonlinear load-deflection relation is observed. These behaviors agree with our observations of the different morphological appearances on the fracture surfaces of laser-heated and control samples. Figure 1.10b shows optical images of the interface between a printed black ABS

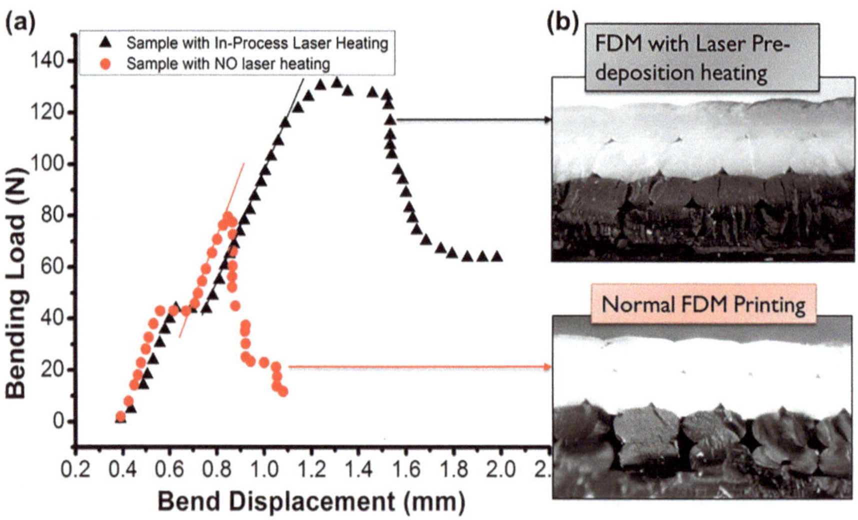

Fig. 1.10 Panel (**a**) is an X-Y chart showing the relationship between bending load (N) and bend displacement (mm) for two samples: one with in-process laser heating (black triangles) and one without laser heating (red circles). The sample with laser heating shows a higher peak bending load. Panel (**b**) includes two images comparing FDM with laser pre-deposition heating and normal FDM printing, highlighting differences in material structure

substrate and a natural ABS layer printed with the laser local heating approach. Discernable differences in the interface geometry are evident in these images between the proposed approach and the intrinsic FDM process. In the laser local preheating approach, the overall more uniform profile suggests reflow of the substrate surface as it is being heated by the laser beam. It is clear that the proposed approach can not only increase the interface temperature and promote increased amount diffusion, but also reduce the defects in between roads of filament and between layers.

Shown in Fig. 1.11 are the flexural strengths of samples at different laser-localized pre-deposition heating powers. Also plotted here are the flexural strengths of samples built with identical raster, layering, and fill parameters but without laser heating, as well as the range of typical flexural strength values of injection-molded ABS. We found that at the chosen set of build parameters, the flexural strength values peaked at 1 W of laser input power. At the peak strength value of 48.2 MPa, a 50% increase in the interlayer bond strength is observed. It also reaches 80% of flexural strength of injection-molded ABS. We attribute this "peak" strength to a local maximum as local heating power increases as a result of two competing mechanisms in the proposed in-process local heating method: (1) the increase in bond strength as the interface temperature increases as a result of increased laser heating

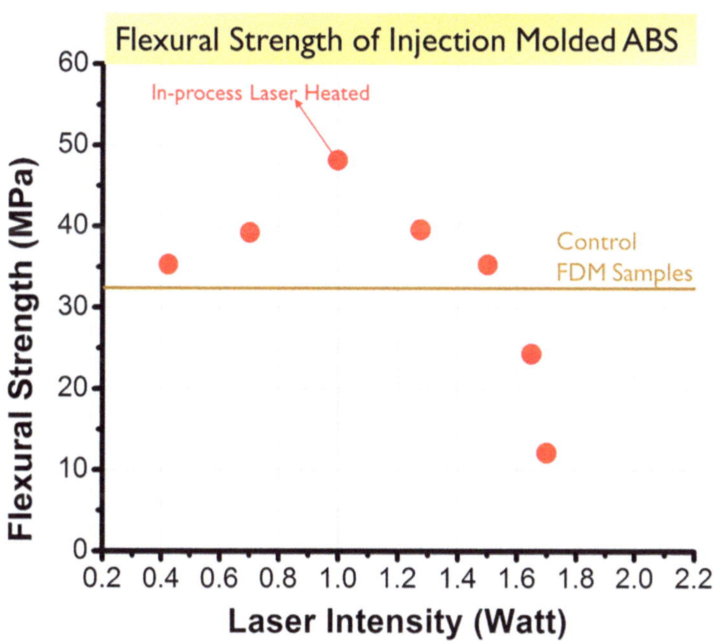

Fig. 1.11 Showing the relationship between laser intensity (Watt) on the x-axis and flexural strength (MPa) on the y-axis. Red data points represent "In-process Laser Heated" samples, indicating an increase in flexural strength with higher laser intensity, peaking around 1.0 Watt. A horizontal line labeled "Control FDM Samples" shows a constant flexural strength of approximately 30 MPa

power and (2) the increase in defects created by the evaporation of material at the high-intensity regions of the Gaussian beam. For the second mechanism, it is similar to what we observed in the bond strength at various nozzle speed studies wherein as the laser power increases, the region of the heated path on the polymer surface where the local temperature reaches the vaporization point of the material also increases. This causes incomplete fill of each road and generation of defects which can later become stress concentrators during the bending test, giving rise to a decrease in flexural strength as laser power increases. On the other hand, as laser power increases, the interlayer interfaces see higher temperature. As the temperature increases, the diffusion across the interface can take place to a larger extent and allow higher interlayer bonds.

The mechanism of the observed increase in interlayer strength as interface temperature goes up can be explained by the polymer interfacial bond formation model proposed [22], where the relation between the interlayer bond strength and the "bond potential" of the interface follows this form:

$$\sigma = \alpha \cdot e^{1/\varphi}$$

where σ is the bond strength, α is a constant, and φ is the bond potential. The bond potential is a quantity that describes the degree to which an interpenetrated bond can be formed on a polymer-polymer interface. It is a function of interface temperature and time:

$$\varphi = \int_0^{\infty} \theta(T) \cdot e^{-k/T} dt$$

$$\theta(T) = \begin{cases} 1 & T \geq T_c \\ 0 & T < T_c \end{cases}$$

The bond potential is zero (or no bonding) when the temperature of the interface is below the critical temperature of the material. In the case of amorphous ABS, the critical temperature is generally agreed to be the glass transition temperature. When the interface temperature rises above the glass transition, the three stages of wetting, diffusion, and randomization [23] can take place and allow an interpenetrated bond to form. The strength of this bond then becomes a function of temperature, as well as the time duration in which the interface stays above the critical temperature.

Based on our experimental results, which agree with the model above, the increase in interlayer bond strength observed in samples built with our proposed laser local pre-deposition heating is, therefore, attributed to the increase in the bond potential on the interlayer interfaces as the laser locally heats up the polymer surface prior to a new filament coming into contact with the heated path on the surface. The preheated surface region allows the interface temperature to stay above the critical temperature of the material after the contact is made and for longer periods of time. As a result of increased time and temperature, the diffusion across the interlayer interfaces increases, allowing the bond strength to increase.

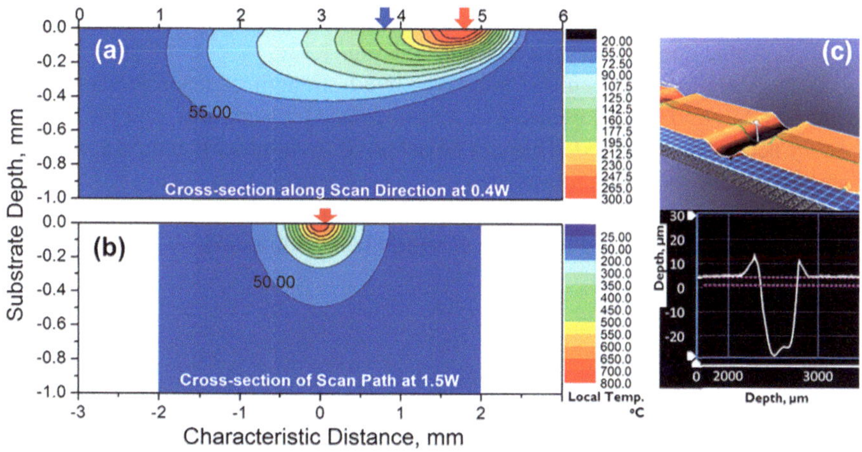

Fig. 1.12 Panel (**a**) is a contour map showing substrate depth in millimeters versus characteristic distance in millimeters, with color gradients representing local temperature in degrees Celsius. It depicts a cross-section along the scan direction at 0.4W, with temperature values ranging from 20 to 300°C. Panel (**b**) is another contour map with similar axes, showing a cross-section of the scan path at 1.5W, with temperature values from 25 to 800°C. Panel (**c**) includes a 3D visualization of a surface with a graph below it, showing depth in micrometers over a distance, highlighting a profile with peaks and valleys

The spatial temperature distribution in the black ABS substrate as a scanning laser beam traverses across its surface is obtained from a transient heat transfer and thermal model established in the commercial finite element modeling package COMSOL. Shown in Fig. 1.12 are temperature profiles in two orthogonal planes intersecting the laser spot at two different laser powers. As can be seen in Fig. 1.12a, at a 10 mm/s scan speed and 0.4 W laser power, at the center of the illuminated spot, the surface of the substrate can be heated to above 250 °C. At 1 mm away from the illuminated spot (where the extruding ABS from the nozzle makes contact with the substrate), the surface temperature, though decreasing, remains above 150 °C. Also seen is that the heat remains at a shallow depth in that at 0.5 mm below the surface, the temperature drops to below 55 °C. The predictions here suggest that the localized laser pre-deposition heating is an effective way to locally raise the substrate–filament interface to promote interdiffusion strengthening without introducing overall heating in the entire work part, causing mechanical integrity issues during the build. In addition, the reduced temperature gradient across the interfaces during printing can reduce the thermal stresses induced due to the shrinkages of different regions of polymer across layers and roads during printing and cooling.

Figure 1.12b depicts the temperature distribution in the cross section of scan at a laser power of 1.5 W and a scan speed of 10 mm/s. At this rate of energy input, an affected zone of several hundred microns in width would rise above the thermal decomposition temperature of ABS, resulting in the ablation of material. In a set of control experiments performed to verify this, we found that the trench created by laser ablation can result in defects as deep as 30 µm and as wide as 500 µm at a scan rate of 5 mm/s and a laser power of 1 W. Optical profilometry of an example laser ablation profile due to slow speed is shown in Fig. 1.12c.

The correlation between the interlayer bond strength of laser pre-deposition heated FDM builds, measurements of actual surface temperatures, and the thermal history need to be obtained. This can be achieved by in-process temperature measurements at the local area illuminated by the laser process beam using through-the-beam optics and IR sensors in the appropriate wavelength range. These are some of the aspects of the proposed approach currently being addressed.

1.3.2 Vat Polymerization

The two key elements of vat polymerization-based processes are photopolymerization resin and a resin exposure system that allow spatial control over polymerization in a vat of monomer resin. There are two primary configurations in this technology: an upright style where the build plate (onto which the desired part is built) is submerged into a vat of resin as the build progresses, and an inverse configuration where resin is contained in a tray and the build plate starts at the bottom of the tray and pulls upward away from the tray as the build continues. The finished part in the upright configuration is completely submerged in the resin vat, while in the inverse configuration, the finished part is completely removed from the resin. Depicted in Fig. 1.13 are examples of those two configurations. In either configuration, the resin exposure system can be either a rastering laser beam or an image projection-based system.

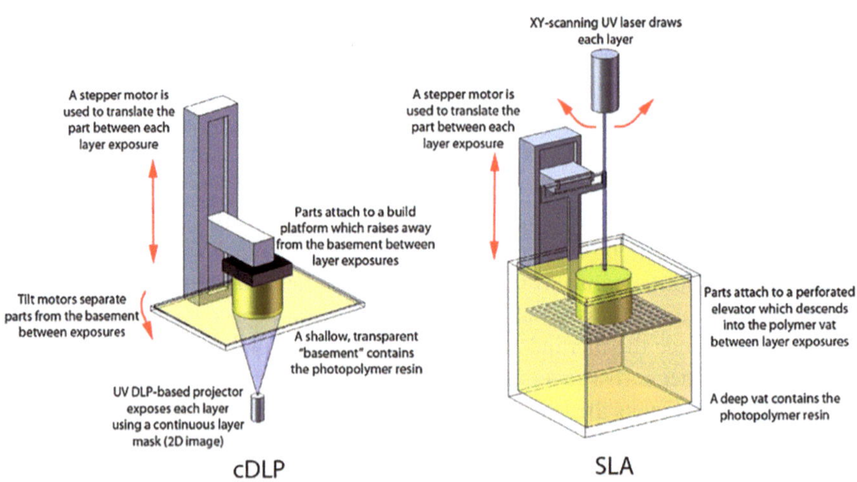

Fig. 1.13 Two configurations of the stereolithography technology. Photo source: J. Wallace et al., Biofabrication, 6, 2014, 015003

In the upright configuration, the vat contains all the available resin in the system and is replenished after each build. The build typically starts with the build plate positioned a few hundred microns just below the resin free surface. The exposure system then polymerizes the layer of resin monomer in between the resin surface and the build plate. Once the entire layer is polymerized, the build plate lowers to allow another layer of fresh resin monomer to form between the resin free surface and the previous layer. The process then repeats until the entire 3D object is completed. Since the viscosity of most resin monomers is typically high, forming a layer of resin monomer of uniform thickness between the build plate in position in the resin and the flow of resin from the edge of an existing layer to fill the layer and settle back down to form a continuous resin monomer layer with just gravity typically takes a long time. In commercial systems, this issue is typically addressed by one of the two approaches: (1) lowering the build plate by an amount larger than the thickness of one layer to allow reduction of time needed for resin monomer to flow and fill in one layer, and then running a leveling blade across the resin free surface to physically create a "flat" resin surface; (2) a resin dispenser is integrated into the leveling blade to allow simultaneous replenish and leveling in between layers.

In the inverse configuration, each layer during the build is defined by the space between the build plate and a solid surface in the vat. Typically, the light path is projected upward into the vat as opposed to the light "shining down" on the resin surface in the upright configuration. Once a layer is completed, separation of the polymerized layer from the solid surface through which the light passes is needed to allow the resin monomer in the vat to flow into the new space formed and allow the exposure and polymerization of the following layer to continue. One distinct advantage of the inverse configuration is that the thickness and geometry of each layer of resin monomer before polymerization are defined by two stiff solid surfaces using a mechanical compression motion. As a result, the time associated with forming and shaping each monomer layer is reduced as compared with the upright configuration.

Commercial systems of both upright and inverse configurations are available, although the inverse configuration is more suitable in an end-user environment. Currently, a major research and development direction in the photopolymerization process-based additive manufacturing approach focuses on material development. In addition to elastic and elastomer polymers for structural applications, higher-temperature ceramic composites and magnetically, electronically, and thermally functional materials are also of great interest.

1.3.3 Material Jetting

Sharing configuration and system-level infrastructure with paper in-jet printing technology, the material jetting additive manufacturing process forms each layer of a 3D article by using a single or large arrays of nozzles to deposit droplets of materials on a surface followed by means to cure the deposited material. Shown in Fig. 1.14

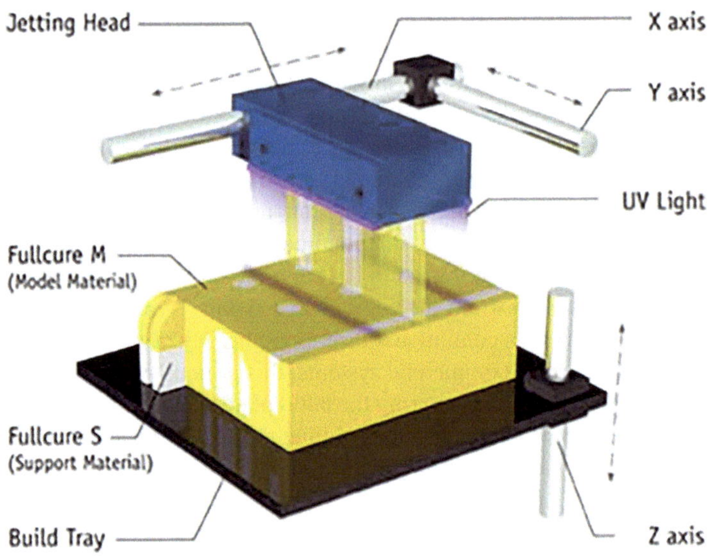

Fig. 1.14 Principle of multi-mateiral jetting process. Image source: www.vt.me.edu

is a demonstration of the concept. Initially commercialized by Objet Geometries, the material jetting technology is the fundamental principle of one of the two main technologies commercially offered by Stratasys. On a commercial system, the same material jetting principle is used for both the model and the support materials required for constructing a 3D component. This technology is widely adopted into prototyping environments as systems based on this technology are often designed for office environments and the build speed, flexibility, and material availability are ranked high on currently available technologies.

Currently, the material jetting technology is the only one that offers voxel-level material property and color tuning. In Stratasys' "connex" series systems, properties of a given voxel can be "digitally" altered and specified by what is essentially mixing of two component model materials of different colors or mechanical properties. A typical example is the ability to print hard plastic type of material, rubbery elastomer, and a range of materials of different elastic/elastomer characteristics by mixing the two.

As compared with the fused deposition modeling and the photopolymerization technologies, the material jetting process can offer higher scalability in productivity, part dimension, and material flexibility owing to the fact that it uses large arrays of nozzles and each can act as an individual process channel (as the laser illumination spot in SLA and the extrusion nozzle in FDM). A typical commercial system can have a number of material jetting nozzles ranging from fifty to several hundreds. Materials available also span a wide range: photopolymers simulating properties of engineering plastics and elastomers, waxes, and conductive inks (Fig. 1.15).

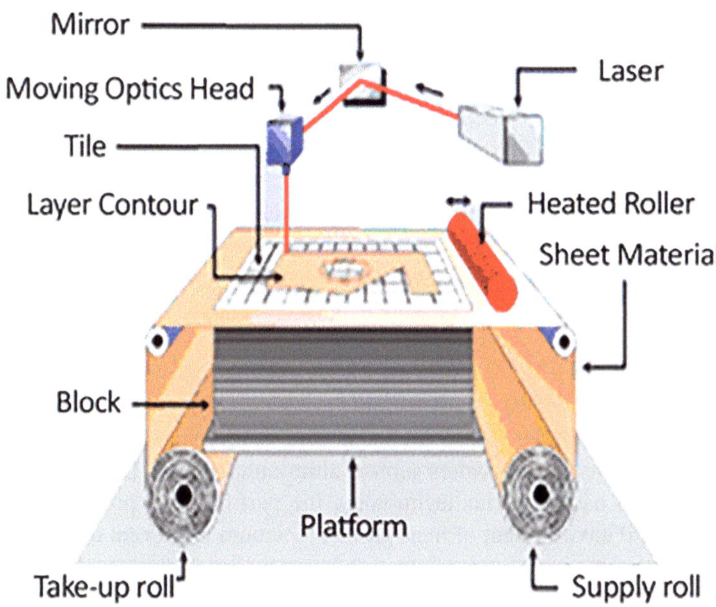

Fig. 1.15 Laminated object manufacturing technology. Image source: topmaxtech.net

1.3.4 Metal Additive Manufacturing Overview

Currently, polymer additive manufacturing, or 3D printing, is accessible and affordable. Systems from $200 do-it-yourself kits to half-a-million-dollar large-scale production printers can be readily acquired. This trend is expected to further develop in the same direction in the near future [26]. Solid metal 3D printing, however, does not see the same development trajectory because of the innate safety concerns and technology and operation costs of existing metal technologies capable of producing solid metal components with densities greater than 95% [27, 28].

Current metal additive manufacturing processes include indirect methods such as binder jet processes and selective laser sintering and direct methods such as selective laser melting, electron beam melting, and laser-engineered net shaping [29]. Indirect methods require post-processing such as hot isostatic pressing to produce parts of density greater than 90%, while direct methods can typically produce parts with more than 90% density with optimized process parameters. In indirect methods, metal powders are partially solid-state sintered together or a low-melting-point binder is used to bind metal particles together to produce a preform. Post-processing operations such as binder removal, sintering, or liquid metal infiltration are used to obtain greater than 90% build density [27]. The process of ultrasonic consolidation was introduced by researchers as a hybrid additive-subtractive process where sheets (or strips) of metal foils are first ultrasonically welded into a stack using a roller sonotrode. A cutting operation (often end milling) is then used to shape the metal stack into the desired layer shape [30]. By alternating between these welding and cutting processes, 3-dimensional objects are constructed. In UC, the 2D shape of layers is obtained by combining a tape or sheet welding process and the subsequent trimming of welded layer to the desired shape. This process is capable of producing pure metal, alloy, and composite material parts with the use of high-power ultrasonics and high mechanical loads [31, 32].

The powder-bed process is now entering a stage of technology maturity and is currently the most common metal 3D printing system for production of engineering components. Current systems of this process use thermal energy to melt and fuse material through manipulation of a melt pool created by laser or electron beam coupled into metal powder as heat. The resulting structures, morphology, and microstructures of printed materials depend highly on the thermal-physical and heat-transfer processes during the micro-welding event [27, 29]. These processes use fine powders as the starting material which can pose health and safety concerns, especially when reactive metal powders such as aluminum are used [28]. Because of the heat melt fusion nature of this technology, the part-building process takes place under controlled environment of inert gases or vacuum to prevent excessive oxidation, beam scattering in the case of electron beam melting, and process hazards [28, 33]. Though high-quality metal parts can be produced, a typical powder-bed metal system starts at $200,000 with an approximately 800 cm^3 build volume, without taking into account the ancillary equipment and facility required to safely handle and process metal powders. For a system with a build volume practical for production of engineering structural components, half a million to a few millions worth of capital investment alone can be expected.

Though in early stages of development, there have been new processes introduced that can allow 3D printing of metals with greater than 95% density without thermal melt fusion. Hu et al. demonstrated a meniscus-confined 3D electrodeposition approach that is capable of producing microscale 3D copper features of more than 99% density [34]. This process showed potential in high-resolution direct printing of fine features with the potential issues of slow deposition rate and limited number of materials available. A number of researchers demonstrated the ability to deposit metals such as copper and silver using a laser-induced chemical reduction process and the ability to have high spatial selectivity and increasing deposition rates [33, 35]. These chemical reduction and deposition-based approaches, however, have common issues in spatial deposition selectivity as the deposited structure increases in height. The process of gas metal arc weld 3D printing is a recently developed inexpensive way of building metal components using an automated articulating welding toolhead [36–38]. This essentially metal welding process allows for a direct writing method where materials are used only when needed (as compared with the powder-bed approach where powder is dispersed across the entire build space at each layer). It is, however, based on melt fusion of metals, and as a result, the thermal physical and thermal mechanical properties are similar to weld joints and to parts made using powder-bed melt fusion processes.

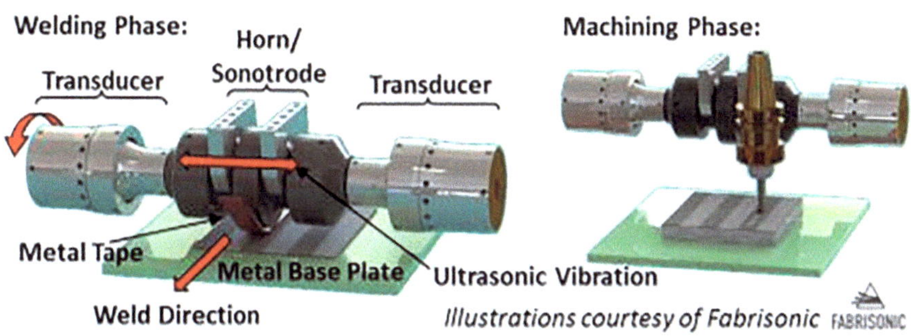

Fig. 1.16 Ultrasonic additive manufacturing technology. Image source: Fabrisonic

1.3.5 Sheet Lamination

Also recognized as laminated objet manufacturing (LOM), the sheet lamination process represents one of the two current commercialized technologies that combines additive and subtractive steps to complete a 3-dimensional article. The process takes place by alternating between steps of bonding sheets of materials (polymer, metals, composites, paper, etc.) in a stack through usually heat- or pressure-activated adhesives and the subsequent cutting and trimming of the material stack to the desired contour of a given layer or stacks of layers. The concept of this process is depicted in Fig. 1.16.

In commercial systems, the additive steps of the process are typically accomplished by applying heat, pressure, or both to chemically or mechanically bond sheets together via adhesive that exists on the sheets using a roller. Following the formation of a stack, mechanical cutting and laser cutting are both possible tools to shape each layer or stack of layers. In the case of metal sheet lamination, a process commercialized by Fabrisonic, instead of using adhesives, sheets of metal are ultrasonically welded together to form stacks before the mechanical milling process is brought in to shape the layers. The bonds formed between layers, or stacks of layers, are metallurgical through the combination of heat as well as ultrasound-induced diffusion processes across interfaces.

1.3.6 Powder-Bed Fusion

There are several variations of the powder-bed fusion technology. Although the process capability, conditions, and part characteristics can vary, all variations share the same working principle. As shown in Fig. 1.17, model material in powder form (mean diameter ranges from a few tens of microns to a few hundred microns) is fed into and spread on a build plate driven by a Z-positioning stage by a blade or a wiper mechanism. The space between the surface of the build plate or a finished layer and

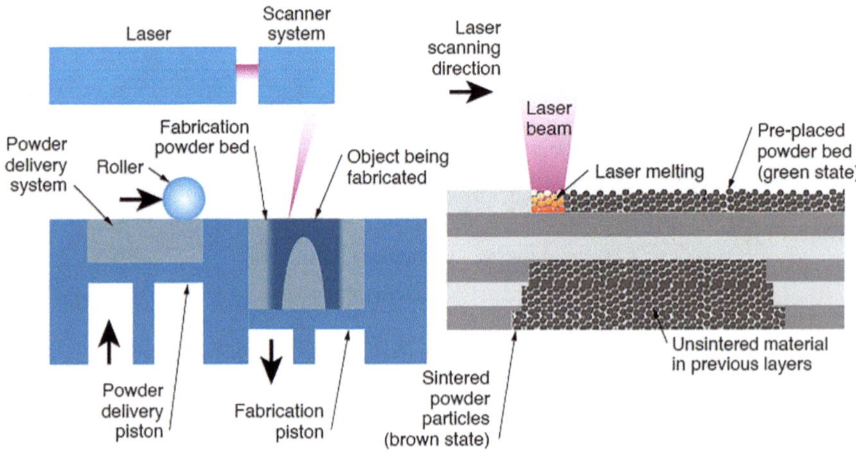

Fig. 1.17 Selective laser melting/sintering technology. Image source: llnl.gov

the bottom edge of this spreading mechanism as it moves across the build area defines the layer thickness and height. The height of a powder layer is typically between a few tens of microns to just below 100 μm in metal systems and between 50 μm and 150 μm in polymer systems. Layer height is an important factor in the powder-bed fusion process and is carefully selected and calibrated against other build parameters and factors in the system such as energy beam geometry, power, as well as powder particle size and size distribution. Once a layer of powder is formed, an energy beam (laser or electron beam) is focused onto the powder bed and rastered across the powder surface in a pattern to fill the area defined by one slice of the desired 3D model. The raster pattern is also a critical factor and has strong effects on the quality, microstructure (in the case of metals), and defect structures of material in the completed part. Once a layer is completed, the powder-bed process shares similar overall process flow with most other technologies where the build platform drops down and the steps for a single layer repeat to complete the following layer. Note that in the powder-bed process, the laser exposure is typically adjusted such that a certain depth into the previous layer is also melted to allow full fusion of each layer into the previous. As a result, the properties in the finished part are less directional as compared with those in, for instance, the FDM technology.

The powder-bed process is capable of processing a wide range of materials including plastics, elastomers, metals, ceramics, and composites. Depending on the materials used, desired property, structure, and limitations of systems used, several variations of the process can be implemented. In thermoplastic polymers, the process follows very closely with what was described earlier, and nearly fully dense materials can be formed. When metals are used, two variations are possible: selective laser sintering (SLS) and selective laser melting (SLM). In SLS, the powder particles do not fully melt but are heated by the energy beam to a high enough temperature to allow solid-state diffusion and bonding of powder particles into a porous layer. Alternatively, a thermoplastic binder solid is mixed into the model material powder to create a mixture and allow the binder material to be heated by the energy beam during rastering to bind the metal powder particles together. This method is typically used in ceramics when remelting of material is not possible.

1.3.7 Binder Jetting

The working principles of the binder jet technology resemble the combination of the powder-bed process and the material jetting process. The powder-form model or base material is laid down on a Z-positioning platform to form a layer of bed of powder with uniform thickness across the layer. A print head with single or multiple nozzles similar to those used in the material jetting processes then runs past the powder bed at a given height above the bed and deposits droplets of binder materials to essentially "glue" the powder particles within areas defined by the boundary of a single slice of a 3D model. This layer formation process is repeated as the Z-stage steps down to complete the entire 3-dimensional part. Fig. 1.18 shows the working principles of this process. Once completed, the finished part is surrounded by loose powder. To harness the printed part, the Z-platform is raised and the loose powder is removed to reveal the completed print. In the multi-nozzle/material environment, the binder can also be combined with colored ink to provide color printing capabilities.

For polymer applications, oftentimes, the part built by this process is immediately usable without any post-processes. However, in metal or ceramic applications, removal of binder material is required and reheating of the printed part at sintering temperatures is carried out to allow solid-state diffusion to take place among the base material particles to achieve higher strength. At this stage of the post-process,

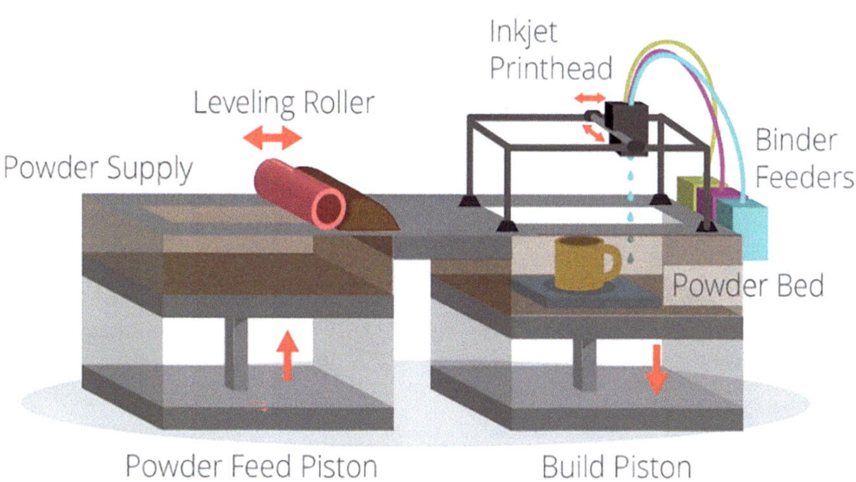

Fig. 1.18 Binder jet binding technology

the printed parts are porous. If high bulk density in materials is required, hot isostatic pressing, HIP, or infiltration of a second material is possible. This process has been applied to the manufacturing of injection molding dies where steel base powder is used to print the molds, and it is followed by a bronze infiltration process to allow full density to be achieved.

1.3.8 Directed Energy Deposition

The directed energy deposition technologies in general refer to processes where the raw material is directed into a spatial location where the energy input and the desired deposition site are co-located. Two processes currently can be categorized under this type: the laser engineered net shaping (LENS), the electron beam free-form fabrication (EBF3), and the wire and arc additive manufacturing (WAAM). These processes are similar in working principles but differ in the energy source used and the form of the raw materials used. Figure 1.19 shows the working principles of these types of processes. In LENS, a laser beam and powder raw material are typically used in an articulating tool head where the powder is injected into a spot on a surface where the laser beam focuses its energy onto. The melt pool formed at this spot allows the material to be metallurgically added into the existing surface and fused into the part being built. By manipulating the spatial locations of the melt pool, a complete 3D article can be built spot by spot, line by line, and then layer by layer. Following the similar working principle, but different raw material form and energy input, the WAAM process feeds a metal wire into the melt pool that is produced by the arc struck between the feed wire and the substrate/existing surface. It is essentially an automated metal inert gas (MIG) welding process in which the weld tool is controlled to follow paths that fill up a 3D part with weld lines. The

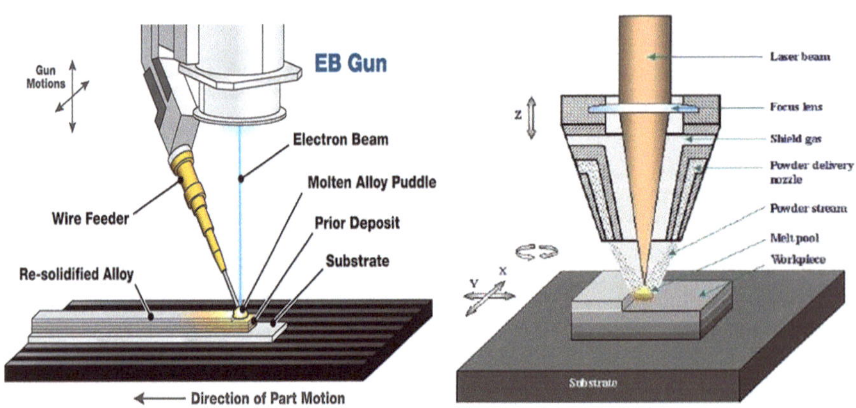

Fig. 1.19 Directed energy deposition processes. Right: Laser-engineered net shaping. Left: Electron beam free-form fabrication. Source: intechopen.com and sciaky.com

EBF3 process was first developed by the NASA and is intrinsically a space-compatible technology. It uses working principles that combine LENS and WAAM in that it uses a wire feed system to introduce materials into a melt pool generated by an energy source. In EBF3, the energy source used is an electron beam and the build environment is under high vacuum to ensure focusing on the operation of the electron beam. The processes in the directed energy deposition category all fall into the category of direct-write technology where the material of a 3D part is introduced locally into the part by continuously directing both energy input and material into the same site.

The directed energy deposition has an intrinsic limitation in surface finish and dimensional tolerances on the built part. Though it can be implemented as a stand-alone 3D printing process, it is typically configured as hybrid additive-subtractive approaches. In this configuration, it is similar to the sheet lamination processes in that the overall process alternates between additive steps and subtractive steps, where a cutting process is brought in to bring tolerances and surface finish to the required range of values. The different processes in this technology are also oftentimes implemented on a 5- or 6-axis articulating system with tool exchange capabilities. It is, therefore, a flexible technology that is well suited for the repair of large mechanical or structural components.

1.3.9 Wire Arc Additive Manufacturing

The working principles of wire arc additive manufacturing (WAAM) are fundamentally the same as metal welding, in the simplest form the standard gas metal arc welding (GMAW), or more commonly referred to as the metal inert gas (MIG) method. The main distinction between the two is that in WAAM, the position of the welding head is precisely controlled by a multi-axis motion system whose motion is synchronized to the amount of the material deposited, while it travels along a prescribed motion path. The process mechanics bear strong similarities to those of the directed energy deposition method described in the previous section, except that in WAAM, the main energy input is through the arc generated between the metal filler rod (electrode as well) and the existing surface—be it the build plate or an existing layer upon which the next layer is to be deposited on—as the electrical current flows through (Figs. 1.20, 1.21, and 1.22).

This form of metal additive manufacturing technique became more accepted recently and found applications in lower-fidelity load-bearing components for prototyping or one-off applications. While this technique is limited in feature resolution capabilities, it usually is implemented in higher-throughput environments in which the material deposition rate can be much higher in comparison with powder-based technologies. Similar to the part characteristics of those of the DED technique, WAAM parts typically require post-processes such as heat treatment to relieve thermal stresses and machining to obtain acceptable geometrical and dimensional tolerances and surface finish. This form of additive manufacturing

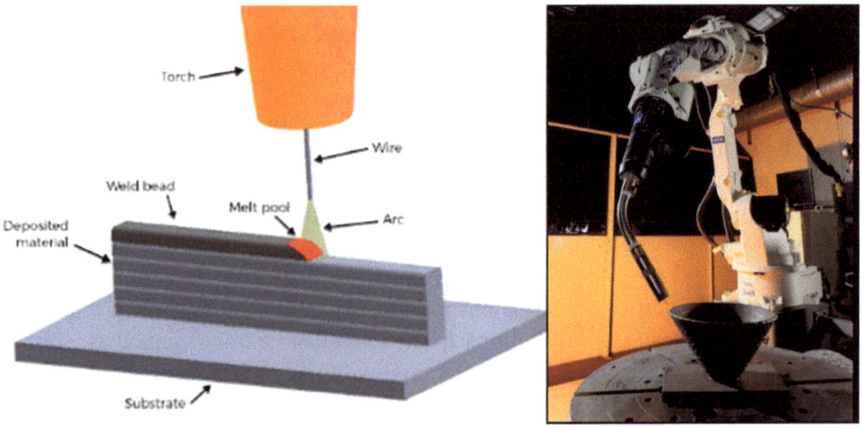

Fig. 1.20 Working principles of WAAM. Source: lortek.com; TWI-global.com

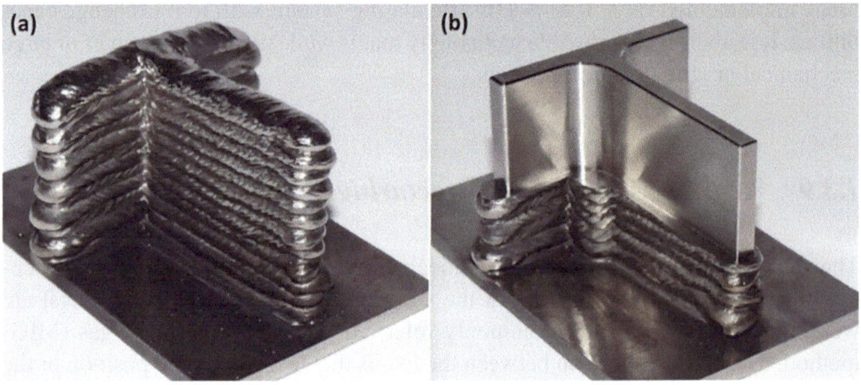

Fig. 1.21 Example WAAM printed part. (**a**) As-deposited material stack. (**b**) After machining, the part with required geometrical, dimensional, and surface finished requirements is obtained. Source: https://www.fronius.com/

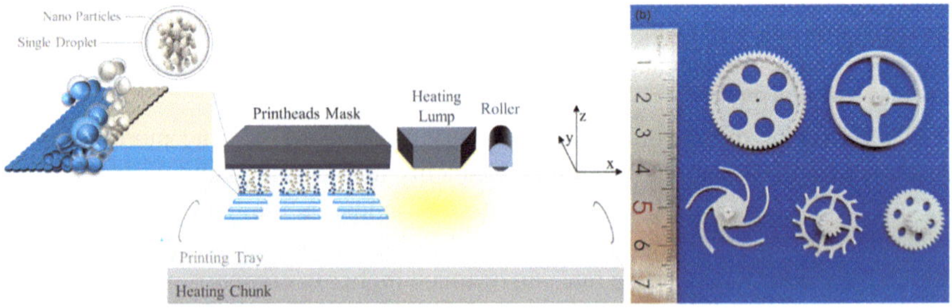

Fig. 1.22 Process principle of nanoparticle jetting additive manufacturing and example parts printed using this technology

technology can also be integrated into multi-axis robotic arm systems as its motion system to accomplish more complex motion required by complex structures or reduction/elimination of support structures during the build process.

1.3.10 Nanoparticle Jetting

The nanoparticle jetting technique works by jetting droplets of fluid droplets carrying high concentrations of nanoscale particles to form "printed" images layer by layer, followed by low-temperature sintering and consolidation. Materials that can be printed using this technology are typically metals and ceramics. In the process, as the droplets land on a surface, the heated print environment allows the carrier fluid to rapidly evaporate, leaving a condensed layer of nanoparticles that are partially sintered. This process is repeated in the print direction to complete a green part. The process is followed by a high-temperature sintering step with temperature settings, ramp rates, holding times, and cooling rate specifically tuned for the material. These post-processes are similar to those for the binder jetting technology. The main difference in this technology is the use of nanoscale particles and the resulting higher apparent density in the as-printed green parts. The apparent density in the green parts in the sintering process has a critical effect on the process itself and on the density of the final parts. Typically, the higher apparent density in green parts produced using nanoparticles results in lower sintering temperature, shorter cycle, and overall higher final part density with less shrinkage or warpage.

1.3.11 Continuous Liquid Interface Production

The continuous liquid interface production (CLIP) method of 3D printing was first demonstrated in 2014 by Tumbleston et al. where a gas-permeable UV-transparent window was used to define layers during the polymerization process. The working principle of this technology is similar to the projection-based vat polymerization processes in the inverted configuration introduced in Sect. 1.3.2. The main difference, which also sets this technology apart from the rest of the photopolymerization-based technologies in terms of build speed, is that the kinetics of photopolymerization of resin is coupled with the oxygen-assisted polymerization inhibition, as well as the continuous motion of the built part. The result is that the separation, recoating, and repositioning steps in the conventional SLA-type approach are completely avoided. The process rate, therefore, can be as high as 100 times higher as compared with other types of photopolymerization technologies. As indicated in Fig. 1.23, the oxygen-permeable window allows an oxygen concentration gradient to be established on the resin side of the window. This concentration gradient allows the rate of photopolymerization of resin monomers to follow an inverse gradient away from the window. At the build platform speed equal to the polymerization rate (measured in

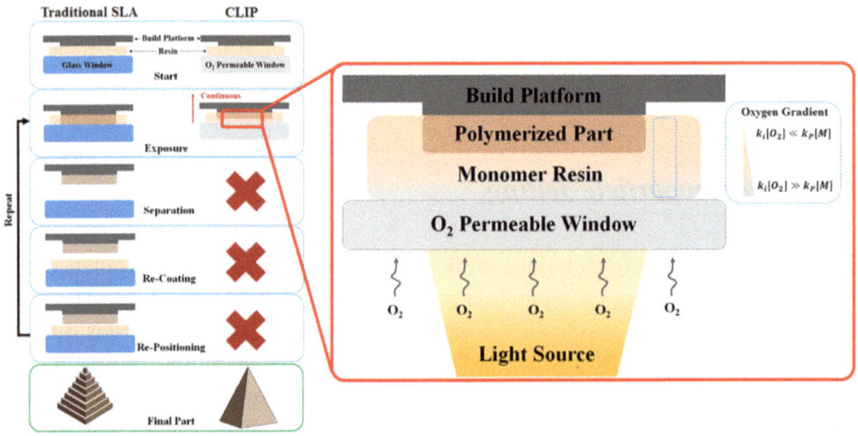

Fig. 1.23 On the left, the traditional SLA process is shown with steps: exposure, separation, re-coating, and re-positioning, each marked with a red cross indicating inefficiency. On the right, the CLIP process is highlighted, showing a continuous build platform with polymerized part, monomer resin, and an oxygen-permeable window. An oxygen gradient is noted, with light source and oxygen molecules depicted below. The CLIP process emphasizes continuous production without the need for repeated steps

increase in the thickness of the formed layer) of resin, the polymerized part can continue to "grow" as the build platform is continuously pulled away from the window.

At 10 μm or less layer height, this technology produces parts with surface virtually without any staircasing (Fig. 1.24). In the work published, a print speed of 500 mm/h was demonstrated. Being a continuous process, this process rate is limited by the resin cure rates and viscosity, not by layer-wise layer formation. With process tuning and at lower resolution, build speeds of greater than 1000 mm/h can be obtained.

1.4 Developmental Additive Manufacturing Technologies

1.4.1 Additive Manufacturing Foundational Notes

Material shaping and joining are the two fundamental processes for many forms of metal manufacturing operations: formative processes such as forging and sheet metal bending, additive processes such as directed energy deposition, and welding operations like metal inert gas (MIG) welding. Fundamentally, metal shaping, or permanent/plastic deformation, begins when the crystalline defects like dislocations in metals move in their slip systems. Joining concerns the exchange of atoms across boundaries or interfaces between adjacent domains of metal. Mass transfer processes such as diffusion or advective mixing enable the healing of interfaces and therefore allow for joining of separate domains of metals. Shaping and joining both

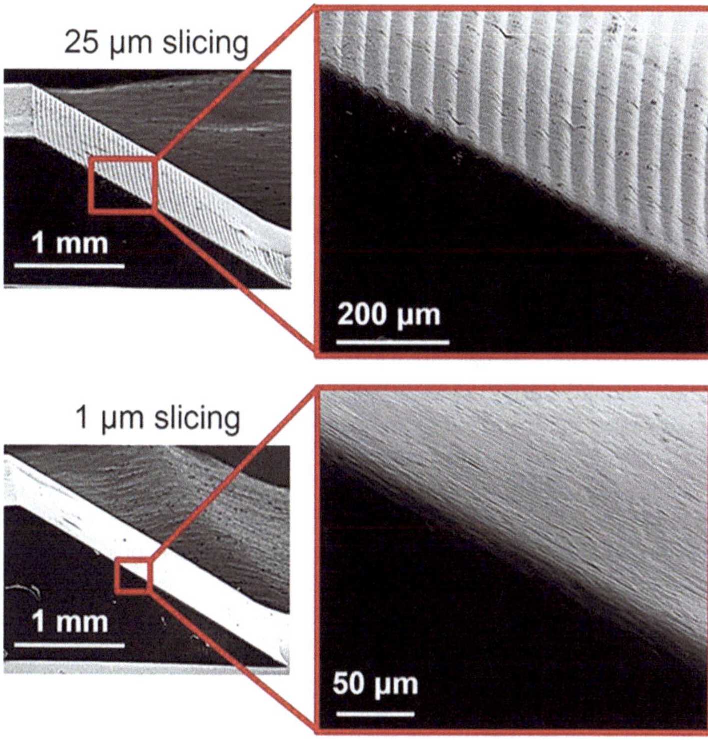

Fig. 1.24 The top panel illustrates a material sliced at 25 micrometers, with a magnified section highlighting surface texture at 200 micrometers scale. The bottom panel shows the same material sliced at 1 micrometer, with a magnified section displaying finer surface details at 50 micrometers scale. Both panels include scale bars for reference

require energy input to overcome the initial energy barrier and to sustain the processes. In addition to energy input to drive these processes, additional energy can be used to assist these processes to enhance these processes by inducing reduction of perceived metals' resistance to plastic deformation or to accelerate the diffusion behavior. Many forms of energy can be used to accomplish such assistance/enhancement to metal shaping and joining. In the case of metal shaping, many operations utilize heat energy to reduce metal's resistance to yield and hardening during plastic deformation and to allow plastic deformation, or shaping, of metals at flow stresses lower than the static yield strength of material. Similarly, in joining, many techniques employ heating and melting to create a molten pool of metals, which is then joined as a result of the mass transfer and mixing happening in the melt pool. These fundamental processes are performed independently or sequentially to fabricate components and parts in manufacturing processes.

Additive manufacturing processes follow the same logic. Shaping and joining are performed either simultaneously or sequentially to produce an additively manufactured part. In commercially available metal AM processes such as those reviewed in the previous sections of this chapter, this is accomplished in a few ways. First, the process can use a high-power point heat source such as a focused laser or an electron beam to create a melt pool from metal powder or wire feedstock which fuses and subsequently solidifies to form a solid unit of material. As the point source moves following a prescribed path, an object can be formed by repeating this process line by line and layer by layer. Examples of technologies that use this process include the selective laser melting (or laser powder-bed fusion) and the directed energy deposition techniques [39]. Another method involves depositing metal particles carried by a polymer binder which is then followed by steps to burn off the binder and sinter the metal particles to obtain a densified part in the follow-on post-print processes. Examples of this type of process include the binder jetting additive manufacturing, the bound metal extrusion, and the metal nanoparticle jetting additive manufacturing technologies [40, 41]. Yet another possible approach to perform metal additive manufacturing is to use localized frictional heat fusion and mechanical strain to shape and join feedstock material into units through local frictional heating and melting, and stirring and mixing of metal in solid state. Examples include the ultrasonic additive manufacturing and the friction stir additive manufacturing technologies [42, 43]. While these processes have proven capabilities and applications, fundamentally the energy efficiency of these processes is limited because they rely on several energy conversion processes between energy input to material shaping and joining. Laser powder-bed fusion, for example, typically uses solid-state fiber lasers which operate at around 20% energy efficiency (from electrical energy input to laser optical energy output). The laser energy incident into the material then goes through another absorption loss in both solid and liquid states. Additionally, the absorbed energy further sees losses into heat transfer from the laser incident area into its surroundings while it serves the purpose of heating and melting the metal feedstock into the melt pool to accomplish shaping the joining. Additionally, the inefficiencies in the electrical energy to optical energy conversion process cause parasitic heating which, in turn, requires yet additional energy to dissipate in order to mitigate lasing issues. For example, a typical fiber laser unit requires kilowatts of additional power to run chillers and cooling systems to remove the parasitic heating in the laser system. As a result, a laser powder-bed fusion AM machine typically consumes 10–20 kW of power during operation, which translates to 300–500 MJ/Kg (or 1 MJ/cm^3) of energy consumption for processing metals like aluminum or copper alloys at the machine level [44, 45]. These energy inefficiency issues are also seen in other thermal or melt fusion-based metal AM technologies.

Fundamentally, other energy conversion processes can be used to accomplish material shaping and joining simultaneously in more energetically efficient ways. For example, when one considers the findings by Langenecker [46], the use of direct mechanical energy coupling into the lattice structures and the defects within crystalline metals to assist deformation represents a significantly more energetically efficient way to accomplish metal shaping and joining. These findings clearly indicate that using small-amplitude, high-frequency repeated strains imposed in metals to assist in shaping results in a significant drop in metal's apparent yield and flow stress. Energetically speaking, using small-amplitude (less than a few tens of micrometers) repeated strains (tens of kHz) is roughly two orders of magnitude more efficient than using heat to cause the same amount of yield or flow stress reduction in metals, in the theoretical conditions with no heat lost or heat transfer from a $1cm^3$ block of aluminum. In reality, the energy efficiency benefits are likely more because of the heat transfer and therefore losses during heating.

Solid-state joining of metal surfaces fundamentally requires physical advective mixing or atomic diffusion across interfaces. Inherently, diffusion rates are higher at metal surfaces or grain boundaries due to the increasing amount of surface diffusion. Conventional approach of using elevated temperature to accelerate solid-state diffusion is an intuitive way to increase diffusion rates since diffusivity of self-diffusion of metals increases with temperature. As compared with how diffusion is enhanced by the presence of crystalline defects such as dislocation pipes, however, mass transport in high-crystalline-defect regions is more effective. As demonstrated in Legro's work where bulk diffusion of aluminum in aluminum is compared with aluminum diffusion in dislocation pipes, increasing the temperature of aluminum from 573 K to 673 K allows for roughly 100 times increase in diffusivity to be observed. In contrast, at 573 K, the diffusivity of aluminum along a lattice line defect like dislocation is nearly 10,000 times higher than bulk diffusion [47]. Conceptually, this means that if one could induce large amounts of lattice defects in the vicinity of metal-metal interfaces with pristine contact, along with a defect concentration gradient across the interface, enhanced diffusion across said interface and therefore joining would be expected. Taken together, the enhancement in the ability to shape metal using energetically more efficient oscillatory vibrations and the increased solid-state mass transfer-based joining can form the basis of a method to simultaneously shape and join small units of metal material without melting or ever heating it. As this method is implemented in a way to gain control of the unit metal shaping and joining process both in space and time, an entirely new 3D printing approach can emerge.

1.4.2 Electrochemical Additive Manufacturing

Fundamentally, the electrochemical additive manufacturing (ECAM) technique shares the same physical principles with electroplating where charged metal ions are moved to targeted areas by an electric field in an electrolyte fluid and reduced to metal atoms through a reduction process. The main distinction is that in ECAM, the area where the metal ions are drawn to is precisely controlled and localized by either a single-point microelectrode or an array of microelectrodes that can be individually turned on or off to introduce area selectivity during deposition of a "layer" with a given shape. An example of development of this type of technology is shown in Fig. 1.25 where a system mechanics similar to that of the projection-based SLA is used. In ECAM, however, the resin is replaced by an electrolyte, and the UV light source is replaced by a micro-cathode array where electrical charge is supplied locally at the micro-cathodes to direct metal ions and for the subsequent reduction and deposition of material.

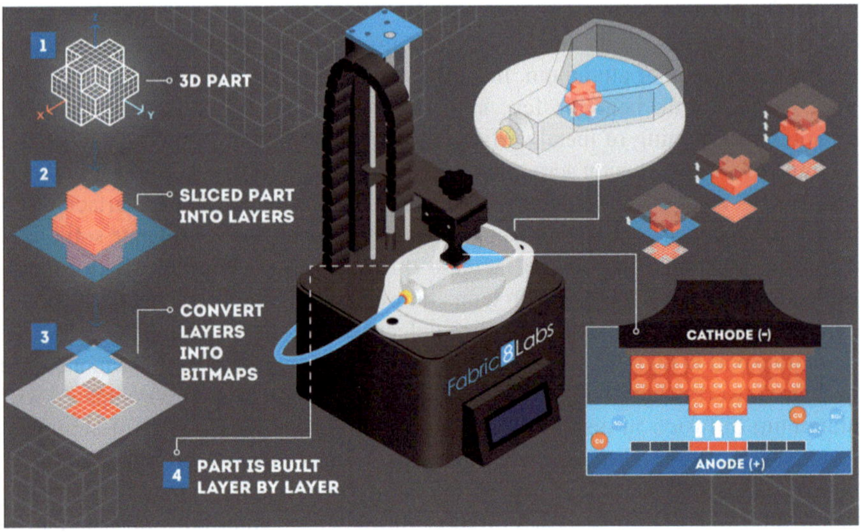

Fig. 1.25 Electrochemical additive manufacturing. Source: imec-int.com and fabric8telabs

1.4.3 Directed Acoustic Energy Deposition

The process directed acoustic energy deposition was first demonstrated by Hsu and Deshpande in 2016 [Patten]. It is currently capable of additively fabricating metal articles of greater than 95% density in ambient conditions at room temperature. The working principle of this process can be thought of as the combination of wire bonding and fused deposition modeling (FDM): a solid metal filament is used as the starting material to form a 3-dimensional object via the metallurgical bonding between the filaments and layers. As shown in Fig. 1.1, the mechanics and tooling configuration of DAED are analogous to the FDM process where a heated thermoplastic extruder directly "writes" the tracks and layers that make up the 3D component. In DAED, a solid metal filament is guided, shaped, and then metallurgically bonded to the substrate (or the previous layer) as well as the adjacent filaments voxel by voxel using a guide tool on a positioning system. The key characteristics of this process are that (1) the mechanical stress (and therefore mechanical energy input) required to "shape" the filament into the desired "track" geometry is drastically reduced (<50%) in the presence of applied acoustic energy as compared to the yield strength of the material, (2) the amount of mass transport across the interfilament and interlayer interfaces to form the metallurgical bonds observed is more than 10,000 times higher [48] than what Fick's diffusion predicts under the observed conditions, and (3) the temperature rise of the DAED process is nearly negligible (5 °C), a reflection of the high coupling efficiency from acoustic energy input into the required plasticity and mass transport. These unique characteristics enable the directed acoustic energy deposition process introduced here to be implemented within a desktop 3D printing environment. Additionally, its unique nature of solid metal 3D printing at room temperature enables simultaneous printing of polymers and metals, a material combination not feasible in melt fuse-based metal additive manufacturing processes.

In the demonstration work, an apparatus based on an ultrasonic energy coupling tool on a 3-axis positioning system was developed to couple acoustic energy into a fully dense aluminum filament, to guide the filament, and to induce the voxel shaping and material fusion required for 3D printing. The ultrasonic vibration source is a piezoelectric crystal-based transducer oscillating in the kHz frequency range. During DAED, the tool delivers the acoustic energy to the interfaces between a solid metal filaments and an existing metal substrate as depicted in Fig. 1.26. As acoustic energy is used to shape the metal filament and allow the metallurgical bond on the metal-metal interface to form, the tool steps down the length of the filament to form a "track" of solid metal. These steps are then repeated to form adjacent tracks that make up one layer, followed by repeating track-wise and layer-wise steps to form a 3-dimensional object. Fig. 1.26 shows scanning electron microscopy images of a two-layer structure built following the described ultrasonic filament modeling approach. These images depict filament track shaping as well as metallurgical bonding with adjacent tracks. In the images shown in the figure, interface regions where native aluminum oxide accumulates can be observed.

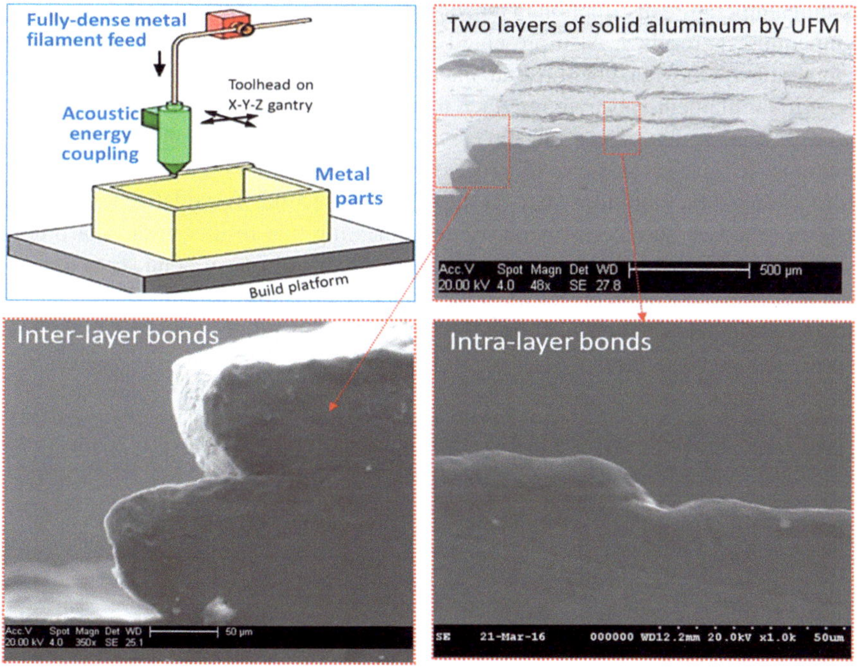

Fig. 1.26 The top left shows a schematic of the process with a toolhead applying acoustic energy to a metal filament on a build platform. The top right image displays two layers of solid aluminum, highlighting the structure at 500 micrometers scale. The bottom images focus on inter-layer and intra-layer bonds, showing detailed surface textures at 50 micrometers scale

In the demonstration work, an L-shaped 3D object and a tensile testing specimen were built following the same procedures. In the case of the L-shaped object, post-process of surface refinishing was used to remove features (less than 200 μm) from the printing process. As shown in Fig. 1.27a–d, the object is 5 mm long, 4 mm wide, and approximately 1.5 mm tall with a layer height of 125 μm. The hybrid additive-subtractive mode of the process was also demonstrated. A 16-layer aluminum structure was printed and machined down to an I-shaped object as photographed and shown in Fig. 1.27c. Preliminary X-ray microtomography results were obtained for the midsection of the sample, and one representative slice is shown in Fig. 1.27f, d. In the micro-CT scans of the DAED printed sample, the interlayer interfaces are discernable, while no inter-filament interfaces are observed. 99% density is observed. The DAED process introduced here can be implemented as a direct 3D printing process or a hybrid additive-subtractive fabrication process.

This process allows room-temperature solid metal 3D printing. Both high-resolution IR imaging and thermocouple probing of surface temperatures were used to quantify the temperature rise in DAED. High-speed IR videography captured by an FLIR thermal imaging camera (whose line of sight is normal to the cross section of the aluminum filament and front surface of tungsten carbide tool) shows that the maximum temperature rise for the formation of one voxel is less than 5°. In Fig. 1.28a, one frame of thermal video captured during voxel formation shows the spatial temperature distribution in the vicinity of the voxel at the time the maximum temperature is reached. Also presented in Fig. 1.28b is the time evolution of the temperature at the filament-substrate interface. The time evolution of temperature at the critical filament-substrate interface indicates that the fusion of a voxel initiates within 30 ms at the power and force setting used while the shaping of the voxel continues to develop as the voxel process time continues. In Fig. 1.28b, the irradiation of ultrasonic vibrations starts at the 50th millisecond. The relative movements between the two surfaces provide frictional heating that results in the sharp temperature rise. Another 30 ms into the process, the metallurgical bond starts to form and the relative movements between the filament and substrate stop. This removes the frictional heat source and allows the interface temperature to drop. A maximum

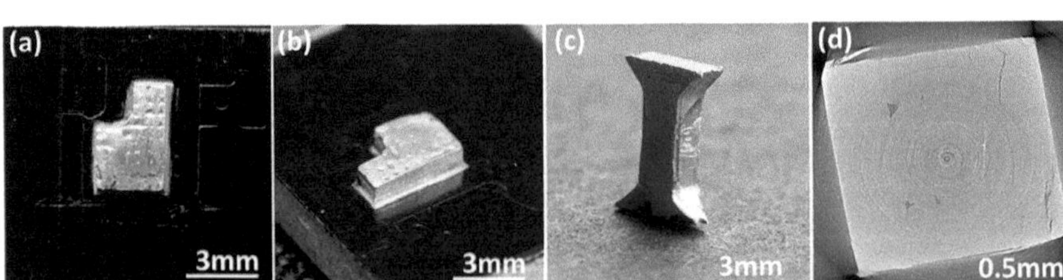

Fig. 1.27 Panel (**a**) displays a rectangular metallic block with a textured surface, labeled with a 3mm scale. Panel (**b**) shows a similar block from a different angle, also with a 3mm scale. Panel (**c**) features a metallic sample with a narrow middle section, resembling an hourglass shape, with a 3mm scale. Panel (**d**) presents a close-up of a cylindrical metallic surface with concentric circular patterns, labeled with a 0.5mm scale

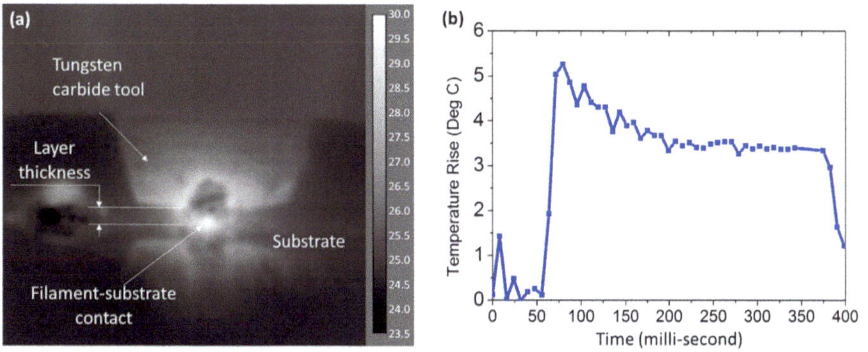

Fig. 1.28 Panel (**a**) is a thermal image showing a tungsten carbide tool, layer thickness, filament-substrate contact, and substrate, with a temperature scale ranging from 23.5 to 30.0 degrees Celsius. Panel (**b**) is an X-Y chart depicting temperature rise in degrees Celsius over time in milliseconds. The chart shows a rapid increase in temperature followed by fluctuations and a gradual decline

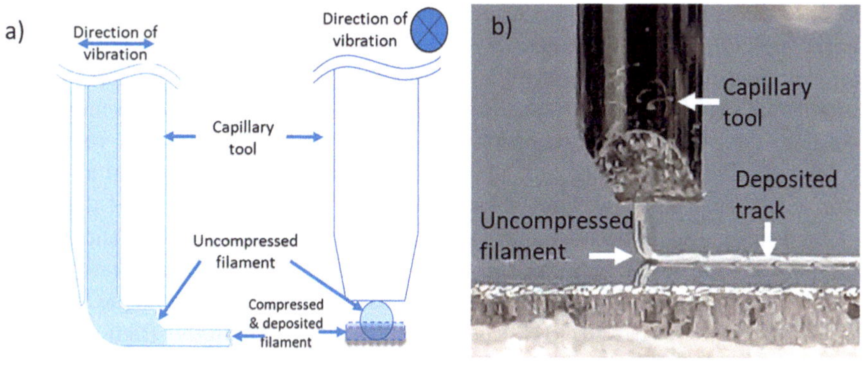

Fig. 1.29 Panel (**a**) shows a schematic of the capillary tool with arrows indicating the direction of vibration. It depicts an uncompressed filament being compressed and deposited. Panel (**b**) is a photograph showing the capillary tool in action, with labels indicating the uncompressed filament and the deposited track

temperature rise of 5° is observed. Another feature in the temporal temperature profile is the sharp reduction at 350 ms where the ultrasonic vibration stops. This indicates the removal of the second heat source in the process: cyclic plastic strain heating due to the high-frequency cyclic shear deformation in the voxel as it forms.

The DAED method introduced here uses two mechanical oscillatory strain-based physical mechanisms to assist shaping and joining. To implement the DAED method into a system to produce a 3D metal object, a mechanical system is set up to use a hollow cylindrical capillary tool to move a solid metal fine wire in space and turn it into "voxels" through "compression cycles" sequentially to form lines, layers, and a complete 3-dimensional part at the end. Shown in Fig. 1.29 is an example of such a setup. In the DAED metal 3D printing process, each compression cycle is the basic step that produces a voxel. In a compression cycle, the feedstock is brought to a desired location and compressed in the direction orthogonal to its axis to shape the cross section of the wire from a circular shape to rectangular while a high-frequency oscillatory shear strain (typically 35–80 kHz at roughly a few micrometers amplitude) along the wire axis direction is simultaneously applied. Simultaneously during shaping of a voxel in each compression cycle, the voxel is joined to all surrounding voxels through crystalline defect-enhanced diffusion near its surfaces. Once at the bottom position of its compression cycle, the tool lifts and moves along the wire axis direction by a prescribed "step size" to begin the next compression cycle of producing the net voxel.

The imposed oscillatory shear strain puts the material in an elevated energy state and thereby mobilizes the lattice dislocations within the voxel. Being in an activated state, these mobilized dislocations move readily when a superimposed static compression strain (in the compression cycle) is applied. In addition, the oscillatory shear strain energy enables faster annihilation of dislocations resulting in reduced hardening and lowering of flow stress required to plastically deform the metal [49, 50]. Oscillatory shear strain imposed on the voxel also causes friction at the feedstock-substrate interface (or feedstock-lower layer interface), but only during the initial few milliseconds. Because the voxel initially has a circular cross section, the area of contact between the voxel and the stiff capillary tool surface applying the oscillatory strain and compressive force is small, allowing for relative movement between the interfaces including the voxel-material interface. As the voxel gets compressed and shaped, the area of contact increases continuously until a point is reached when the relative movement between the surfaces stops, and the voxel experiences purely oscillatory shear strain. It is at this point that the frictional heat on interfaces ceases to be generated. The frictional heat that was generated in the initial roughly 50 ms is not sufficient to cause a significant increase in the temperature of the voxel (Fig. 1.28) [51]. However, the combination of this friction and the increase in surface area because of the shape change allows breaking and redistribution of native oxides on the surface of the wire feedstock and brings about nascent metal-to-metal contact at the interface.

Simultaneously as the voxel is being compressed and shaped, the second crystalline defect and oscillatory shear strain-based phenomenon promote atomic diffusion across the interface, which allows joining of a voxel to neighboring ones. This athermal diffusion enhancement phenomenon has also been employed in the wire bonding process for semiconductor device packaging applications [52]. Since the metal voxel is in an elevated energy state due to the imposed oscillatory shear strain, the energy barrier to atomic diffusion and mass transfer is subsequently lowered. Additionally, the chemical potential that is also required for atomic diffusion is raised across voxel-voxel boundaries because of the crystalline defect density gradient across the same interfaces. Combined with inherently high crystalline defect density near interfaces, the mass transfer across voxel interfaces, and therefore joining, is orders of magnitude higher than what would be expected due to thermal diffusion at temperatures at which the process occurs. This translates to either increased diffusion length across the interfaces or reduced time needed for joining to occur.

As the compression cycles repeat along a prescribed linear motion, a line of metal is deposited and welded onto the previous layer (or an existing surface). The lines can be of any pattern to either encompass the boundaries of an enclosed area or tracks with a given width to fill up the interior space of the enclosed area until a layer of material is completed. Much like conventional 3D printing processes, this layer-by-layer process repeats in a given direction, typically referred to as the print direction, to form a 3-dimensional object. Shown in Fig. 1.30 are examples of objects constructed using this approach. As seen in 30b and c, top surface of printed metal parts (surface orthogonal to the print direction) has different surface morphology as compared with the side surfaces parallel to the build direction. The top surface morphology is dictated by the tooling surface in contact with the metal voxel during each compression cycle, as well as the material flow and track-to-track overlaps in the tool path. The bottom surface morphology of the build is determined by the joint strength between the first layer of the printed part and the build plate and how it is separated from the build plate. The side surfaces have a characteristic "scalloping" effect similar to what is typically seen in the fused filament fabrication polymer 3D printing method as large amounts of material strain are experienced by the feedstock material during the printing process. In the DAED process, the ratio of voxel height (also layer height) to feedstock diameter is between 0.28 and 0.43. While the exact ratio is dictated by the amount of compression required in each compression cycle to establish full space filling between adjacent tracks and proper joining of interfaces, this ratio also determines the extent of the scallops on the side surfaces. The larger the layer height, the higher the waviness on the sidewalls. Within the feasible parameter range, the resulting roughness values of top surfaces

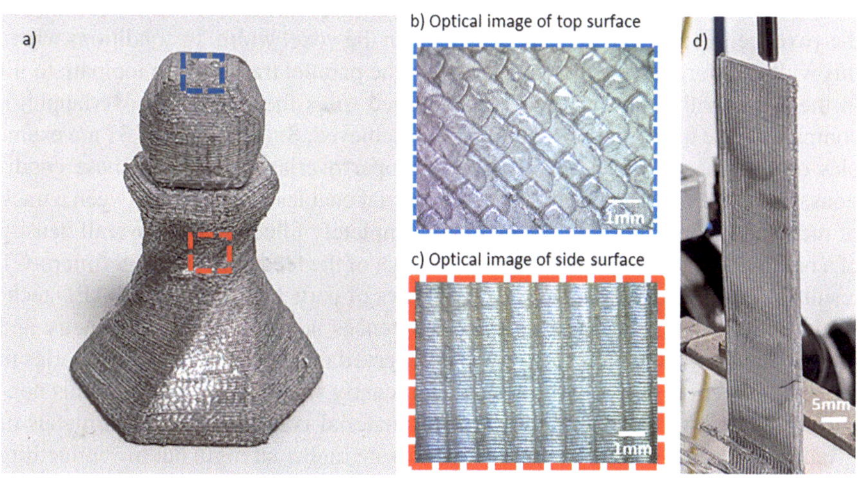

Fig. 1.30 Panel (**a**) displays the full object, resembling a chess piece with a textured surface. Panel (**b**) is an optical image of the top surface, highlighting a pattern of overlapping layers. Panel (**c**) shows the side surface with visible horizontal lines. Panel (**d**) presents a vertical view of a similar object, emphasizing its height and layered construction. Scale bars indicate 1mm and 5mm for reference

are typically around 10–20 μm Ra, while the side surfaces are in the range of 15–25 μm Ra. The as-printed surfaces of a part can also be processed using conventional surface process tools such as sanding, bead blasting, and tumbling. Depending on the accuracy and repeatability of the motion system into which the tooling is integrated, the tolerances of features in a built part can vary. On a typical linear bearing and ball screw-based Cartesian motion system, the tolerance of a printed part can fall within ±50–100 μm given that the toolpaths are generated with precisions down to 1 μm or less. Additionally, the DAED technique offers a unique capability of printing articles with very high aspect ratio as shown in Fig. 1.30d. This capability can be attributed to the absence of residual stresses in the printed parts.

Parts printed using the DAED technique typically have finer grains in comparison with the feedstock. Unsurprisingly, the large amount of strain experienced by the metal in each compression cycle during the process causes grain refinement similar to what is typically seen in deformation-based manufacturing processes such as forming. One unique feature, however, in the RAD-printed parts is that the grain refinement is not nearly as much as that seen in processes like cold rolling/drawing, but rather, similar to typical warm or hot working conditions. This behavior is linked to the observed reduction in the apparent flow stress of material during each compression cycle and can be rationalized by considering the increased dislocation annihilation in the material which counteracts the dislocation multiplication process that refines the grain sizes. The effect of this behavior on the overall grain texture and the corresponding process parameter are described elsewhere [49, 53]. We found that it is possible to locally control the grain size down to each voxel by altering the rate of compression strain or the amplitude of oscillatory movements of the tool as the grain refinement process is confined locally to each compression cycle.

In each compression cycle, the feedstock is compressed in the build direction to the voxel height as it expands laterally to form the voxel width. In conditions where this width is more than the spacing between the parallel tracks in the toolpath to fill in the space within an enclosed area (referred to as the inter-track overlapping), complete space filling of the cross section is achieved. Shown in Fig. 1.31 are examples of cross-sectional space filling with proper overlapping. Under these conditions, we observe that the lateral flow of material enables the spaces between a track of metal and its neighboring tracks to be completely filled, and the overall density of a build part reaches 99.95% or greater of that of the feedstock material (micro-CT results in Fig. 1.31c–e). Mechanical properties of parts built using the DAED technique reflect directional and locational differences in both the grain geometry and size distribution in the built material. It is observed that near the voxel boundaries in the build direction, the grain sizes are significantly smaller. As a result of this non-uniform grain size distribution, the overall material is spatially made of metals of two different properties: a finer grain region with higher strength but lower ductility and a coarser grain region with lower strength but high ductility, where the finer grain regions are sandwiched between layers of coarser grain regions. Effectively, this is similar to alternating layers of reinforcing agent and a softer matrix phase in composite materials. The overall result is that the mechanical property of a DAED part is directionally dependent on the relative orientation between the load and the stronger phase, or the finer grain regions, in the built material.

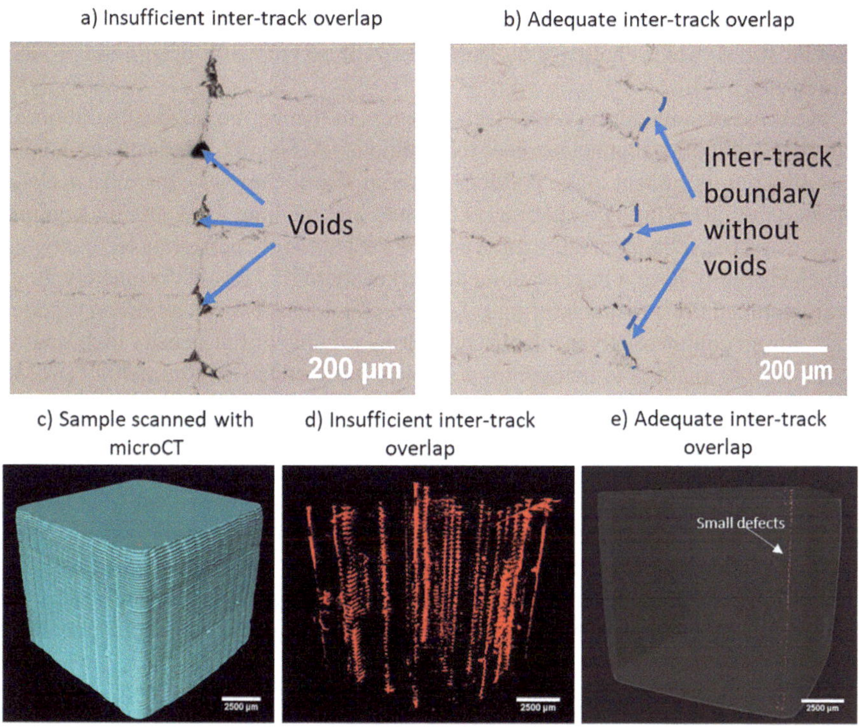

Fig. 1.31 Subfigure (**a**) shows a microscopic view labeled "Insufficient inter-track overlap" with visible voids marked by arrows. Subfigure (**b**) displays "Adequate inter-track overlap" with inter-track boundaries without voids, also marked by arrows. Subfigure (**c**) is a 3D model labeled "Sample scanned with microCT," showing a green block structure. Subfigure (**d**) shows a red pattern labeled "Insufficient inter-track overlap," highlighting gaps. Subfigure (**e**) presents a 3D model labeled "Adequate inter-track overlap" with small defects indicated by an arrow. Each subfigure includes a scale bar for reference

1.4.4 Additive Manufacturing Innovation to Address Market Needs

Accessible 3D printing has truly deep impact on many layers of human society: from affordable medical products and health care to transforming STEM education and to point-of-use design and manufacturing of goods. Accessibility, however, is the key to reaching these potentials. Take the extrusion-based polymer 3D printing technology as an example. The explosion of accessible extrusion-based 3D printing systems not only initiated a shift in how polymer goods can be designed and made, but also enabled the community to reach their impact by drastically lowering the barrier to access design-to-manufacturing freedom. In the industrial landscape, the American industry is dominated by small- and medium-scale firms which, though with limited resources, are agile and ready to contribute to making an impact. For these firms, however, changing customer requirements has been identified as a major "pain point" [54–56]. Customers are increasingly demanding services that require shorter product design and development cycles and customized designs and specifications, which challenges existing means of executing the design and production cycle.

Additive manufacturing (AM) technologies, in theory, are an ideal solution for this pain point of changing customer requirements. A sign of the industry moving in this direction is evident in the polymer material space. Currently, the most accessible polymer 3D printing technologies to small and medium firms and the community are the fused filament fabrication and the vat photopolymerization technologies, which currently have a total equipment market share of 71%. Unfortunately, however, metal additive manufacturing (MAM) technologies exist in a different model. The current commercially available technology capable of full-density high-quality parts without major post-fabrication processes is the selective laser melting (SLM), especially for aluminum alloys [57–60]. Even though it has the highest total equipment market share of all MAM technologies, it only occupies 8% of the market. As the broader market needs for metal and polymer materials have roughly equal shares, there is a large gap between market needs for metal parts and what current MAM technologies can provide. The key issue is, again, accessibility. The accessibility issue lies not only in the high capital, operational, and maintenance costs, but also in the safety, environmental, and health concerns associated with the use of powder feedstock, high-power lasers, and high temperatures in the SLM technology.

References

1. Ravi A, Deshpande A, Hsu K (2016) An in-process laser localized pre-deposition heating approach to inter-layer bond strengthening in extrusion based polymer additive manufacturing. J Manuf Processes 24(7):179–185
2. Anitha R, Arunachalam S, Radhakrishnan P (2001) Critical parameters influencing the quality of prototypes in fused deposition modelling. J Mater Process Technol 118(1–3):385–388
3. Nancharaiah T, Raju DR, Raju VR (2010) An experimental investigation on surface quality and dimensional accuracy of FDM components. Int J Emerg Technol 1(2):106–111
4. Thrimurthulu K, Pandey PM, Reddy NV (2004) Optimum part deposition orientation in fused deposition modeling. Int J Mach Tools Manuf 44(6):585–594
5. Horvath D, Noorani R, Mendelson M (2007) Improvement of surface roughness on ABS 400 polymer using design of experiments (DOE). Mater Sci Forum 561:2389–2392
6. Wang CC, Lin TW, Hu SS (2007) Optimizing the rapid prototyping process by integrating the Taguchi method with the gray relational analysis. Rapid Prototyp J 13(5):304–315
7. Sood AK, Ohdar R, Mahapatra S (2009) Improving dimensional accuracy of fused deposition modelling processed part using grey Taguchi method. Mater Des 30(10):4243–4252
8. Zhang JW, Peng AH (2012) Process-parameter optimization for fused deposition modeling based on Taguchi method. Adv Mater Res 538:444–447
9. Sahu RK, Mahapatra S, Sood AK (2013) A study on dimensional accuracy of fused deposition modeling (FDM) processed parts using fuzzy logic. J Manuf Sci Prod 13(3):183–197
10. Lee B, Abdullah J, Khan Z (2005) Optimization of rapid prototyping parameters for production of flexible ABS object. J Mater Process Technol 169(1):54–61
11. Laeng J, Khan ZA, Khu SY (2006) Optimizing flexible behavior of bow prototype using Taguchi approach. J Appl Sci 6:622–630

12. Zhang Y, Chou K (2008) A parametric study of part distortions in fused deposition modelling using three-dimensional finite element analysis. Proc Inst Mech Eng Part B 222(8):959–968
13. Nancharaiah T (2011) Optimization of process parameters in FDM process using design of experiments. Int J Emerg Technol 2(1):100–102
14. Kumar GP, Regalla SP (2012) Optimization of support material and build time in fused deposition modeling (FDM). Appl Mech Mater 110:2245–2251
15. Ahn SH, Montero M, Odell D et al (2002) Anisotropic material properties of fused deposition modeling ABS. Rapid Prototyp J 8(4):248–257
16. Ang KC, Leong KF, Chua CK et al (2006) Investigation of the mechanical properties and porosity relationships in fused deposition modelling-fabricated porous structures. Rapid Prototyp J 12(2):100–105
17. Sood AK, Ohdar RK, Mahapatra SS (2010) Parametric appraisal of mechanical property of fused deposition modelling processed parts. Mater Des 31(1):287–295
18. Percoco G, Lavecchia F, Galantucci LM (2012) Compressive properties of FDM rapid prototypes treated with a low cost chemical finishing. Res J Appl Sci Eng Technol 4(19):3838–3842
19. Rayegani F, Onwubolu GC (2014) Fused deposition modelling (FDM) process parameter prediction and optimization using group method for data handling (GMDH) and differential evolution (DE). Int J Adv Manuf Technol 73(1–4):509–519
20. Masood SH, Mau K, Song WQ (2010) Tensile properties of processed FDM polycarbonate material. Mater Sci Forum 654:2556–2559
21. Ahn S-H, Montero M, Odell D, Roundy S, Wright PK (2002) Anisotropic material properties of fused deposition modeling ABS. Rapid Prototyp 8(4):PP248
22. Rodriguez JF, Thomas JP, Renaud JE (1999) Tailoring the mechanical properties of fused-deposition manufactured components. In: Proc. rapid prototyping and manufacturing '99, vol 3. Society of Manufacturing Engineers, Dearborn, MI, pp 629–643
23. Yan Y, Zhang R, Guodong H, Yuan X (2000) Research on the bonding of material paths in melted extrusion modeling. Mater Des 21:93–99
24. Partain S (2007) Fused deposition modeling with localized pre-deposition heating using forced air. Master's Thesis, Montana State University
25. De Gennes PG (1971) Reptation of a polymer chain in the presense of fixed obstacles. J Chem Phys 572
26. Stansbury J, Idacavage M (2016) 3D printing with polymers: challenges among expanding options and opportunities. Dent Mater 32(1):54–64
27. Gu D, Meiners W, Wissenbach K, Poprawe R (2012) Laser additive manufacturing of metallic components: materials, processes and mechanisms. Int Mater Rev 57(3):133–164
28. Wolff I (2/8/2016) Breathing safely around metal 3D printers. Advanced manufacturing.org
29. Guo N, Leu MC (2013) Additive manufacturing: technology, applications and research needs. Front Mech Eng 8(3):215–243
30. White D (2003) Ultrasonic consolidation of aluminum tooling. Adv Mater Process 161:64–65
31. Ram G, Robinson C, Yang Y, Stucker B (2007) Use of ultrasonic consolidation for fabrication of multi-material structures. Rapid Prototyp J 13(4):226–235
32. Sriraman M et al (2011) Thermal transients during processing of materials by very high power ultrasonic additive manufacturing. J Mater Process Technol 211:1650–1657
33. Hu J, Yu M (2010) Meniscus-confined three-dimensional electrodeposition for direct writing of wire bonds. Science 329(16):313–316
34. Thomas D, Gilbert S (2014) Costs and cost effectiveness of additive manufacturing. NIST Special Publication 1176
35. Semenok D (2014) LCLD Laser processing technology for microelectronics printed circuit boards of new generation. In: Proceedings of the 7th International Conference Interdisciplinarity in Engineering, vol 12, pp 277–282
36. Kochemirovsky VA et al (2014) Laser-induced copper deposition with weak reducing agents. Int J Electrochem Sci 9:644–658
37. Anzalone G et al (2013) A low-cost open-source metal 3D printer. IEEE Access 1:803–810

38. Haselhuhn A et al (2015) In situ formation of substrate release mechanisms for gas metal arc weld metal 3D printing. J Mater Pocess Technol 226:50–59
39. Vafadar A, Guzzomi F, Rassau A, Hayward K (2021) Advances in metal additive manufacturing: a review of common processes, industrial applications, and current challenges. Appl Sci 11(3):1213
40. Suwanpreecha C, Manonukul A (2022) A review on material extrusion additive manufacturing of metal and how it compares with metal injection moulding. Meta 12(3):429
41. Galati M, Minetola P (2019) Analysis of density, roughness, and accuracy of the atomic diffusion additive manufacturing (ADAM) process for metal parts. Materials 12(24):4122
42. White DR (2003) Ultrasonic consolidation of aluminum tooling. Adv Mater Processes 161(1):64–66
43. Griffiths RJ et al (2019) A perspective on solid-state additive manufacturing of aluminum matrix composites using MELD. J Mater Eng Perform 28(2):648–656
44. Liu Z, Li C, Fang X, Guo Y (2018) Energy consumption in additive manufacturing of metal parts. Procedia Manuf 26:834–845
45. Priarone PC, Lunetto V, Atzeni E, Salmi A (2018) Laser powder bed fusion (L-PBF) additive manufacturing: on the correlation between design choices and process sustainability. Procedia Cirp 78:85–90
46. Langenecker B (1966) Effects of ultrasound on deformation characteristics of metals. IEEE Trans Sonics Ultrason 13(1):1–8
47. Legros M, Dehm G, Arzt E, Balk TJ (2008) Observation of giant diffusivity along dislocation cores. Science 319(5870):1646–1649
48. Haselhuhn A et al (2016) Structure-property relationships of common aluminum weld alloys utilized as feedstock for GMAW-based 3D metal printing. Mater Sci Eng A 673:511–523
49. Deshpande A, Hsu K (2018) Acoustic energy enabled dynamic recovery in aluminium and its effects on stress evolution and post-deformation microstructure. Mater Sci Eng A 711:62–68
50. Yao Z et al (2012) Acoustic softening and residual hardening in aluminum: modeling and experiments. Int J Plast 39:75–87
51. Deshpande A, Hsu K (2018) Acoustoplastic metal direct-write: towards solid aluminum 3D printing in ambient conditions. Addit Manuf 19:73–80
52. Harman G (2010) Wire bonding in microelectronics. McGraw-Hill Education
53. Deshpande A, Tofangchi A, Hsu K (2019) Microstructure evolution of Al6061 and copper during ultrasonic energy assisted compression. Mater Charact 153:240–250
54. Intelligence M (2019) 3d printing market-growth, trends, and forecasts. Mordor Intelligence
55. May B (2017) 2017: the year of small to medium manufacturing
56. Timmons J, Gold S, McNelley J (2011) Facts about manufacturing. National Association of Manufacturers
57. Sanders WW, Mcdowell KA (1978) Fatigue behavior of 5000-series aluminum-alloy weldments in marine-environment. Weld Res Counc Bull (in English) 242:1–14. <Go to ISI>://WOS:A1978FT38700001
58. Mossman MM, Lippold JC (2002) Weldability testing of dissimilar combinations of 5000- and 6000-series aluminum alloys. Weld J (in English) 81(9):188s–194s. <Go to ISI>://WOS:000186767400017
59. Nishida T, Ogura T, Fujimoto M, Hirose A (2011) Microstructure and mechanical property of 5000 series aluminum stud joint with zinc insert using friction welding. Mater Trans (in English) 52(5):960–966. https://doi.org/10.2320/matertrans.L-MZ201124
60. Park DH, Choi SW, Kim JH, Lee JM (2015) Cryogenic mechanical behavior of 5000-and 6000-series aluminum alloys: issues on application to offshore plants. Cryogenics (in English) 68:44–58. https://doi.org/10.1016/j.cryogenics.2015.02.001

Chapter 2
Additive Manufacturing Process Chain

A series of steps go into the process chain required to generate a useful physical part from the concept of the same part using additive manufacturing processes. Depending on the technology and, at times, the machines and components, the process chain is mainly made up of six steps:

- Generation of CAD model of the design
- Conversion of CAD model into AM machine-acceptable format
- CAD model preparation
- Machine setup
- Part removal
- Post-processing

These steps can be grouped or broken down and can look different from case to case, but overall, the process chain from one technology remains similar to that of a different technology. The process chain is also constantly evolving and can change as the existing technologies develop and new technologies surface. In this text, the focus will be on the powder-bed metal technology. Therefore, the process chain for this technology will be discussed in detail, while others will be roughly mentioned.

2.1 Generation of Computer-Aided Design Model of Design

In any product design process, the first step is to imagine and conceptualize the function and appearance of the product. This can take the form of textual descriptions, sketches, to 3-dimensional computer models. In terms of process chain, the first enabler of AM technologies is 3D digital computer-aided design (CAD) models where the conceptualized product exists in a "computer" space and the values of its geometry, material, and properties are stored in digital form and are readily retrievable (Fig. 2.1).

K. Hsu et al., *Additive Manufacturing of Metals*, Springer Series in Advanced Manufacturing, https://doi.org/10.1007/978-3-032-04961-2_2

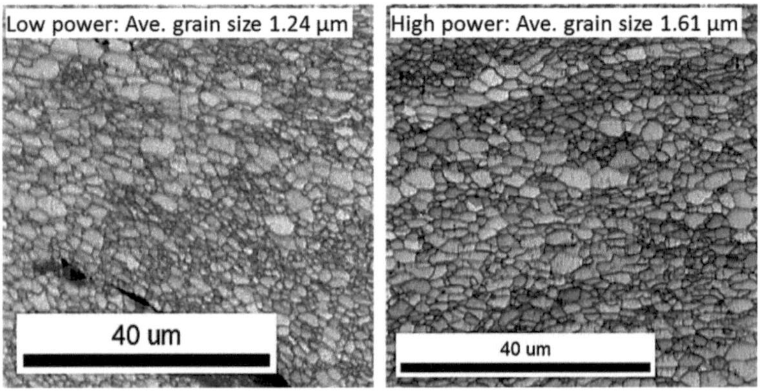

Fig. 2.1 The left image, labeled "Low power: Ave. grain size 1.24 µm," shows a pattern of grains with an average size of 1.24 micrometers. The right image, labeled "High power: Ave. grain size 1.61 µm," displays a similar pattern with slightly larger grains, averaging 1.61 micrometers. Both images include a scale bar indicating 40 micrometers

In general, the AM process chains start with 3D CAD modeling. The process of producing a 3D CAD model from an idea in the designer's mind can take on many forms, but all require CAD software programs. The details of these programs and the technology behind them are outside the scope of this text, but these programs are a critical enabler of a designer's ability to generate a 3D CAD model that can serve as the start of an AM process chain. There are a large number of CAD programs with different modeling principles, capabilities, accessibilities, and cost. Some examples include Autodesk Inventor, SolidWorks, Creo, NX, etc. The examples here are, by no means, representatives of CAD technologies, but are intended to provide a few key words with which a search on topics on CAD can provide specific information a reader needs.

Once a 3D CAD model is produced, the steps in the AM process chain can take place. Though the process chain typically progresses in one direction that starts with CAD modeling and ends with a finished part or prototype, it is often an iterative process where changes to the CAD model and design are made to reflect feedback from each step of the process chain. Specific to the metal powder-bed technology, critical feedback can come from geometry and property anisotropy on parts due to build orientation, distortion of part or features due to thermal history of build, issues in generating and removal of support structures, etc. Issues like these may arise in the AM process chain and may call for design changes and revisions. Similar to the topic of "design for manufacturability" in the conventional manufacturing space, design for additive manufacturing is critical and is developing in parallel with the technologies themselves.

2.2 Conversion of CAD Model into AM Machine-Acceptable Format

Almost all AM technology available today uses the STL file format. Shown in Fig. 2.2 is an example part in its STL format. The stereolithography (STL) format of a 3D CAD model captures all surfaces of the 3D model by means of stitching triangles of various sizes on its surfaces. The spatial locations of the vertices of each triangle and the vectors are normal to each triangle. When combined, these features allow AM pre-process programs to determine the spatial locations of surfaces of the part in a build envelope and on which side of the surface is the interior of the part.

Although the STL format has been considered the de facto standard, it has limitations intrinsic to the fact that only geometry information is stored in these files while all other information that a CAD model can contain is eliminated. Information such as unit, color, and material can play a critical role in the functionality of the built part that is lost through the file translation process. As such, it places limitations on the functionality of the finished parts. The "AMF" format was developed specifically to address these issues and limitations and is now the ASTM/ISO standard format. Beyond geometry information, it also contains dimensions, color, material, and additional information that is also possible with this file format. However, currently, the predominate format of file used by AM systems and supported by CAD modeling programs is still the STL format. An increasing number of CAD program companies, including several major programs, have included support of AMF file formats. Currently, actual use of the information stored in the AMF file is still limited due to the capabilities of current AM systems and the state of current technology development.

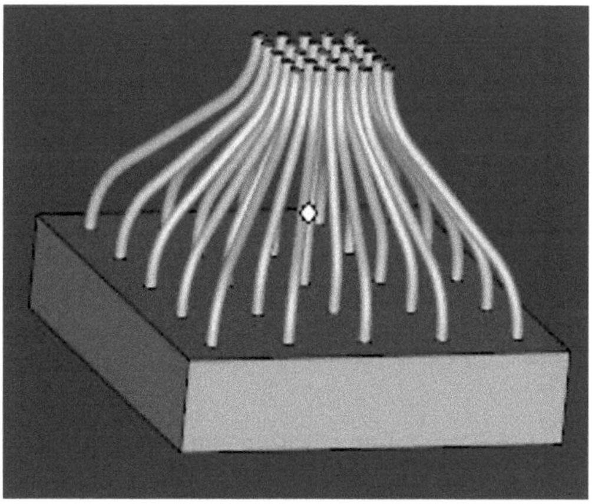

Fig. 2.2 Fine-wire type of 3D structure fabricated by laser powder-bed fusion AM

2.2.1 CAD Model Preparation

Once a correct STL file is available, a series of steps are required to generate the information an AM system needs to start the build process. The needed information varies depending on the technology, but in general, these steps start with repairing any errors within the STL file. Typical errors can be gaps between surface triangle facets, inverted normal where the "wrong side" of a triangle facet is identified as the interior of the part. Once the errors have been repaired, a proper orientation of the 3D model with respect to the build platform/envelope is then decided. Following the orientation, the geometry, density, and geometry of support structures are decided and generated in 3D model space and assigned to the part model. The process then progresses to slicing the 3D model defined by the STL as well as the support structure into a given number of layers of a desired height, each representing a slice of the part and support models. Within each slice, the cross-sectional geometry is kept constant. Once 2D slices are obtained, a "unit" area representing the smallest material placement is then used, combined with a set of strategies, to fill the area within each layer that is enclosed by the surface of the part and the support. For the SLA process, the unit area is related to the laser spot size and its intensity distribution as well as absorption of monomer, and the strategies of filling the enclosed area in one layer are the path in which the laser would raster the resin surface. At this point, the surface data that are originally in the STL file have been processed and machine-specific information to allow placement of the material unit into the desired location in a controlled manner to construct the physical model layer by layer.

Specifically for metal powder-bed processes, an STL file is first imported into a software that allows repairing and manipulating of the file, as well as the generation of support and the slicing of the part and support models. The sliced data are then transferred into the AM system machine for build preparation and the start of the building process. There are a number of software programs that allow these tasks to be carried out. Magics, for example, by Materialise is one such software program that is capable of integrating all CAD model preparation steps into one program and generating data files directly accepted by powder-bed machine systems. In the sections below, additional details of model preparation specific to the powder-bed metal process are described.

2.2.2 STL File Preparation

The CAD model preparation starts with importing an STL, or other compatible file formats, into the pre-process software program (e.g., Magics). Once imported, the dimensions can be modified if needed. Once the model is in desired dimensions, a series of steps are carried out to correct possible errors in the model file. These errors can include missing triangles, inverted or double triangles, transverse triangles, open edges and contours, and shells. Each type of error can cause issues in the building process or result in incorrect parts and geometries. While some errors such as shells and double triangles are noncritical and can sometimes be tolerated, errors such as inverted triangles and open contours can cause critical issue in the building process and need to be resolved during STL preparation.

2.2.3 Support Generation

For metal powder-bed process, the primary function of the "support" structure is to extract heat from the model and to provide anchor surfaces and features to the build plate to avoid warpage due to thermal stresses during and after the build. It does not "support" the part against gravity that causes overhanging or angled features to fail to build.

Generation of support structures in powder-bed processes can be accomplished in a few different ways. Also applied to any other AM processes, the first way is to generate the support structures during CAD modeling and design the support to be features of the geometry of the part. Alternatively, the support structures can be generated in the STL pre-process software program. This second approach provides much more flexibility in terms of being able to tailor the structures based on the detailed needs. For example, since the support structure is only used during the build process and is removed during post-process, the amount of material that goes into it needs to be minimized. However, since the primary function of the support is to conduct heat and provide mechanical anchor, a minimum amount of cross-sectional area in the support is needed for it to be functional. Optimization of the volume, geometry, location, and part-support interface geometry is important and part dependent. Therefore, careful design of support structure plays a critical role in the success of a build process. Figure 2.3 shows examples of support generated in CAD model space.

2.2.4 Build File Preparation

Once the CAD models of part and support are generated and prepared in the pre-process software program, a slicer program is used to divide the models into layers in the build direction based on the desired layer thickness. For typical metal powder-bed systems, the layer thickness can be anywhere from 25 μm to close to 100 μm. Typical thicknesses used are 25 μm for high-resolution builds and 50–75 μm or higher for high-rate applications. Layer thickness is also correlated to the powder dimension and distribution. Ideally, the layer thickness would be slightly larger than the mean diameter of powder to achieve high coupling of laser energy input into the absorption, heating, and melting of powder and remelting of the previous layer. At larger thicknesses, scattering of optical energy input may not be sufficient to allow uniform optical energy absorption and heating, resulting in partial melting.

Fig. 2.3 Support structure generated on the model

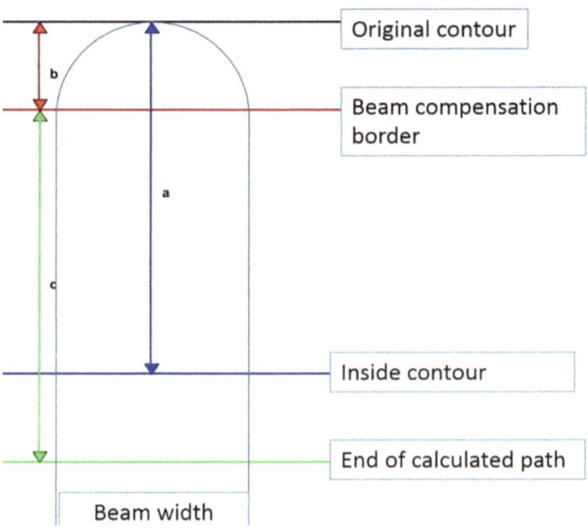

Fig. 2.4 Beam compensation to determine correct beam paths

Within each slice, the slicer software also determines the rastering path that the energy beam takes to fully melt the entire layer. How a melt pool created by an optical beam can move in an enclosed area to ensure that every part of it is covered is analogous to how an end-milling tool does a pocket milling operation to create an enclosed 2D area with a depth on a surface. Figure 2.4 shows how the use of contour scan coupled with in-fill fully covers an area. As seen in the illustration, an important factor is the beam diameter, which is captured by the beam compensation parameter. This value is typically chosen such that sufficient overlapping of adjacent paths occurs and partial melting is avoided.

Other important parameters include "islands" within each layer, and shifting of scan directions, offset of scan lines from layer to layer as build progresses. The use of islands is important in that the rapid heat generation and cooling in the build part are a result of absorbing a fast-moving optical energy input source (~1000 mm/s). Heat input into small "islands" of randomized locations within a layer allows the heat to be evenly introduced spatially and reduces the chance of deformation and failures caused by uneven warpage during a build. Changing scan directions and offsetting scan lines from layer to layer allow for uniform distribution of scanline-wise thermal and microstructural anisotropy in the entire 3D space within a build part. It allows for the overall isotropy of material properties.

In addition to rastering strategy, critical parameters that determine energy input into the powder bed to achieve controlled melting are beam power (laser or electron beam), scan speed, and focus move (if the hardware of the systems allows). Overall, these parameters determine the amount of energy incident onto the powder bed per unit time and directly relate to the heating, melting, and cooling of the material. Each factor, however, also plays a role in different characteristics of the build. High power at faster scan rate produces different thermal histories as compared with lower power at slower scan rates, even if the total power input is the same. As a result of the differences in thermal history, the microstructures and properties of parts can differ. In addition, the advective flow of molten material in the melt pool can also have different behaviors when the melt pool travels at different rates, resulting in also differences in the solidification process.

Once the slice information is generated, it is transferred into the interface program that runs on AM systems. The interface program serves as the interface between information of the build and machine controls that carries out the actual build process.

2.3 Machine Setup

Following software preparation steps in the AM process chain, machine preparation is the next step before a build can start. Machine preparation can roughly be divided into two groups of tasks: machine hardware setup and process control. Hardware setup entails cleaning of the build chamber from the previous build, loading of powder material, a routine check of all critical build settings, and process controls such as gas pressure, flow rate, and oxygen sensors. Details of how each task in this group is carried out can vary from one system to another, but overall, once the machine hardware setup is complete, the AM system is ready to accept the build files (slices generated from the previous step) and start the build.

The tasks in the process control task group allow an AM system to accept and process the build files, start the build, interrupt the build at any given time if desired or required, and prepare the machine for finished part extraction and unloading of material. This first task is usually importing and positioning of build parts in the area defined by the build plate. In this step, some capabilities of scaling and basic manipulation of build parts are usually provided to account for changes needed at this step. Once the physical locations of parts are decided upon, it is followed by a series of steps of defining the (1) build process parameters, (2) material parameters, and (3) part parameters.

The build process parameter controls machine-level parameters that are applied to the entire build. Examples of these parameters include gas injection processes, material recoater motions, and ventilation processes. These parameters define the basic-level machine operation to enable a build environment. Material parameters typically control powder dosing behaviors and chamber environment control through inert gas injection. A critical parameter such as "dose factor" determines the amount by which the dose chamber plate rises as compared with the drop in the build chamber build plate. A 100% dose factor means that the rise in the dose chamber plate is the same as the drop in the build chamber plate. Typically, higher values (150–200%) are desired at the beginning of the process to ensure full coverage of a new layer with powder material. This factor needs to have values greater than 100% because, as the powder is melted and fused into a solid layer, the volume occupied by the material in a solid layer form is smaller than that of the powder form, due to elimination of spaces between powder particles. Therefore, the surface of melted areas is lower than the rest of the powder bed. As the portion of the powder-bed surface occupied by the build part increases, the dosing factor also needs to increase accordingly. This factor is typically adjustable anytime during a build process as well to provide adjustments to the build process as needed.

Inert gases such as nitrogen or argon are typically used in AM system to control the build chamber environment and maintain low oxygen concentration. Oxygen concentration inside the build chamber is of critical importance to not only the build quality, but also to the success of a build process. Typically, the oxygen concentration is maintained below 1–2%. Upper and lower limits of concentrations are often set to allow the gas injection and ventilation system to maintain the oxygen content. Above threshold values, AM systems will shut down the build process. In reactive material (such as aluminum and titanium) powder-bed processes, oxygen content control is particularly important due to safety reasons, and the inert gas injection typically remains on even after the build process ends.

Part parameters are assigned to each and every component/part to be built. Multiple sets of part parameters can be used in the same build on different parts. These parameters are taken into account in the slicing process that takes place in the previous step of the process chain. As a result, the parameters chosen in the slicer have to correspond to the actual parameters selected on the AM system. Once part parameters are selected, the building process starts and is controlled and monitored by the AM system itself. Some feedback and in-process monitoring systems are possible. Most current systems are outfitted with basic in-process diagnostic tools such as melt pool monitoring, where a diagnostic beam coaxial to the process beam monitors the intensity of emission of thermal radiation from the melt pool and does basic analysis of melt pool size and radiation intensity spatial distribution. A quality index can be extracted from the results of monitoring to provide an indication of part quality. Another type of in-process feedback tool available to some of the current systems is related to the powder recoating process. The tool typically takes an optical image of each and every layer and uses the reflectivity information within the optical image to determine where a full coating on a completed layer is achieved. In some cases, this value is used to pause the process to prevent failure of builds due to insufficient powder coating and overheating.

2.4 Build Removal

The build time of the powder-bed process depends on a number of factors. Of them, the height of the entire build has the largest effect on the total time. It can take anywhere from minutes to days. Nevertheless, once the build completes, the laser metal powder-bed technology allows for immediate unpacking of the build chamber and retrieval of the finished part, because the process does not maintain the build platform at elevated temperatures (as opposed to laser powder bed for polymers and electron beam-based powder-bed processes). The unpacking process typically involves raising the platform in the build chamber and removing loose powder at the same time. The loose powder from one process can be reused and has to go through a series of sieving steps to remove contaminates and unwanted particulates. Figure 2.5 shows an example of the process. Once the loose powder is removed from the finished part, the build is ready for post-process. The finished parts in metal powder-bed AM at this point are welded onto the build plate via support structures. Removal of the finished part from the build plate typically involves the use of cutting tools such as band saws or wire EDM for higher fidelity and flexibility.

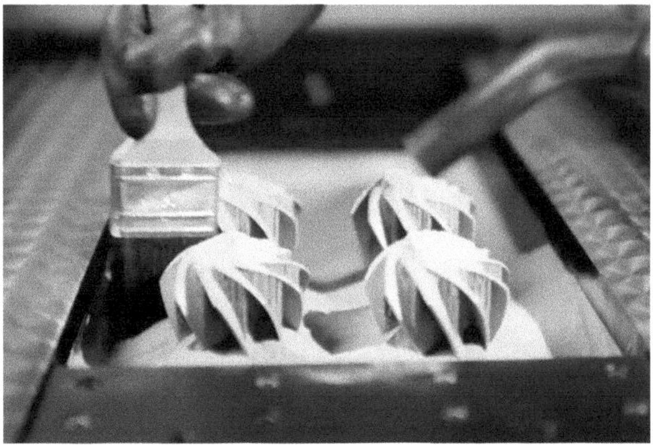

Fig. 2.5 SLM part being extracted from build chamber. Image source: topmaxtech.net

2.5 Post-processing

Depending on the AM technology used to create the part, purpose, and requirements of the finished part, the post-fabrication processes can vary in a wide range. It can require anything from no post-process to several additional steps of processing to change the surface, dimensions, and/or material properties of the built part. Shown in Fig. 2.6 is an example of the unique surface features on powder-bed AM parts where partially melted particles are bound to the surfaces of built parts. These features, along with the weld lines rastering the melt pool in different directions, result in a new type of surface finish that is distinct from any existing manufacturing processes. In metal powder-bed AM systems, the minimum required processing is removal of the built part from the build plate and the removal of support structures from the built part. Removal of support structures can be as simple as manually breaking the supports from the surface of the part, but it can also be a process that utilizes CNC tools to not only remove the support but also achieve desired surface finish and/or dimension tolerance. In addition, the metal powder-bed AM systems can introduce large amounts of thermal stresses into the built part. Under these conditions, the support structure serves as the mechanical "tie-downs" to hold the built part in place and keep it in its intended geometry. If the supports were to be removed from the part, warpage in the part will occur. A thermal annealing process can be used to relieve the thermal stresses in the part before it is removed from the build plate, to prevent part warpage upon removal from the build plate.

Hot isostatic pressing, HIP, is a process where a component is subjected to elevated temperature and isostatic pressure in a pressure vessel. At a temperature of

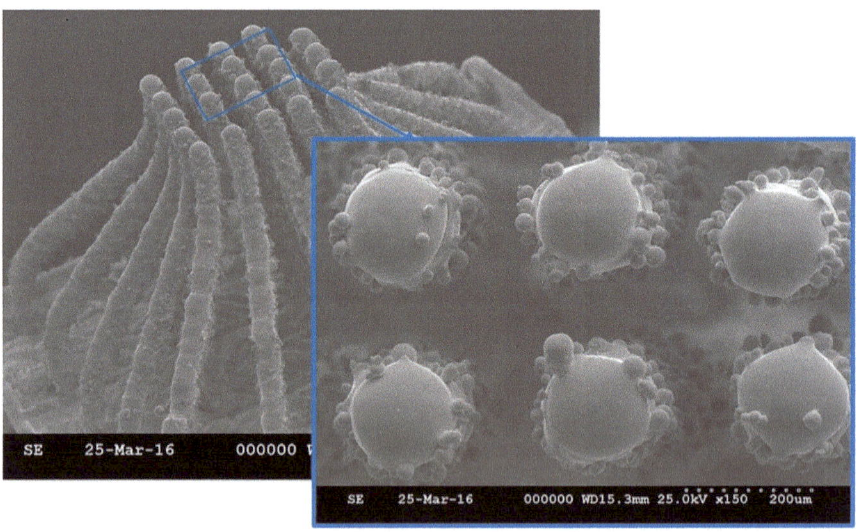

Fig. 2.6 SEM images of SLM parts showing unique surface features unlike any other current manufacturing processes

over 50% of the melting point of materials and pressures above 100 MPa (can be as high as 300 MPa), the voids and porosity inside a powder-bed metal AM part can be greatly reduced. Once processed by HP, the final bulk density can reach more than 95% of the true density of material. Under these extreme conditions of pressure and temperature, not only the material in a component being treated undergoes localized plastic deformation, but also the processes of creep and solid-state diffusion bonding are allowed to take place. This enables the required shaping and mass transport around internal defects to happen and to "heal" them, increasing the bulk density of the component.

Chapter 3
Design for AM (DfAM) and Design with AM (DwAM)

For most applications, the intended functionality or properties are determined by not only intrinsic material properties but also geometry designs. It is well known that during the product design process, it is important to incorporate sufficient consideration of factors related to the manufacturing. In general, these additional consideration factors impose certain restrictions to the designs. There exist extensive guidelines and theories for modifying and optimizing geometry designs of structures and materials, which is generally termed as design for manufacturing (DFM). In a more extended context, DFM also encompasses material selection, process design, and quality inspection as additional factors. In general, the goal of DFM practice aims to reduce the cost and manufacturing-induced defects of a product. Due to the significant impact of design in the overall cost and supply chain structure of a product, ideally DFM should be implemented as an integral concept throughout the design process. However, in practice, this is often difficult to realize due to various reasons such as the fragmented supply chain, management structure, cost restriction, and cultural barrier. Consequently, DFM is often more narrowly interpreted as geometry modification practice after the primary form of the design is generated, potentially limiting its effectiveness.

The same argument also applies to AM when it comes to product design utilizing the principles. With digitally driven process flow and the layer-wise process characteristics, AM exhibits significantly higher capability for realizing complex geometries. However, despite the popular notion, AM also imposes various restrictions to the design of geometry, material, and process, and thus the idea of design optimization for reduced cost and improved manufacturing quality is still a legitimate argument. With many applications, the existing product designs are often already optimized for traditional manufacturing processes such as machining, casting, or forging, and therefore when adopting AM as a substitutive manufacturing process, substantial design modification is likely needed. One contemporary example of such AM adoption scenario is the replacement part manufacturing in aerospace and

K. Hsu et al., *Additive Manufacturing of Metals*, Springer Series in Advanced Manufacturing, https://doi.org/10.1007/978-3-032-04961-2_3

automotive industries. With the primary form/geometry of the component fixed, the need for design typically focuses on the modification of secondary features such as side wall draft angles, corner fillets, minimum feature sizes, as well as setup of orientation with the components. Alternatively, the new designs could also deviate significantly from the original forms, which leads to more disruptive product innovations, but also possibly substantial design validation and certification works as well as disruption to the existing supply chain. One example is the GE90 jet engine fuel nozzle tip by GE Aviation, which not only has a simplified single-body integrated structure compared to the previous design, but also features a more efficient geometry with reduced material use [1]. On the other hand, the development of the new nozzle tip product involved significant engineering efforts, as well as substantial investment for the setup of entirely new facilities [2]. Regardless of the design approach, the following questions are generally addressed:

- How manufacturable is the design using certain AM processes?
- How optimal is the design for intended performance when processed with AM?
- Could AM be utilized to achieve optimized design for improved performance?

These design questions/goals also reflect different levels of complexity with the design. Manufacturability can be considered as the minimum design requirement that would ensure that AM processes and materials could be successfully adopted for the design with adequate qualities. On the other hand, the manufacturability of a design does not warrant process optimality. As with AM, the material properties are closely associated with both the geometry and the process setting, in order to further ensure that the material characteristics of fabricated structures could satisfy the performance requirements, more comprehensive analysis that often involves a suite of tools such as physics-based process simulations and empirical database. When the form of the design is not subjected to significant limitations, further design optimization could be carried out, which utilizes the unique capability of AM to generate complex geometries and specific microstructures. Such approach requires a more comprehensive knowledge with the geometry-process-material (GPM) characteristics of the AM processes and often results in radically different geometries and process strategies with the new designs. In the typical context, the design for AM (DfAM) theories and database focus on the first two design questions, whereas the last design question represents a more paradigm-shifting design approach that might be more properly defined as design with AM (DwAM). Some of the recent design concepts such as metamaterials, heterogeneous materials, and shape-morphing materials that are uniquely enabled by AM utilize the AM capabilities to conceive novel material/structure architectures and designs previously considered infeasible. Although the real-world applications of these concepts are currently scarce, they represent some of the most promising directions of AM applications. As the knowledge with AM continues to evolve, it is reasonable to anticipate that the design of AM product will also gradually evolve from being driven by AM capabilities into one that incorporates AM concept as an integral part of the performance-driven design optimization problem.

3.1 Design for AM (DfAM) Overview

In general, the DfAM framework bears significant similarity to the traditional DFM. For the typical product development workflow as illustrated in Fig. 3.1, the design requirements should be defined during the initial discovery stage, which could include various design aspects such as performance and functionality, quality, cost, supply chain, manufacturability, and other customer/user needs. It is generally desirable that more comprehensive knowledge across the entire product life cycle can be utilized for the identification of design requirements. However, with the fragmented supply chain and highly hierarchical infrastructure commonly seen in contemporary manufacturing enterprise, it is often difficult to achieve this. Also, depending on the application, the form of the product (e.g., geometry, material type) of the supply chain might be subjected to additional design restriction. As a result, the design of a product related to manufacturability might be effectively reduced to a "patching" effort, severely limiting its effectiveness.

With AM, another notable characteristic with the design is that a significant portion of the product life cycle management could be converted into digital information. Often considered as an integral piece of the digital manufacturing and Industrial 4.0 concept, AM allows for an unprecedented level of digitalization during the design stage. Various design tasks such as model creation and manipulation, performance analysis, and process/machine setting are routinely carried out in digital environment for most current AM systems. Many AM systems are also equipped with in-process monitoring capabilities via system-standard sensors/monitors and/or additional custom-configured sensors, thus potentially enabling digital process validation and qualification. In addition, there also exist extensive and coordinated efforts in the research and development of various protocols and guidelines for digitalization of AM and management of AM data [3–7].

Despite the potential, currently the utilization of digital information for the design is generally limited in scope. This is caused by various factors. Firstly, the different design steps are typically carried out in different tools/software, which is

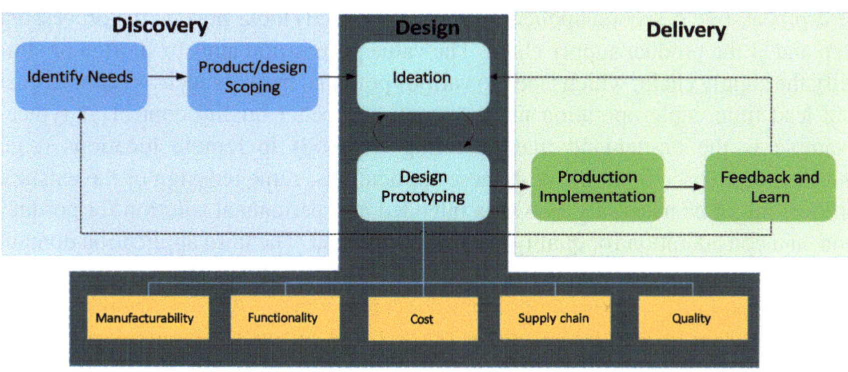

Fig. 3.1 Typical product design workflow

cumbersome to say the least. In many cases, due to the inconsistent data structures, design information might not be fully retained during the transfer of modeling information between software/tools, and the organization and utilization of certain data features often require extensive manual interference by the users. This also prevents effective design optimization based on heterogeneous data types. Secondly, for many AM processes, the high-fidelity process simulation tools are still both computationally expensive and scarce. Consider that with these processes, there usually exhibit a complex geometry-process-material (GPM) characteristic; the lacking of such tool often makes comprehensive design impractical. Thirdly, the understanding of the potentials of AM in creating new structures and metamaterials is still under intense research development and can be generally considered to be immature for practical applications. On the other hand, with the development of more advanced process hardware (system, sensor) and software (e.g., integrated AM product management), as well as the utilization of advanced data analytics tools such as machine learning, it is widely expected that AM design is rapidly becoming more informed and thus more optimizable. Ultimately, it is not unreasonable to speculate that AM could become one of the first manufacturing technologies to be deeply integrated with artificial intelligence, which combines advanced data modeling with decision optimization and thus is potentially able to largely replace the role of current designers.

Coykendall et al. categorized the AM adoption in the defense and aerospace industries into four categories depending on its disruption levels on both the supply chain and overall product concept [8], as illustrated in Fig. 3.2. Applications with lowest impact on the existing supply chain and product concept (Path I) focus on utilizing AM as a more effective or convenient replacement to the existing traditional manufacturing solutions and therefore introduce relatively small technological and business barriers for implementation. A typical example is rapid design prototyping. The use of AM could often accelerate the iterative design evaluation process by allowing for local or in-house prototyping directly from the digital model. With these applications, the DfAM consideration mainly focuses on ensuring that the intended design could be adequately represented by the AM prototype. This typically includes the minimization of geometrical inaccuracy and the ease of post-processing. A second application domain (Path II) looks beyond the processing step and at the product supply chain. The value proposition aims to shorten or simplify the supply chain, which leads to various potential benefits such as reduced cost and lead time, agile operation management, and better quality control. A typical example is the on-demand manufacturing of goods in remote locations (e.g., International Space Station). With these applications, some redesign of the existing product might be necessary as AM is intended as a permanent solution for production, and consideration for quality becomes important. The third application domain (Path III) explores more into the geometry free-form capability of AM to create new designs based on the existing product concepts. Various AM lightweight design concepts such as bicycle frame, engine bracket, and custom shoe midsole are examples of such an adoption strategy. Extensive DfAM considerations are necessary for these design innovations, as the interplay between geometry design and quality of

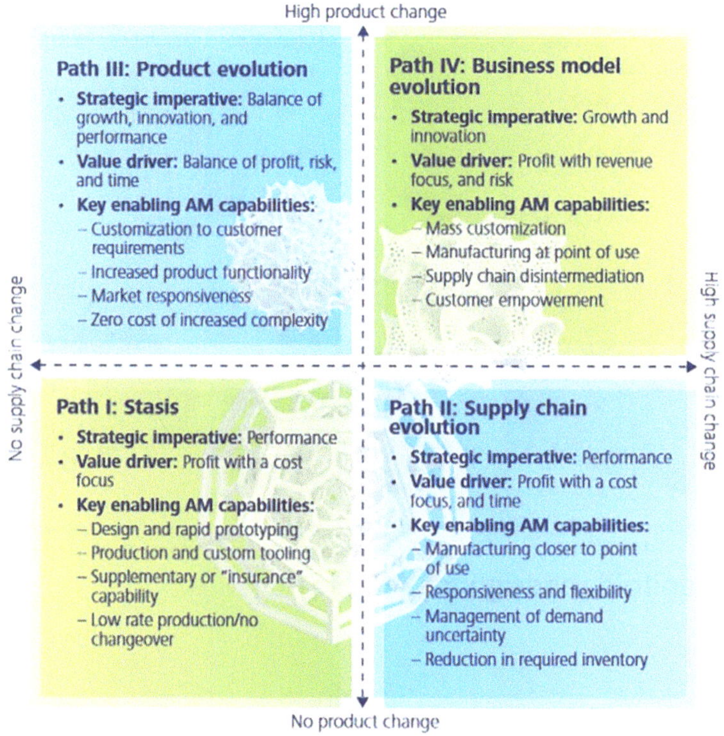

Fig. 3.2 AM adoption by supply chain and product disruptiveness level [8]

the structure becomes complex. Lastly, with the fourth application domain (Path IV), new business models that involve the redesign of not only product geometry but also the supply chain system will be created. A close example is the crowdsourced fully customized car concept proposed by Local Motors in its earlier year's AM adventure [10]. This particular concept took advantage of AM's digitalized design workflow to utilize crowdsourcing for customer-defined car frame design, which would eventually be printed via a big-area additive manufacturing (BAAM) system. DfAM and even DwAM concept is integral for the realization of such AM-centric business concept, and comprehensive knowledge of AM characteristics must be incorporated into the entire design process.

Figure 3.3 shows two design examples utilizing DfAM, which are also representative of two major types of DfAM design scenarios, including redesign for processing/post-processing (Fig. 3.3a) and new design demonstration (Fig. 3.3b). In the first scenario, the design process is often defined as multiple successive steps, with possible iterations. In the example shown in Fig. 3.3a, the original design of the manifold was optimized for machining, which results in some compromise with both material utilization and functionality. Correspondingly, the first step of redesign focuses on improving the primary functionality after removing the constraints of orthogonal flow pathway. The redesign is subsequently further refined to improve

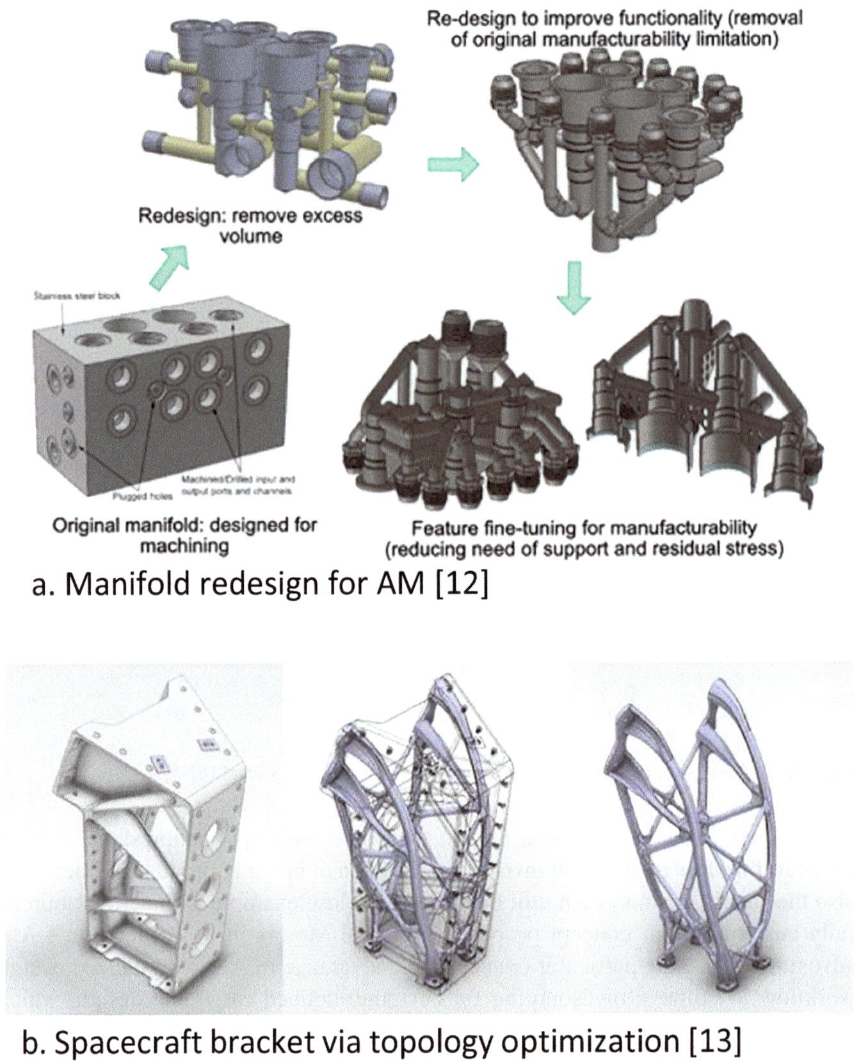

Fig. 3.3 Typical examples of DfAM applications. (**a**) Manifold redesign for AM [11]. (**b**) Spacecraft bracket via topology optimization [12]

its AM manufacturability, such as introducing features to reduce the need of support structures and the thin-wall dimension adjustment to reduce thermal residual stress. In comparison, in the second scenario, less constraints are imposed to the general form of the geometry, and instead the design is more informed by certain performance optimization algorithms or tools. In the example shown in Fig. 3.3b, the bracket geometry is generated by topology optimization based on the required performance criteria (e.g., rigidity, strength, weight) as well as prescribed boundary conditions (e.g., assembly features and their locations), and the resulting design not only is significantly different from the original form but also exhibits highly organic characteristics that are difficult if not feasible to traditional manufacturing. The generated geometry might still need to be subjected to additional scrutiny for AM manufacturability and often end up being used as design reference instead of the final design [9]. However, the utilization of the design algorithm is generally considered to be highly effective in providing more mass-efficient design optimization alternatives and thus reduces the manual design efforts. Some researchers also referred to the first approach as "restrictive approach" as it emphasizes on the design restrictions, as well as the second approach could be termed as "opportunistic approach," due to its more open and explorative nature [10].

Figure 3.4 illustrates the typical workflow of a DfAM process, once the product concept and user needs are fully determined. The first step is the identification of AM value proposition, which is the most critical step of the process. Not all product designs require the use of AM, and in many cases, AM could actually result in worse outcomes compared to the non-AM solutions. One widely known example is the fasteners, which are both highly standardized and mass produced. There exist a number of typical scenarios for AM value proposition, including the following:

- **Rapid prototyping**: Product that requires iterative design prototyping, which often results in long lead time and expensive tooling.
- **Complex geometry**: **Complex** geometries that are difficult to realize via traditional manufacturing, such as internal features, organic geometries, and conformal channels.
- **Tool-less fabrication**: **Products** that require extensive tooling efforts, which results in issues such as long lead time, high cost, geometry limitation, and rigid product design.
- **Simplified supply chain**: **Product** that has supply chain issues such as high complexity, fragility, long lead time, or high cost.
- **Design customization**: Highly customized products and designs that favor agile manufacturing systems.

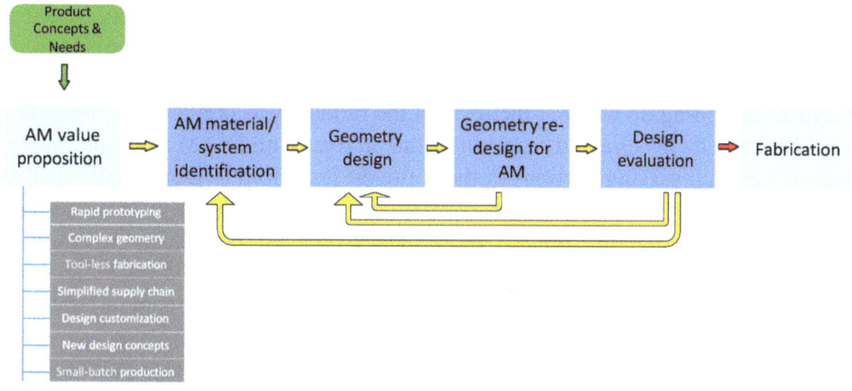

Fig. 3.4 DfAM typical design workflow

- **New design concepts**: Disruptive **products** with new material and structure designs infeasible or difficult with traditional process technologies.
- **Small-batch production**: **Products** with very small demand but that possess good values.

There exist certain overlaps between individual value proposition scenarios, and in many cases, the AM adoption is justified by multiple scenarios. It is also important to note that the justification of AM value proposition does not correspond to an "AM-only" solution. Often, a hybrid solution that utilizes both AM and traditional manufacturing technologies is more optimal.

If AM adoption is justified, detailed design process often starts from preliminary identification of AM material and process system. With traditional DFM theories, the material and process selections are often determined sequentially with the order dependent on the application cases. However, with most current AM systems, the material options are relatively limited, and correspondingly, the selections of material and process systems are closely coupled. It is worth noting that here the material option does not indicate system material capability. Many AM processes are compatible with a broad range of materials; however, typically, only a small number of options are designated by system OEMs. This is largely due to the material quality control and process reliability considerations with the "nonstandard" materials. Take powder-bed fusion (PBF) as an example; although in theory, the energy beam has sufficient energy density to introduce melting to any feedstock, the resulting integrity and quality of the materials can vary significantly. For ceramics, due to high melting point and susceptibility to thermal cracking, the PBF processing often results in part cracking or low toughness with the materials. Another example is pure Cu, which exhibits very low absorptivity to the Nd:YAG laser spectrum in solid state, therefore necessitating large processing energy density. However, once melted, liquid Cu exhibits significantly higher absorptivity, which would cause rapid overheating of the melt pool and consequently defect generation [13]. Due to the highly dynamic nature of the PBF process, the material quality is also highly sensitive to various input setting conditions such as powder feedstock characteristics, system layout configuration, and part geometry, further increasing the challenge of material quality control. Although highly variable, the cost of developing relatively robust or comprehensive process setting for a new PBF metal material can easily exceed million-dollar mark, preventing users from easily undertaking such tasks.

Next, depending on whether the form of the product was previously defined to a certain extent, the design and redesign of the product with DfAM rules and guidelines can take up varying portions of the entire design effort. Regardless of whether extensive change of form design takes place, the detailed geometry design usually follows an iterative functionality/performance-oriented design optimization—restrictive AM manufacturability design modification loop. Additional analysis tools such as thermomechanical simulation or virtual support-generation check are often employed to facilitate the design assessment in the iterative process. Also, physical experimentation is often utilized, which provides necessary validation or calibration to the design iterations.

It should be noted that for both the DFM and DfAM, the design guidelines and algorithms are intended for providing statistically robust outcomes. In another word, the guidelines provide conservative and restrictive design recommendations that might not be necessary but that ensure a certain confidence level with the manufacturability. As AM technologies are still rapidly evolving, new materials and process systems that exhibit new characteristics might not be adequately accounted for by the existing DfAM guidelines. To further complicate this argument, with most AM process, the understanding of the GPM characteristics is rather incomprehensive; thus, the DfAM guidelines might not even fully account for the potential effects of specific geometry designs. One such example is with the lightweight structure designs such as cellular structures and topology optimization structures. Due to their small feature cross-sectional dimensions, the processing characteristics of these structures are often significantly different from those of bulk geometries, resulting in different material properties and qualities. However, currently there exist very limited guidelines specifically for the material and quality design of these geometries.

Despite the limitations, current DfAM knowledge base can still be considered as a helpful tool for the designers. For those who are unfamiliar with AM, the DfAM knowledge base provides an efficient guideline for understanding and overcoming some of the most common design issues of AM, as well as some quick tips to accelerate the detailed design process. For designers with AM knowledge, DfAM can serve as useful references for identifying critical design elements. In the following sections, some specific aspects of the DfAM will be discussed in more details. Instead of focusing on the specific guidelines, the discussions intend to provide some insights into the technological backgrounds of different design aspects.

3.2 General Design Consideration for AM

Although there exist significant differences among the different AM technologies, they also share various common characteristics of relevance to the design.

Currently, all the AM processes are layer-wise. This brings about several implications: (a) The resulting characteristics of the material/structure exhibit directional heterogeneity. For example, for most AM processes, the material exhibits directional anisotropy between the in-plane directions (xy plane) and the build direction (z-direction) [14]. Also, in-process characteristics such as thermal expansion/shrinkage and heat transfer also exhibit directional anisotropy. (b) There exist staircase effects on the non-vertical exterior and interior surfaces. Generally, this results in surface textures with characteristic length similar to the layer thickness (e.g., waviness), as well as surface defects such as micro/macro-cracking between the layers. (c) The processing routing pathways (e.g., toolpath) are generated in a layer-wise manner, which brings about various complications for different processes such as nonuniform heating/cooling, differential material expansion/shrinkage, and boundary toolpath termination. The various defects corresponding to these characteristics have been some of the most challenging issues for AM process design. (d) Continuous phase along the build direction (z-direction) could only be achieved in discrete layer-by-layer manner. Continuous reinforcement phases such as fiber could be added within a layer for layered reinforcement but not across the layers.

Another profound characteristic of AM is that the geometry and the material properties are generated simultaneously with most process types except for binder jetting. This means that as the process proceeds, the process conditions, such as overall structural mass, total surface areas and volumes, structural strength, and temperature distribution, are constantly changing. A direct consequence of such characteristic is that the process generally exhibits heterogeneity even under constant process settings. The process heterogeneity manifests in both local and structural levels. At local level, the newly added/deposited material might be subjected to varying boundary conditions, which potentially result in local property variabilities of defects. At structural level, the cumulative effects from the process, such as heat accumulation, geometry distortion, and diminished substrate thermal effects, could also result in gradual change of material properties along the build direction. Table 3.1 summarizes some of such local and structural heterogeneity for the different types of AM processes. It should also be noted that in addition to these process heterogeneities, there also exist stochastic process variability, which is typically labeled as process quality variability and treated by a different approach.

Geometry dependency with material properties also creates challenges for the qualification of materials and products. For traditional manufacturing processes that also exhibit geometry-dependent material characteristics such as forging and casting, the qualification of materials and processes generally requires statistically significant numbers of test coupons with representative dimensions [15, 16]. For structures with various characteristic dimensions (e.g., thickness), the qualification can be sectioned into simpler geometries of more consistent dimensions [17]. The more variability the dimensions are within a single component, the more test coupons with representative dimensions are required. Consequently, the cost and lead time for qualification are substantial. For products with large production quantity, such approach might be more easily justifiable. However, this could be a challenge for highly customized or small-batch products. In addition, for traditional manufacturing processes, it is often assumed that the bulk of the material exhibits relatively consistent microstructure and properties. However, this assumption holds very limited validity for AM processes [18, 19]. With such argument, some of the existing material databases for AM materials are established based on standard coupons (e.g., ASTM E8) and thus should not be considered as reliable design data for all the products. Instead, these databases provide references on typical achievable material properties with the AM materials, which can potentially serve as useful benchmark for part-specific material qualification analysis.

A rather rough strategy for incorporating geometry features into material qualification of AM product is to decompose the geometry into "bulk" and potentially critical features and then characterize these features individually. Instead of simply following the classic CAD geometry classification system, the identification of the

Table 3.1 Process heterogeneity for AM

AM process type	Local process heterogeneity	Structural process heterogeneity
Material extrusion	Location within the layer (e.g., boundary vs. interior, toolpath turning point), neighboring track status (temperature, distance)	Heat accumulation, structural rigidity and strength (extruder compression and shear), cumulative geometry distortion
Vat photopolymerization	Location within the layer (e.g., boundary vs. interior, toolpath turning point if not digital projection-based), neighboring track status (distance, level of photopolymerization)	Cumulative geometry distortion, structural strength (liquid feedstock refilling/ recoating)
Powder-bed fusion	Surrounding solid/powder (heat transfer), sequence of melting/heating, powder particle size/shape variability	Heat accumulation, structural thermal mass (heat dissipation), thermal distortion accumulation, structural rigidity and strength (powder recoating)
Binder jetting	Location within the layer (boundary vs. interior) (binder permeation)	Build height (binder permeation and bleeding), structural strength, and rigidity (powder recoating)
Material jetting	Neighboring structure/support materials	Cumulative geometry distortion
Directed energy deposition	Neighboring solid distribution, location in the toolpath	Heat accumulation, cumulative geometry distortion
Sheet lamination	Presence of internal features and boundaries	Structural rigidity and strength (e.g., processing head/ sonotrode pressure)

critical features should be carried out with the consideration of the scan toolpath [20, 21]. Some of these potentially critical features are briefly discussed further below.

Surface region: For many AM processes, the contour + hatch scan toolpath is adopted, and therefore it is reasonable to treat the surfaces and near-surface regions of the parts (surface region) separately. The surface regions typically exhibit higher defect occurrence due to (a) the transition of local boundary conditions to exterior boundary and (b) the transition between contour and hatch toolpaths. In addition, the surface quality characteristics such as surface roughness are also determined by this region. Another often relevant factor is the post-printing processing for surface and geometry. With many AM products, the AM components will be subjected to additional machining or surface treatment to improve dimensional accuracy and surface quality, which removes some of the defects in the surface region or the region itself. In these scenarios, the qualification requirement of the surface region is reduced.

Corners and sharp features: Although the definition of these features is not robust, they can be described as generally sharp-angle features. Sharp-angle features represent significant feature shape and dimension variation in local areas. As illustrated in Fig. 3.5a, with the contour + hatch scan strategy, the contour scan path is offset inwards from the nominal contour in order to account for the energy beam diameter or deposition track width. This results in loss of the sharp feature tip. The sharp turning of the contour scan path results in local stalling effect and the corresponding over-deposition of heat or material. Further, the hatch line generation in the near-tip region is also highly inconsistent between layers (when the typical rotating or alternating strategy is utilized). This results in inconsistent heating/material deposition at different layers, further exacerbating the process heterogeneity in that region.

Feature-type transition: Even for features with more gradual dimensional transition, significant material property change and defect generation could potentially take place when the feature changes from one type to another. Although there is no robust theory or knowledge base for the definition of the different feature types, a conservative approach based on simple geometry primitives (triangle, rectangle, circle, ellipse, and their variations with curved edges) can often be considered. As illustrated in Fig. 3.5a, the transition from a triangle feature to a semicircle feature could correspond to change of trend (consider the zero point of the second derivative of a function as a mathematical analogy) of the hatch scan vector line length. With the typical hatch strategy, there exist certain layers periodically that such trend change would manifest, which might translate to microstructure heterogeneity or defect occurrence. The size of such transition zone is rather difficult to quantify, as it could be influenced by various process setting conditions and material properties. On the other hand, for specific products, it might be possible to utilize high-fidelity process simulation to more efficiently establish qualification strategy under this guidance.

Feature aspect ratio change: As illustrated in Fig. 3.5b, when features of similar types exhibit different aspect ratios, the rotating/alternating hatch scan patterns could exhibit significant difference between them. For low aspect ratio features, the

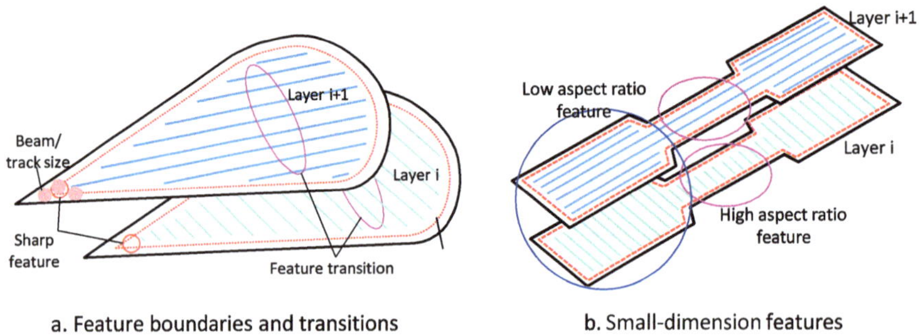

Fig. 3.5 Illustration of hatch scan toolpath in some AM processes. (**a**) Feature boundaries and transitions. (**b**) Small-dimension features

vector lengths of the hatch scan pattern tend to be less heterogeneous. In addition, the different feature aspect ratios also indicate different exterior boundary conditions. Combined, these factors likely bring about heterogeneous microstructure and defect characteristics.

Overhanging feature: While the previous feature types can be labeled as "in-layer," overhanging feature is defined as a cross-layer feature. The implications of overhanging features and the corresponding guidelines for both the geometry design and process design have been extensively studied by the AM community. The impact of overhanging feature on the need of support structure is discussed elsewhere, and for the purpose of qualification, the defects generated around the down-facing surface regions are of most interest. Due to the lack of solid substrate underneath, the downward-facing surface exhibits various defects. As illustrated in Fig. 3.6, for metal PBF, various defects such as over-dimension, rough surface texture, and extensive porosity could occur, which often makes these features most critical in the quality control of the structure. Many AM systems address this by allowing for different process settings for overhanging feature or downward-facing surface. Alternatively, post-printing processes could be considered to remove or reduce these defects.

Lightweight feature: With many advanced AM design concepts, the lightweight designs such as cellular structure and topology optimization are employed. These structures are characterized by features that exhibit very small dimensions in one or multiple directions, such as thin walls and thin struts. Correspondingly, the hatch vector scan length is short in at least one of the directions. The frequent turning of the scan trajectory results in significant deviation of the process from the "stable" conditions, which implies altered microstructure and other quality characteristics. Further, as the dimensions of these features often approach those of the theoretical process resolution limit, the significance of the stochastic process variability also becomes more pronounced. This type of quality variability is inherited directly from the process instability and is influenced by various hardware configuration and process setting parameters. For PBF, the process stochasticity is significantly influenced by the powder feedstock stochasticity and the variation of local process physical conditions (e.g., air curtain flow turbulence, output energy fluctuation). The formation and evolution of the local melt pool are influenced by various highly

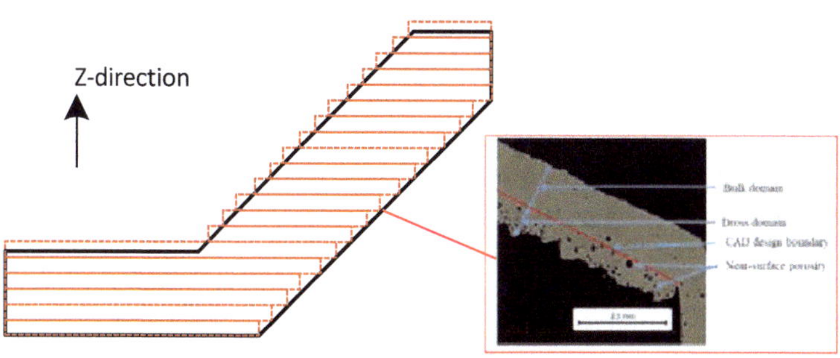

Fig. 3.6 Downward-facing surface from overhanging feature

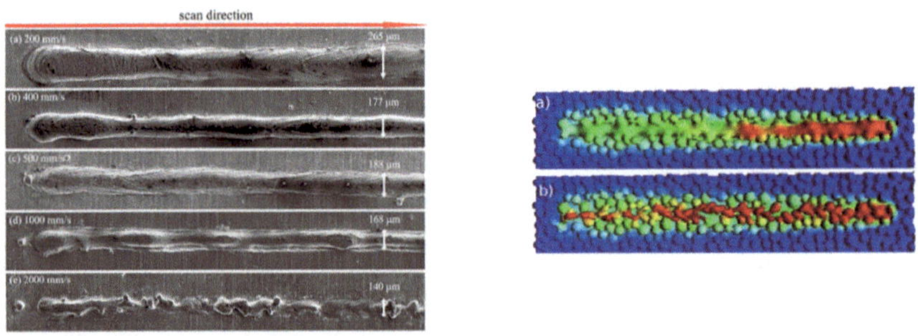

a. Experimental study (laser PBF, 304 stainless) [30]

b. Simulated single track (electron beam PBF, Ti6Al4V) [31]

Fig. 3.7 Single track illustration of process stochasticity of PBF. (**a**) Experimental study (laser PBF, 304 stainless) [29]. (**b**) Simulated single track (electron beam PBF, Ti6Al4V) [30]

dynamic process phenomena such as evaporation, Marangoni convection, surface convective heat transfer, and spattering and exhibit high sensitivity to the material phase status, substrate condition, local airflow, surrounding solids/powder, and scan history [22–27]. As a result, various local defects such as lack of fusion, denudation, balling, inclusion, and keyholes could take place, as can be clearly observed with experimental and simulation studies with single tracks (Fig. 3.7). For lightweight features such as thin walls and thin struts, such process and material variability will be accumulated over layers, resulting in significantly different overall material properties and qualities compared to bulk features [28]. Some representative observations are illustrated in Fig. 3.8, which shows the various quality characteristics such as overall internal porosity level, porosity distribution, yield strength, elastic modulus, and internal thermal residual stress. With lightweight features, the material properties and quality often exhibit clear trends as a function of their characteristic dimensions. This provides a potential direction for the qualification of these structures, which involves the inclusion of feature dimension as an explicit input variable.

It should be noted that although much of the discussions above utilized PBF as exemplary process, the general design analysis approach can be similarly applied to most other AM processes, even though the specific technical details might differ. For example, for material extrusion (ME), similar contour + hatch deposition strategy is also employed, and therefore there also exist similar quality design challenges caused by the inhomogeneous scan pathways, such as end-of-vector turning, hatch pattern inconsistency across different layers, and transition from surface to interior. For another example, for binder jetting, although the concept of toolpath is not applicable, there still exists some process heterogeneity between the exterior boundary regions and the interiors due to the different binder liquid permeation boundary conditions (e.g., whether one side of the porous media is saturated with liquid phase). Although the binder phase will ideally be completely removed from the final material phase, its permeation characteristics during the green part printing could still impact both the geometrical and mechanical qualities of the final part

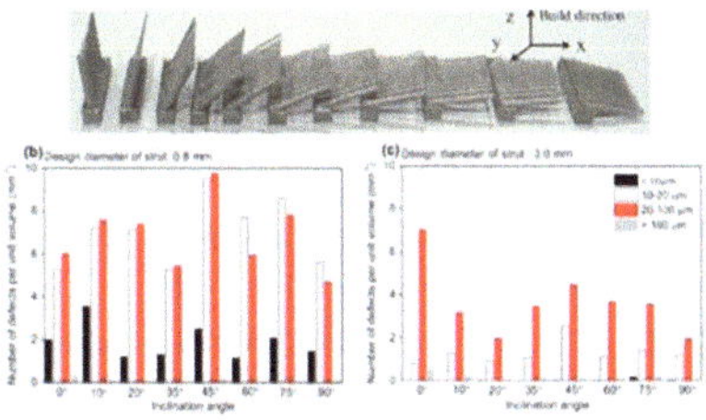

a. Diameter-dependent internal porosity for Ti6Al4V struts [32]

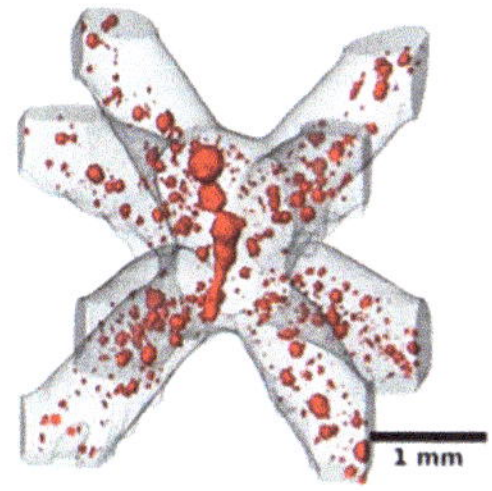

b. Internal defect distribution within a AlSi10Mg cellular structure joint [33]

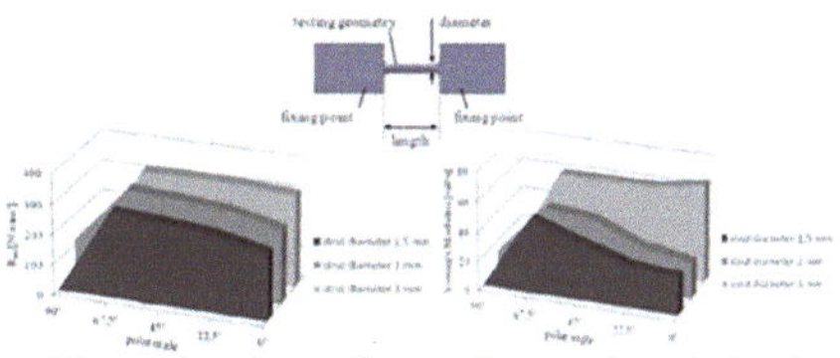

c. Dimension-dependency of strength and elastic modulus of AlSi12 thin struts [34]

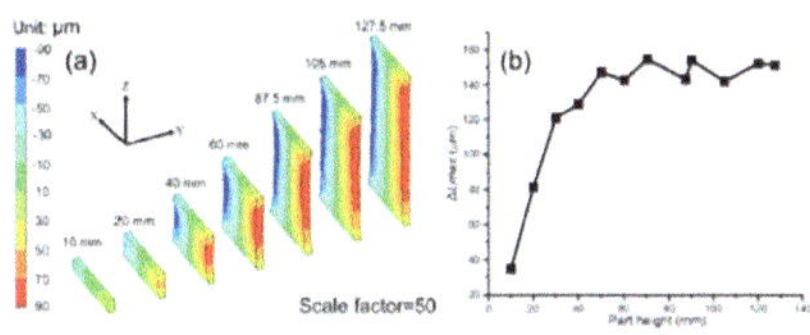

d. Geometry-dependency of thermal distortion for Ti6Al4V thin wall [35]

Fig. 3.8 Unique dimension dependency of quality characteristics for PBF lightweight features. (**a**) Diameter-dependent internal porosity for Ti6Al4V struts [31]. (**b**) Internal defect distribution within an AlSi10Mg cellular structure joint [32]. (**c**) Dimension dependency of strength and elastic modulus of AlSi12 thin struts [33]. (**d**) Geometry dependency of thermal distortion for Ti6Al4V thin wall [34]

[35]. In addition, for sharp corner features, binder jetting printing will also lose some geometrical definition at the tip due to the shape and size of the printing voxel (primitive from binder droplet permeation) [36].

Another process aspect closely related to the processing of lightweight features is the process discretization resolution. While the implication of discretized layer slicing is more acknowledged by the designers, the in-layer discretization is often much less understood. For different AM processes, the in-layer discretization manifests in different ways. The concept of minimum voxel could be employed to help understanding of the discretization of AM process. For melting-based processes such as PBF and DED, the in-layer process voxel can be effectively defined by the melt pool. The voxel continuously moves during the processes to create "trajectories" of solids and ultimately forms the in-layer geometries. A similar approach can be employed for material extrusion, although in this case, the material voxel is generated by the printhead nozzle extrusion instead of melting. With material extrusion, the voxel can be defined as the minimally extrudable volume, which is in turn determined by both the nozzle diameter and the step size of the filament feeding motor system. In comparison, for printhead and digital projection-based processes, the in-layer discretization can be defined by the minimum material primitives generated by the voxelized process control. For example, for the binder jetting process, the voxel can be defined as the size of the droplet-powder bed interaction zone, which is in turn determined by several factors including binder droplet size and velocity,

Table 3.2 Discretization of AM processes

Process	In-layer discretization "voxel"	Layer-wise discretization
Material extrusion	Nozzle head diameter and extrusion motor step size	Deposition track (by height)
Vat photopolymerization	Minimum UV light spot/voxel size	Layer thickness, curing penetration depth
Powder-bed fusion	Melt pool size (influenced by powder feedstock and energy beam size)	Layer thickness (restricted by powder feedstock size), melt pool depth
Directed energy deposition	Melt pool size	Layer thickness, melt pool depth
Binder jetting	Size of the primitive binder droplet permeation primitive in the powder bed (influenced by printhead nozzle size and powder feedstock characteristics)	Layer thickness (restricted by power feedstock size), droplet permeation primitive height
Material jetting	Deposited droplet size (influenced by the printhead nozzle size)	Droplet deposited height
Sheet lamination	Sonotrode contact area (during material addition); machining motor resolution (during machining)	Sheet thickness

powder-bed porosity, and binder-powder wettability binder chemistry. For material extrusion, as the new material volumes are added through droplets, the droplet sizes can be effectively considered as the discretization voxel. For sheet lamination, the interpretation of in-plane voxel is different from the other AM process types, since the geometry generation is accomplished by a cutting/machining operation. Consequently, the minimum feature sizes in the x-y plane are mostly defined by the minimum achievable size of the material removal mechanism (e.g., CNC cutter). Table 3.2 summarizes the definition of process discretization for the major AM process types.

The knowledge about process discretization resolution is useful when designing for lightweight features in both positive space (e.g., thin wall, extrusion) and negative space (e.g., slot, hole). In addition, even when the process discretization resolution is significantly smaller than the dimensions of the geometrical features, it is still a relevant design factor of consideration for the material microstructure control. One of the most intriguing design scenarios of AM is the design of microstructure, which has been preliminary demonstrated in some works [37, 38]. One example is 316 stainless steel processed by L-PBF, as illustrated in Fig. 3.9 [37]. In addition to refined columnar grain, the sub-grain cellular structure and nanoparticle intercellular precipitate contribute significantly to the enhanced strength and ductility of the L-PBF 316 stainless steel. Another example illustrated in Fig. 3.10 with the EB-PBF of IN718 implemented a custom point offset-type scan strategy, with the beam sequentially scan patterns of locations with fixed offset for each subsequent pattern. With the altered scan strategy, the process demonstrated the altering of both microstructural grain morphology (columnar ↔ equiaxed) and crystallographic texture via energy density input change [38]. Although in many cases the evolution of the

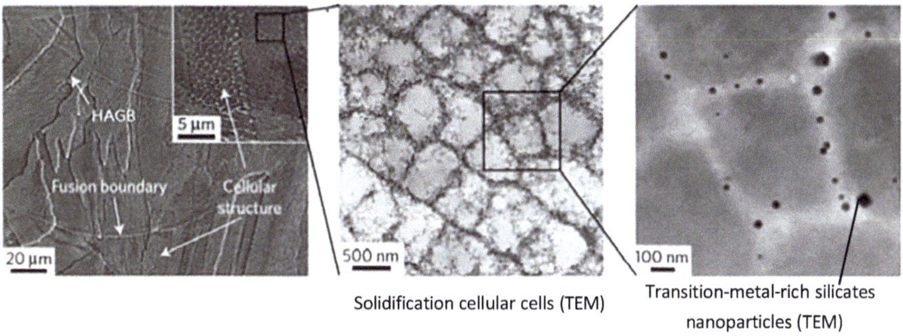

Fig. 3.9 Hierarchical microstructural morphology from L-PBF of 316 stainless steel using typical process setting, adopted from [37]

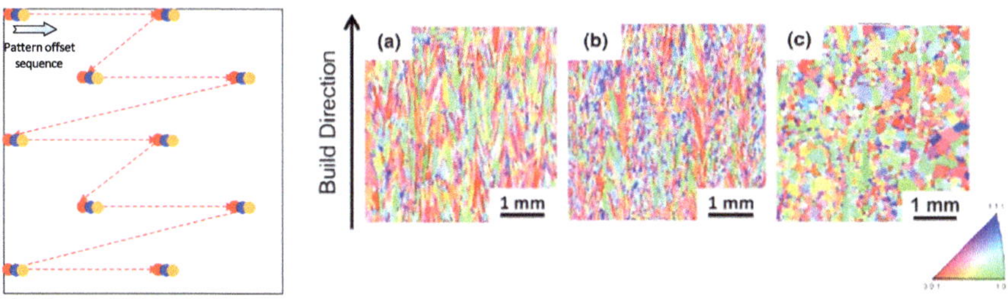

Fig. 3.10 Control of microstructural grain texture of IN718 via scan strategy and beam energy using EB-PBF (*a*—120 J/mm^3, *b*—225 J/mm^3, *c*—300 J/mm^3); on the left, the same-color points indicate locations scanned in sequence, and then the next sequence of location pattern is determined by offsetting the current one by certain distance [38]

microstructure is spatially continuous, the process discretization resolution is still indicative of threshold resolution level of microstructure control.

Last but not least, the support structure is also a universal issue with the AM design. For many AM process types, the design consideration for support is integral to the design methodology. In general, support structure is necessary for many geometry types while also introducing additional design constraints due to the challenges for post-processing.

Support structures are needed for various reasons. For vat photopolymerization, supports are needed due to the existence of the buoyancy force and the shrinkage-induced distortion effect during the photopolymerization processes. For material extrusion processes, both thermal residual stresses and gravity necessitate the need for support structures. For material jetting process, support structures are needed to both enable track deposition and counter the gravity effect. For PBF and directed energy deposition processes, the main purpose of the support structure is to counter the thermal residual stresses generated during the melting-solidification processes. For metal PBF processes, especially the laser PBF, the large temperature gradients generated during the fabrication process usually necessitate extensive support structures. In comparison, for binder jetting and sheet lamination processes, the material

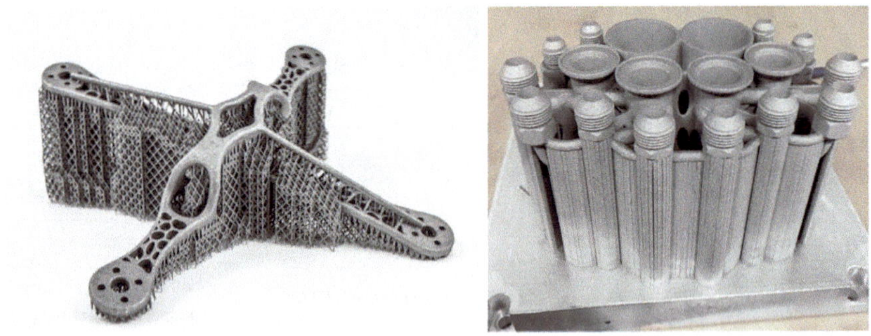

a. Drone frame wih support [40] b. Hydraulic manifold with support [41]

Fig. 3.11 Examples of metal AM parts with support structure designs. (**a**) Drone frame with support [39]. (**b**) Hydraulic manifold with support [40]

feedstock usually provides adequate support and therefore can be considered support free.

Although important for most AM processes, support structures are also generally undesirable for post-processing purposes, as they eventually need to be removed. The removal of support structures from the parts is often considered to be the most constraining factors for the design. For structure designs such as the examples illustrated in Fig. 3.11, the geometry complexity of the part structures often translates directly to the complexity of the support structures and consequently increased difficulty for support removal. As support removal must be carried out by using non-AM means, the process effectively becomes a traditional manufacturing design issue that would impose various manufacturability design constraints.

Some of the manufacturability constraints with the support structure removal are more straightforward for consideration. For example, if wire EDM is employed for support removal, then the support-part interface should be designed to allow for 2D planar cutting movement, as illustrated in Fig. 3.12 with a random octopus bracket part (from Thingiverse). Undercut and internal support structures might be difficult for a regular three-axis CNC machining operation, as illustrated in Fig. 3.13. Further, custom work holding solutions might become necessary, which might require additional designs of fixtures, as illustrated in Fig. 3.14.

On the other hand, some of the support removal and post-printing treatment processes require more complicated design considerations. For example, for the design of bonding between the support and the functional materials, local process modification and geometry design might be needed. Due to the potential surface damage from support removal, additional surface finishing treatments might become needed. This is particularly the case for processes that generate large amount of in-process stress, such as laser PBF. Due to the need of strong support strength, the post-printing support removal process is often plagued with surface damage that becomes defects that require additional attention. Also, the introduction of support structures alters the distribution of the internal process-induced stress, necessitating iterative design approach in order to achieve the desired geometry quality level.

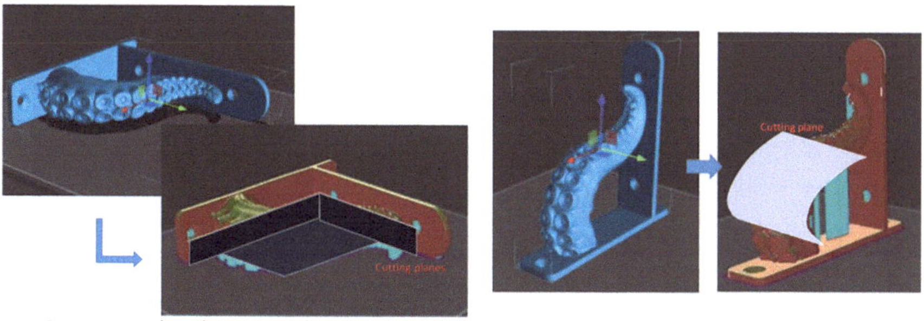

a. Support needs to be removed via half-planes (infeasible for wire-EDM)

b. Support can be removed via planar cutting (feasible for wire-EDM)

Fig. 3.12 Support cutoff using wire EDM: planar cutting movement requirement. (**a**) Support needs to be removed via half-planes (infeasible for wire EDM). (**b**) Support can be removed via planar cutting (feasible for wire EDM)

Fig. 3.13 An aerospace tubular part with internal support features that are difficult for three-axis CNC machining [41]

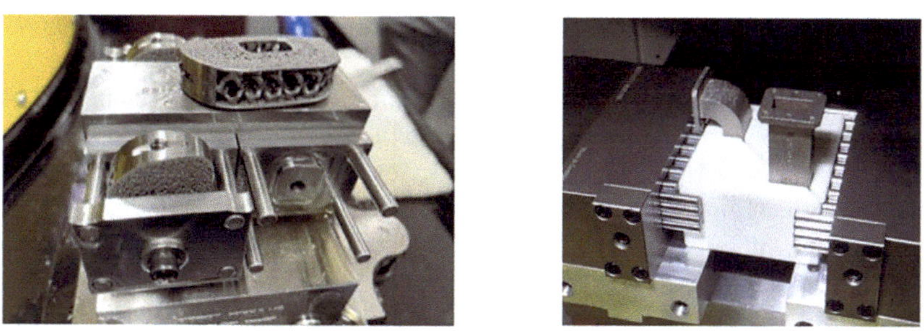

a. Custom part holder for metal implant [43]

b. Custom 3D printed part holder for a satellite waveguide [44]

Fig. 3.14 Custom workpiece holding for custom AM components. (**a**) Custom part holder for metal implant [42]. (**b**) Custom 3D printed part holder for a satellite waveguide [43]

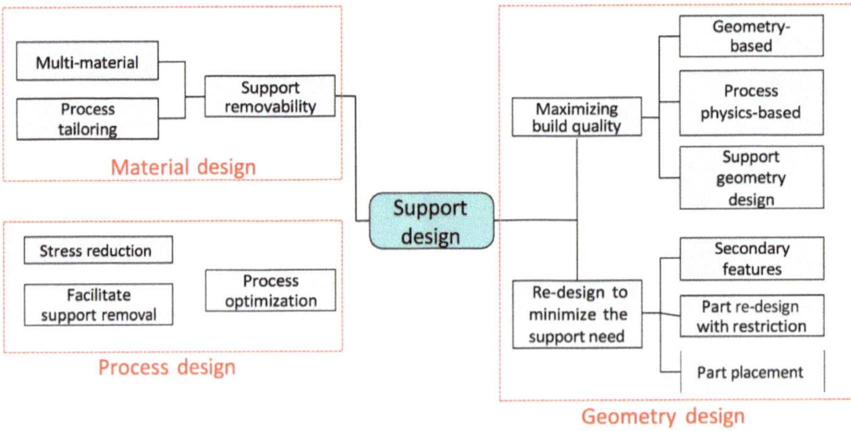

Fig. 3.15 General support design approaches

The design challenges associated with the support structures have prompted extensive efforts that follow different approaches, which can be categorized into geometry design, material design, and process design, as well as the combined use of them. Figure 3.15 illustrates a summarization of the different design approaches for AM support structures.

Geometry design: This approach generally focuses on the geometry aspects for addressing the support issues and can be further categorized into two types of approaches based on the design objective, including maximizing the build quality and minimizing the need of support structures. With maximizing build quality, the design attempts to introduce limited geometry change with the functional structures and instead focuses on the design of the support structure for ensuring best build quality. In contrast, support minimization approach focuses more on the design modification with the functional structures in order to minimize the need of supports. While this approach might require certain design trade-off with either functionality or other economic aspects (e.g., build time and costs), it might be preferable when the downstream post-printing support removal is considered challenging.

The design of support structures can be further categorized into geometry-based and process physics-based approaches.

Geometry-based support generation methods are commonly employed to determine the geometrical characteristics where the support structures are needed, as illustrated in Fig. 3.16. For many AM processes, a general guideline is that features with overhanging angle of over 45° require support structures (surfaces C and D in Fig. 3.16a), although this condition can be relaxed under certain circumstances. For example, when it is acceptable to have rougher surface finish for the down-facing surfaces, overhanging angle of up to 70° is still realizable if the process parameters are set up properly [44]. In addition, features that are unsupported (cup handle in Fig. 3.16b) and overhanging features with very large projected areas (Fig. 3.16c) generally require support structures. The algorithms for the support

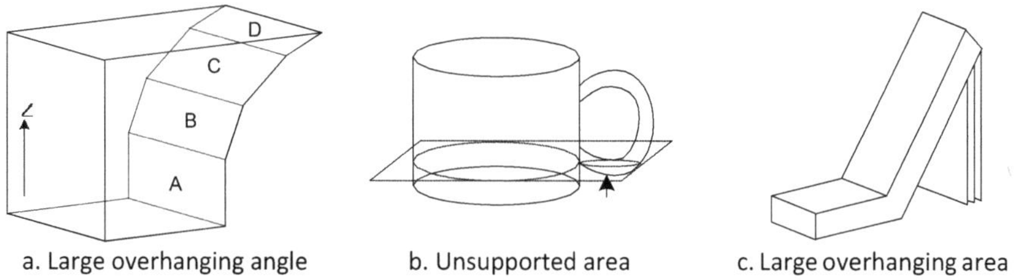

Fig. 3.16 Typical geometries that require support structures. (**a**) Large overhanging angle. (**b**) Unsupported area. (**c**) Large overhanging area

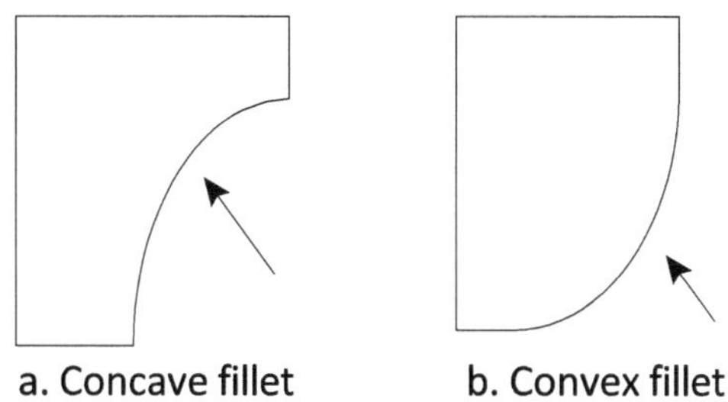

Fig. 3.17 Overhanging features with transition that may be self-supportive. (**a**) Concave fillet. (**b**) Convex fillet

structure generation calculate the unsupported projected areas of each layer and determine the need for support structures based on various conditions such as the total areas, area change over the layers, and interference with the parts. For some special geometries, such as the concave and convex fillet features shown in Fig. 3.17, even though the features possess large overhanging angle and projected overhanging areas, due to the gradual transition of features, the overhanging can be adequately self-supported with proper process designs [45].

There exists an obvious limitation to the geometry-based support generation method in that it does not take into consideration the mechanisms that necessitate the need of support structures, such as process pressures and internal residual stresses. As a result, the use of such support generation approach often relies heavily on the user's experience levels with the processes, materials, and designs, which can be highly empirical.

Process physics-based support generation methods can potentially overcome this issue. In the past several years, thermomechanical modeling has rapidly become a practical tool for process design with PBF [46]. With the emphasis on the fidelity of thermal residual stress and thermal distortion calculation, these software and tools often employ simplified melt pool models that ignore dynamic phenomena such as melt pool convection, evaporation, spattering, and melt pool-powder bed interaction, as well as the detailed microstructural evolution during solidification. The simplification modeling treatment significantly reduces the computational cost of the problem, which is considered critical for the practicality of the simulation tools. The thermal evolution of the PBF process is significantly influenced by both the part geometry and the process scan strategy, and therefore a full-scale model

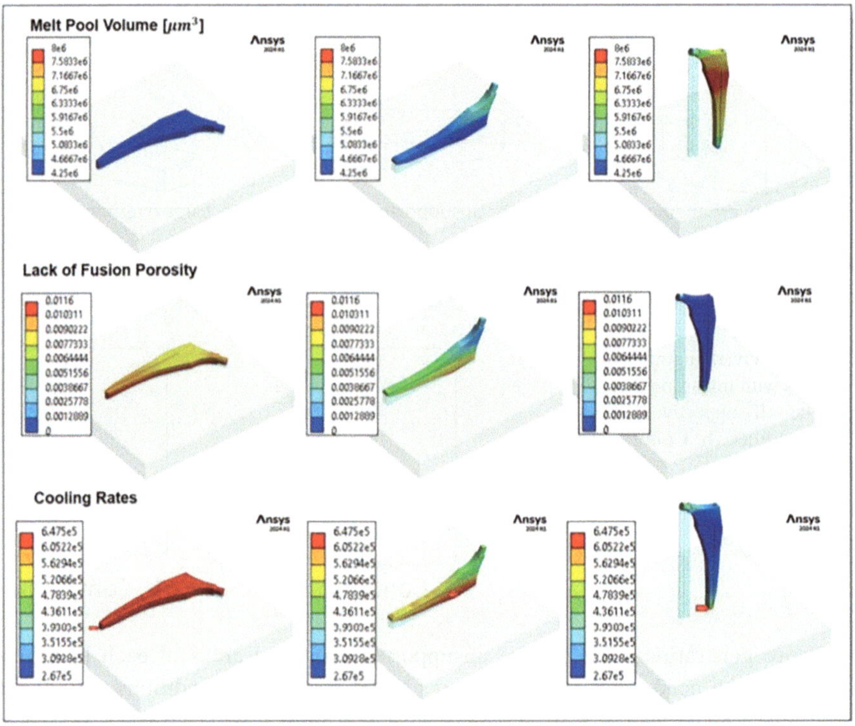

Fig. 3.18 An example of physics-based modeling for support generation and part quality analysis (ANSYS) [47]

must be utilized. Without proper simplification, it could take excessively large amount of time for the simulation, rendering it ineffective for real-world process development. On the other hand, due to its computational treatment approach, the thermomechanical simulation tools typically do not provide accurate information of the other important material/part characteristics, such as internal porosity defect occurrence, surface topography, and microstructure. When used in combination with the other physics-based modeling tools, the thermomechanical thermal residual stress simulation tools could be employed to evaluate different support generation options for trade-off analysis between different criteria, before a design and process decision is finally made, as illustrated in Fig. 3.18 [47].

Support geometry design: There also exist additional design considerations with the detailed geometry design of the support structures. Considering the need of balance between support strength and post-printing support removal, the support structures are often designed as lightweight features such as thin walls and thin struts. Figure 3.19 illustrates several types of support structure geometry design options provided by commercial support design software. These types of support structure designs are commonly employed for the preparation of metal PBF processes. For higher support stress levels, denser support geometries, i.e., higher

Fig. 3.19 Typical support geometries for metal powder-bed fusion AM [48]. (**a**) Block. (**b**) Point. (**c**) Web. (**d**) Contour. (**e**) Line

Fig. 3.20 Block support structure for an electron beam PBF. Left: default support; right: variable-density support [49]

numbers of walls per unit support area or larger wall thickness, are considered. In addition, some of the support "unit cell" geometries (e.g., Fig. 3.19b and c) exhibit different directional symmetry and therefore might be more convenient for certain support areas.

Figure 3.20 illustrates an example of the support wall density design based on the local residual stress levels for a part made by metal PBF. Block support geometry was employed, and in the regions where the thermal residual stress with the original part (without any support) is high, higher-density block support is used. With the support designs, two competing design factors are considered: (1) As the support structures ultimately need to be removed, it is desirable that thinner walls and/or fewer wall counts are designed. (2) There exist manufacturability limitations for the thin walls depending on the process types and individual systems, such as the minimum achievable wall thickness, minimum wall spacing, and property/quality variability with the thin walls. Additionally, due to the rigidity of the support structures, the internal residual stress distribution of the functional structures could be altered after support structures are introduced, necessitating iterative design process approaches. Figure 3.21 illustrates one example of the iterative support design, which employs fractal subdivision to increase the local support structure density and to reduce the size of the high-strain areas of the functional parts [48].

Although usually not exploited by users, additional design variables are available for the thin wall-type support structures. As shown in Fig. 3.22, for individual support bars, teeth structures are usually introduced for the contact/connection between the support and the part. Various design variables such as tooth height, tooth spacing, offset, base interval, base length, and top length can be considered. Top length

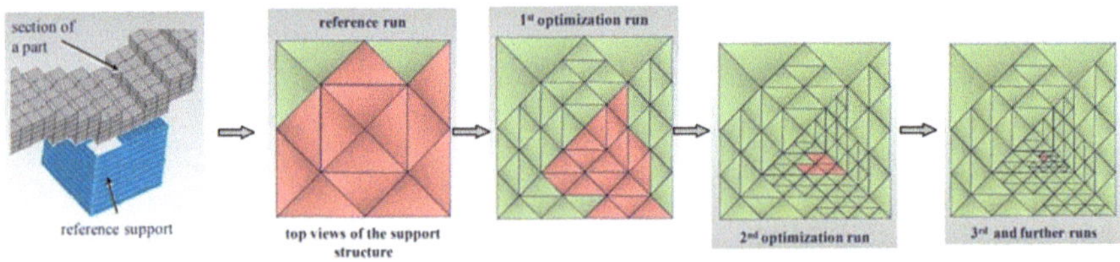

Fig. 3.21 Iterative support structure design optimization using a fractal subdivision method to selectively increase support structure density, resulting in increasingly diminished high-strain sections (in red) [48]

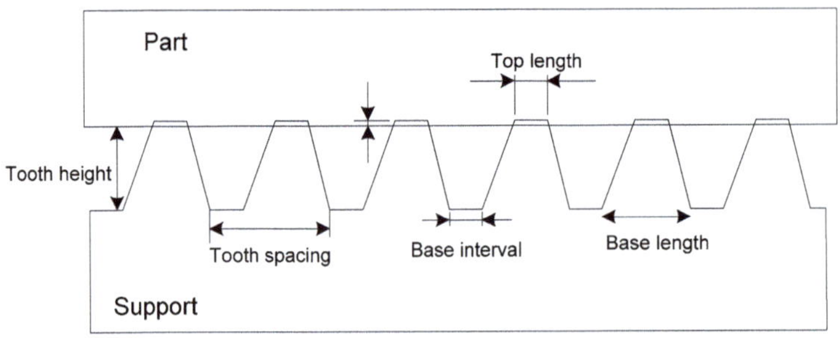

Fig. 3.22 Support-part contact design

was shown to be a significant variable, and larger top length generally corresponds to stronger support effects [50]. Larger offset could also help improve the strength of the support attachment, although this effect diminishes with larger top length. Often, since support structures and parts are fabricated with different process parameters, the overlapped regions resulted from the offset setting are more prone to microstructural defects, potentially facilitating support removal. Additionally, smaller tooth spacing tends to strengthen the support effect since it corresponds to increased number of teeth.

In addition to the geometry design aspects, with the support structures, the process aspects also need to be custom-designed. For example, typical laser PBF utilizes contour + hatch scan strategy for in-layer scanning toolpath generation, which would effectively result in minimum feature thickness of at least 2x the beam diameter when default process beam offset is in place. Instead, in order to generate minimum wall thickness, single-pass scanning would be needed, as illustrated in Fig. 3.23. Moreover, with typical PBF process settings, the scan pathways also exhibit changing patterns between successive layers, e.g., alternating or rotation. As illustrated in Fig. 3.24a, for some layers, the hatch scan paths extend through the length of the thin walls, while for other layers, the hatch scan paths might be reduced to very short vectors or even become completely absent. Consequently, for these features, both the geometrical and mechanical characteristics might exhibit unintended variability across the z-direction. For more consistent wall thickness control,

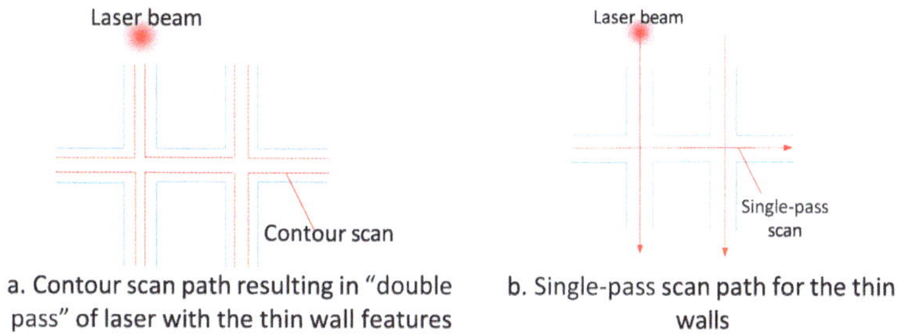

Fig. 3.23 Scan strategy for thin wall-type support structures. (**a**) Contour scan path resulting in "double pass" of laser with the thin-wall features. (**b**) Single-pass scan path for the thin walls

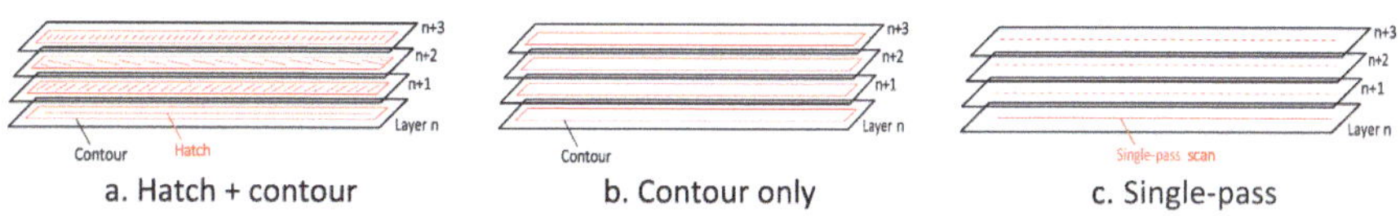

Fig. 3.24 Different multilayer scan strategies for L-PBF. (**a**) Hatch + contour. (**b**) Contour only. (**c**). Single pass

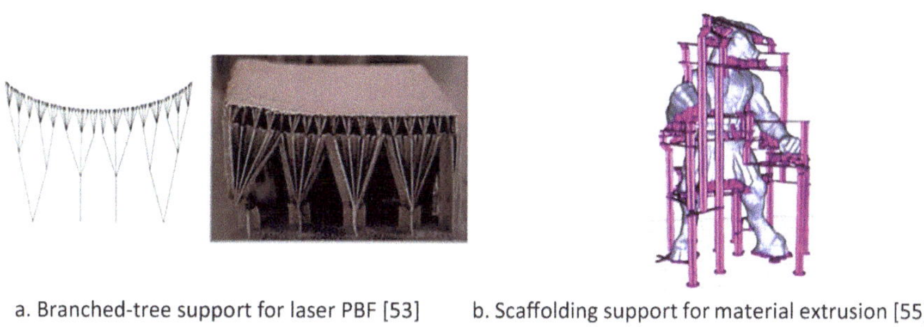

Fig. 3.25 Some sparse truss-based support structure designs. (**a**) Branched-tree support for laser PBF [51]. (**b**) Scaffolding support for material extrusion [53]

either a contour-only scan or a single-pass scan might be needed, as illustrated in Fig. 3.24b and c.

Beside thin walls, other types of support structures such as branched trees, scaffolding, and cellular structures have also been proposed and investigated. A branched tree-type support is illustrated in Fig. 3.25a. Generally, small tree branches that are in direct contact with the overhanging part surfaces are first generated based on the overhanging geometries, which are then consolidated to increasingly larger and sparser branches at lower z-heights. The tree branch support structures could also be optimized based on the residual stress analysis with the functional parts [51, 52], although such algorithms have not yet been broadly implemented in commercially available tree branch support generation tools. Figure 3.25b illustrates a scaffolding

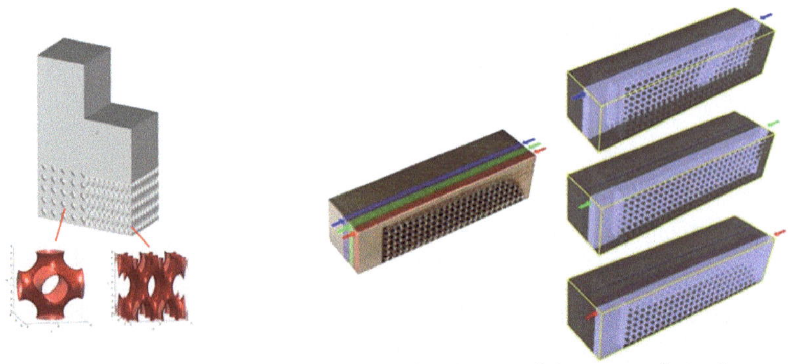

a. Cellular support using Schwartz-P with different relative densities [56]

b. Variable-density cellular support design for a cantilever structure [57]

Fig. 3.26 Cellular structure-based support structure designs. (**a**) Cellular support using Schwartz-P with different relative densities [54]. (**b**) Variable-density cellular support design for a cantilever structure [55]

support design, which also employs similar design approach as the tree branch supports by utilizing thin truss structures in order to facilitate support volume reduction and subsequent ease of support removal.

Cellular structure-based support design approach also gained substantial interests in recent years, due to the designability and broad range of achievable properties of cellular structures. For cellular structures, the mechanical and physical properties can be readily designed at broad ranges via adjusting the relative density with the designs. In addition, with different cellular topology designs, different elastic modulus and strength combinations can be further tailored. Figure 3.26a illustrates a support design example utilizing the Schwartz-P triply periodic minimal surface (TPMS) cellular structure of two different relative densities (via adjustment of lateral periodicity) [54]. Figure 3.26b illustrates a variable-density cellular support structure design based on thermomechanical stress analysis with a cantilever part [55]. With the design of cellular supports, the topology-relative density-property relationships for cellular unit cell designs are commonly established via asymptotic homogenization method, which effectively treats the cellular unit cells as voxels of continuum material with equivalent properties. This allows for the application of various well-established computational mechanics tools for efficient design of support structures, while at the same time subjects the designs to potential inaccuracies due to the deviation of the boundary conditions from asymptotic assumptions. When cellular support structures are designed, the support structures are typically designed as solid designs alongside the intended functional parts, which means that they are effectively treated as "parts" in the build setup. There exist various cellular structure design software and tools, and additional details about the design of cellular structures will be discussed later in this chapter.

There also exist some unconventional support design concepts that have been demonstrated in literatures. Figure 3.27a illustrates a contact-free support structure

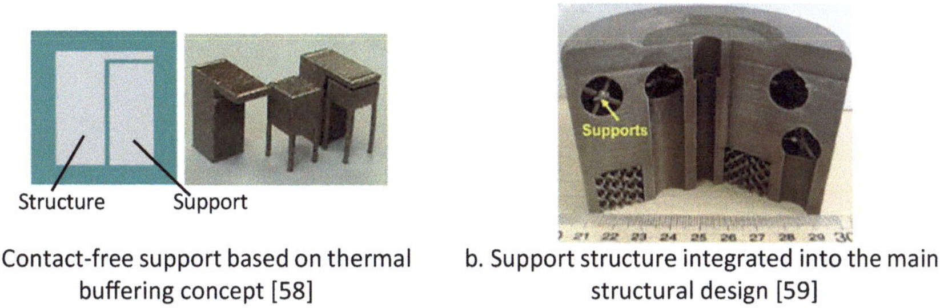

Fig. 3.27 Some nonconventional support design concepts from literatures. (**a**) Contact-free support based on thermal buffering concept [56]. (**b**) Support structure integrated into the main structural design [57]

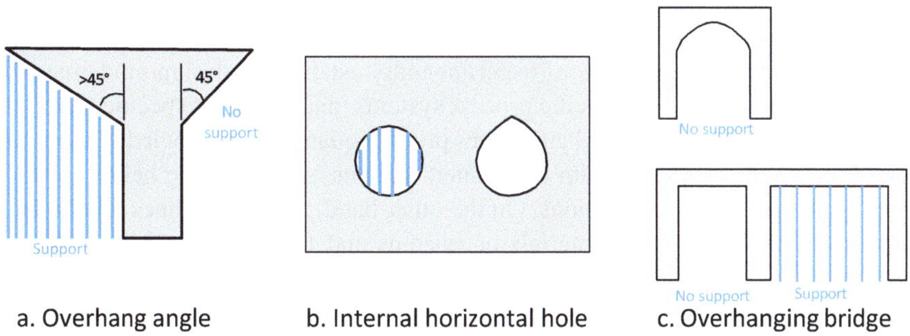

Fig. 3.28 Some generic guidelines for minimizing support structures. (**a**) Overhang angle. (**b**) Internal horizontal hole. (**c**) Overhanging bridge

design concept employed in EB-PBF process [56]. The support structures were designed with a small spacing from the overhang feature surfaces in order to provide sufficient thermal conditions to minimize the thermal residual stress on the functional structures. While still subjected to limitations such as geometry complexity, overhang dimensions, and process conditions, it still has the potential to reduce support-part contact while providing supporting effects. Figure 3.27b demonstrates another approach that integrates support structures into the functional part design [57]. By optimizing the lattice support geometries, both the geometrical quality and the fluid flow capabilities of the internal cooling channels of the mold satisfy the functional requirements.

Part redesign for support reduction: Instead of designing for the support structures, the functional parts could be redesigned or optimized to reduce the need of support structures. There exist extensive design guidelines for part feature redesigns that aim at this purpose. For example, the filleted feature transition illustrated in Fig. 3.17 is often recommended for cornered overhanging features for reducing support needs. Figure 3.28 illustrates some additional general guidelines for feature redesign that generally reduces support needs. Most of these guidelines are empirical and usually do not aim to provide optimized solutions. Instead, they provide

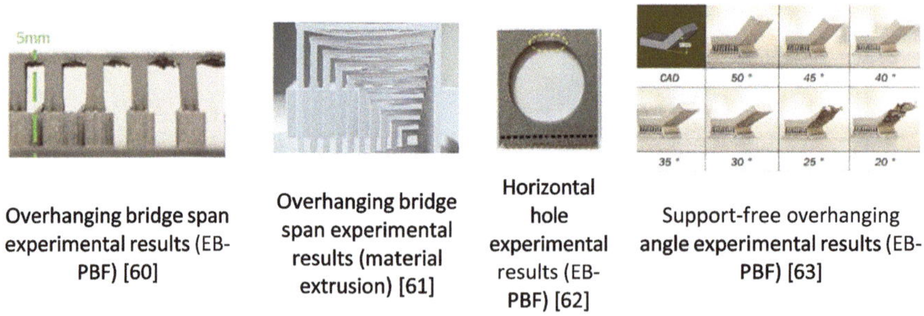

Fig. 3.29 Some experimentally established support-free design guidelines. (**a**) Overhanging bridge span experimental results (EB-PBF) [58]. (**b**) Overhanging bridge span experimental results (material extrusion) [59]. (**c**) Horizontal hole experimental results (EB-PBF) [60]. (**d**) Support-free overhanging angle experimental results (EB-PBF) [61]

reasonable preliminary design ideas that could be further improved. Figure 3.29 further illustrates some results of experimentally established design guidelines for support-free structures for specific process systems (and likely for specific materials as well). Often, these empirical guidelines provide quantitative knowledge about the loss of geometrical quality with the printed features, which can be helpful toward making design trade-off decisions. On the other hand, these guidelines tend to lack transferability to different materials or systems and therefore should be adopted only as references. In this case, additional experimental benchmark studies could be carried out to obtain more accurate design information for specific system/material combinations.

Another commonly employed strategy for reducing support structure is part orientation adjustment. Based on the geometries of the fabricated structures, it is often possible to reduce the amount of support structures required via the adjustment of part orientation. However, the orientations that result in minimum amount of support also often result in very long fabrication times, since under these cases, the largest dimensions of the parts are more or less aligned to the build direction, such as the example shown in Fig. 3.30. Also, as the removal of support might damage the surface, critical surfaces with requirement of surface integrity should avoid support structures. As a result, for complex geometries such as illustrated in Fig. 3.30, the optimal part orientations for support removal are often neither horizontal nor vertical, but instead intermediate angles in between.

3.3 Geometry Design Optimization

Besides the considerations for support structures, there also exist other geometry design optimization aspects when adopting AM.

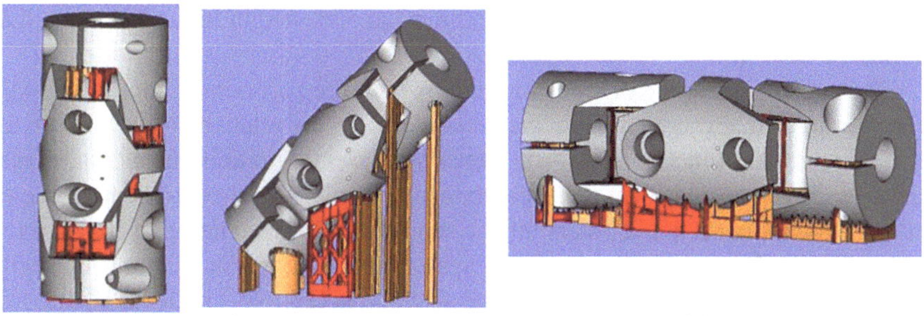

a. Fewest support b. Easy-to-remove support c. Shortest fabrication time

Fig. 3.30 Support generation (Materialise Magics) for a universal joint. (**a**) Fewest support. (**b**) Easy-to-remove support. (**c**) Shortest fabrication time

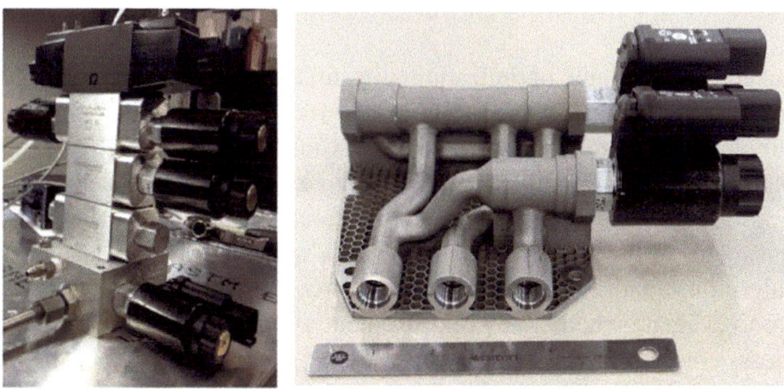

a. Original design b. Consolidated redesign

Fig. 3.31 An aerospace hydraulic manifold redesign with part consolidation [62]. (**a**) Original design. (**b**) Consolidated redesign

3.3.1 Part Consolidation

One of the most notable aspects is the redesign for part consolidation. Due to the capability to realize complex geometry designs, AM can often allow for multiple parts to be consolidated into fewer numbers of parts. During this redesign process, some geometry design optimizations become helpful in both reducing the design redundancy and improving the performance efficiency of the parts. There exist various potential advantages for part consolidation, including cost reduction, simplification of supply chains and production, performance enhancement, and improved reliability and sustainability. Meanwhile, the part consolidation process is usually more than just the merging of multiple existing components into a single one. Instead, it involves various geometry design modifications, such as the reduction/removal of assembly features between consolidated components, addition of transition features (e.g., fillets), and lightweighting. Figure 3.31 illustrates a case study of an aerospace hydraulic manifold part consolidation redesign for AM adoption [62].

The redesign achieved 60% weight reduction and 53% height reduction, as well as improved fluid flow.

There also exist some obvious limitations with the part consolidation redesign. For example, if the components are required to move relative to each other for achieving certain functionality, or if there exist assembly requirements that necessitate multipart design, then these components should not be consolidated. Multimaterial components could potentially be subjected to consolidation restrictions too depending on the material type and AM system capabilities. For example, currently most PBF systems could only handle one single material for a single build, and few AM systems could process different material types (metals, ceramics, polymers).

3.3.2 Topology Optimization

The concept of topology optimization was first introduced in early 1900s and received further research intention in the 1980s [63]. In a nutshell, topology optimization employs optimization algorithms to maximize design objectives via the optimization of spatial material distributions within a given design space, usually with certain design constraints such as minimum safety factors and allowable compliance. During the optimization, individual voxels were selectively added or removed by the algorithm, and the resulting structure would be checked against the target objective to evaluate whether any improvement was made. This iterative optimization process continues until either a satisfactory result is achieved or the improvement diminishes below the predefined threshold. There exist various design strategy variants based on the topology optimization concept. Depending on the level of geometry variable design freedom, the design optimization could be categorized into sizing optimization, shape optimization, and topology optimization, as illustrated in Fig. 3.32. Sizing optimization is often carried out on designs with discretely definable features such as truss structures [64]. With sizing optimization, the shape/construction of the structures is known a priori, and the optimization is carried out with the cross-sectional sizes of the feature elements. With shape optimization, the initial structure topology was defined, which becomes further optimized within the established space (domain). With shape optimization, the topology design of the structures can be considered as somewhat predefined. The per-defined design could potentially be determined from prior designs or other design information. Finally, with topology optimization, such predefined design inputs are further reduced, which expands the design freedom. Topology optimization aims to identify an optimal layout of material within the designed space (domain), and the solutions are often not unique (i.e., local optimum instead of global optimum). In topology optimization, shape and size are not treated topologically as with the geometry modeling algorithms, but instead as distribution functions defined within the design domains. As such, the optimized solutions are sometimes not physically tractable. An example of topology optimization is shown in Fig. 3.33, in which a minimum

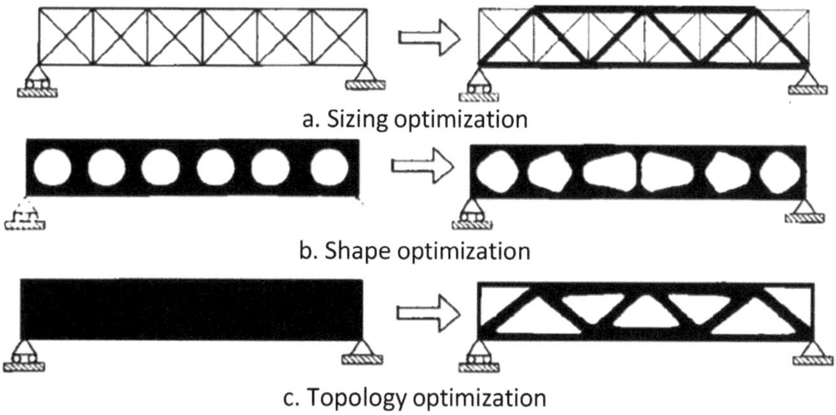

Fig. 3.32 Comparison of different optimization algorithms [64]. (**a**) Sizing optimization. (**b**) Shape optimization. (**c**) Topology optimization

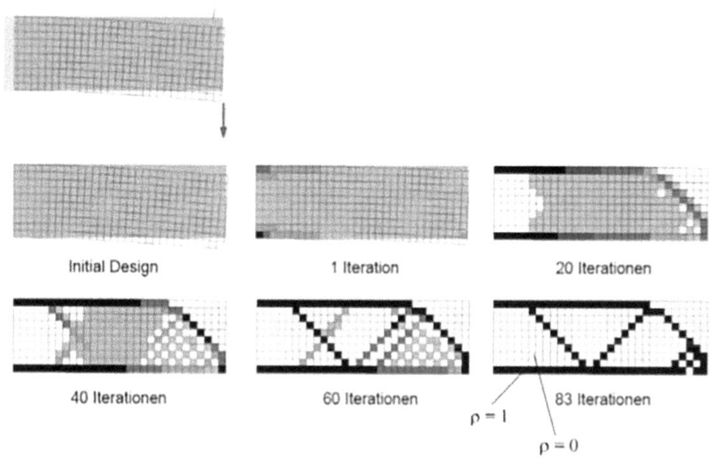

Fig. 3.33 An example of a 2D topology optimization problem [65]

compliance/deflection optimization was performed on an overhanging beam with one fully constrained end [65].

Various topology optimization algorithms have been employed for the exploration of the design potentials utilizing the free-form AM processes, especially with voxel-based processes (e.g., material jetting). In general, these algorithms can be categorized into density-based (e.g., bidirectional evolutionary structural optimization (BESO), solid isotropic microstructure with penalization (SIMP)) and level set-based frameworks. In short, the density-based framework focuses on the design of the pseudo-density variables at discrete voxel level, with the pseudo-density variables taking either discrete or continuous values. Discrete variable values of 0 and 1 are commonly used to represent the material status within the design voxel, i.e., "0" represents no material and "1" represents fully filled material. Such design scenario is considered reasonable with manufacturing processes including most AM technologies, in which "partial density" design is not feasible. On the other hand, with some AM processes such as material jetting, variable material property over a semi-continuous range is feasible via the digital material concept, i.e., blending of two material ingredients with different properties. Correspondingly, continuous pseudo-density levels from 0 to 1 could be employed for the design, which could potentially lead to more optimized solutions [66, 67]. There exist some notable mathematical challenges for the density-based framework, such as the checkboard solution and mesh sensitivity [68]. In comparison, with the level set-based framework, the design domain is expressed by level set functions, which express topological evolution in more natural ways and alleviate the issue with infeasible intermediate densities that often arise from density-based frameworks. On the flip side, level set-based approach is subjected to more constraints from initial design such as internal holes and existing borders [69].

The concept of metamaterial has also been applied in the topology optimization, which exploits the potentials of AM in realizing free-form geometries for the generation of metamaterial voxels as the material design voxels. Cellular structure is frequently employed for the design of representative volume elements (RVEs), as there exists extensive knowledge on the geometry-property design of cellular structures from literature. The cellular unit cells could be either used as pre-characterized equivalent material voxels, generated during the optimization through the introduction of shape/topology rules within each design voxels, or simply treated as periodic porosity [70–72]. Figure 3.34 illustrates two examples of such approach. Although potentially subjected to certain additional design restrictions such as topology freedom, minimum feature size, and orientation ranges, the combination of metamaterial geometry design and topology optimization appears to provide a promising approach toward achieving a balance between optimized performance and process/material quality control, as will be elaborated later in this chapter during the introduction of cellular structures.

As topology optimization algorithms deal with mathematically defined problems, it is necessary to translate design constraints and performance objectives into mathematical forms. This has been shown to pose various challenges for the application of the algorithms to AM designs. Manufacturability constraints such as minimum feature dimensions and support-free overhanging angles could generally be clearly defined geometrically. However, complications arise when these criteria are

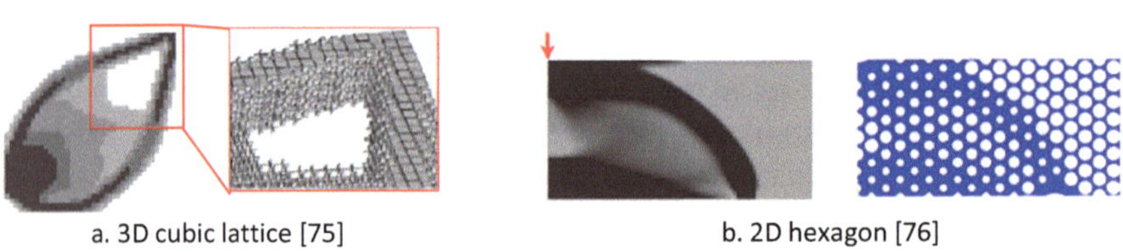

Fig. 3.34 Examples of utilizing cellular structures in topology optimization for variable densities. (**a**) 3D cubic lattice [73]. (**b**) 2D hexagon [74]

expressed in mathematical terms. For example, the measure of minimum feature dimensions is direction dependent and thus involves the counting of the minimum numbers of neighboring elements, which typically requires additional filter in the algorithm [75]. Similarly, overhanging angle must be identified with reference to the geometry boundary and overall feature orientation, which often involves mathematical capture of spatial geometry features that could not be easily accommodated [76]. In cases of computational difficulties that result in infeasible solutions, the feasibility could only be imposed via mathematical means instead of physical principles, which are nonintuitive. There also exist other manufacturability aspects that could not be easily defined, such as the post-processing for support removal or surface finishing.

Some of the examples for post-printing machining include internal holes, undercut features, and tooling clearance [77]. Further, in many multi-objective design problems, the trade-off relationships among different performance criteria are often not clearly quantifiable, while simultaneous optimization of multiple objectives might not be attainable [78]. For formulatable multi-objective design problems, the Pareto front-based approach could be utilized for solutions generated by topology optimization, which usually involve human decisions by the designers. A multi-objective topology optimization design example is shown in Fig. 3.35, which clearly illustrates the divergent characteristics of the optimized geometries under different objective trade-offs [78].

A well-noted challenge for the adoption of topology optimization in AM is the non-simplistic material/process property and quality. Various process characteristics such as layer-wise geometry generation, scan strategy/toolpath, process voxel

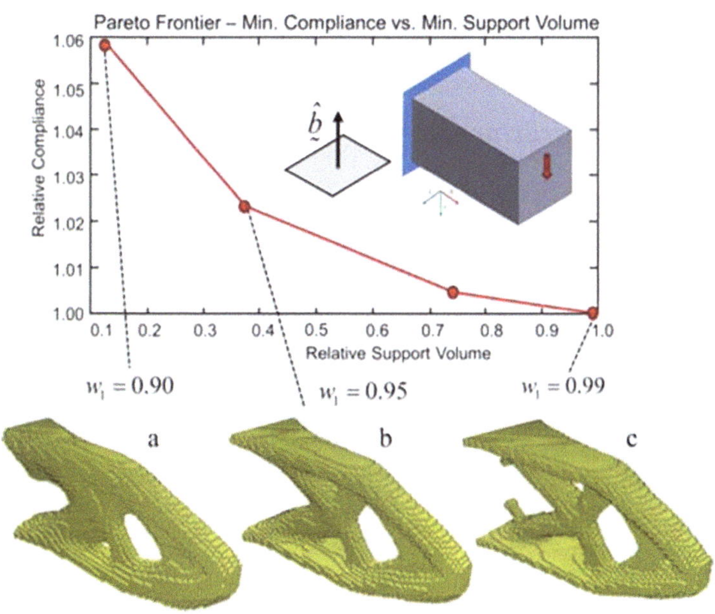

Fig. 3.35 A Pareto front of topology optimization design for a cantilever beam, with stiffness/compliance and support volume as objectives [78]

resolution, surface roughness, and feedstock stochasticity, as previously discussed, result in generally non-isotropic, nonhomogeneous material properties and stochastic quality fluctuation with most AM processes [68, 69, 75, 77]. In addition, some of these characteristics also result in additional design constraints that limit the feasibility of the eventual design solutions [79]. Many of these characteristics exhibit nontrivial dependency on both process setting and geometry design, necessitating the needs of cost-inhibitive experimental process-material-property database development. Further, as the material properties and qualities of many AM processes are also dependent on the geometry of the features, the material-geometry coupling could not be easily handled by the current topology optimization framework. As such, for applications that require part reliability or have tight design performance tolerance, topology optimization designs are currently not yet practical.

On the other hand, topology optimization could efficiently provide design references that serve to inform or inspire designers for further design refinements. As shown in Fig. 3.36, topology optimization was carried out on the baseline design of a bracket, which generated a voxelized design that was subsequently used as the reference for additional redesign by the designers. The interpretation of the design concepts and treatments from topology-optimized references is often a challenging practice, due to the "black box" nature of the iterative design optimization approach. On the positive side, the reference designs have the potential to significantly disrupt the current product design concepts, helping the designers to overcome the initial design barrier more efficiently. Meanwhile, additional design for AM manufacturability knowledge from the designers could be more easily applied to the design process, which is currently still rather under-structured and tends to be highly empirical and isolated within individual organizations.

3.3.3 Cellular Structure Design

Another major class of approach for advanced AM design concepts is cellular structure-based design. Cellular structures are generally defined as networks of interconnected solid beams or walls with included voids. Classical theory for

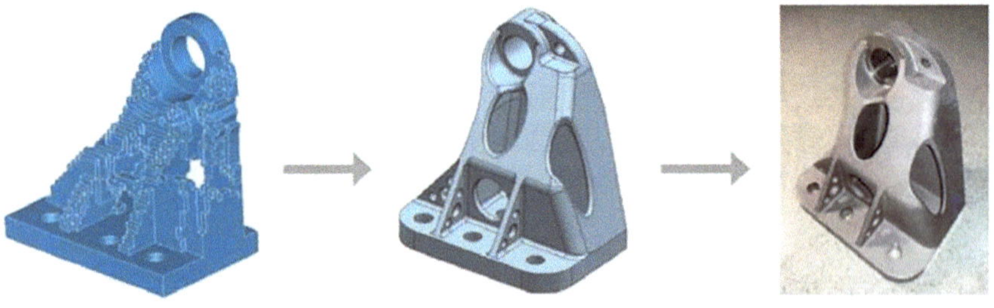

Fig. 3.36 Utilization of topology optimization as references for the redesign of a bracket [42]

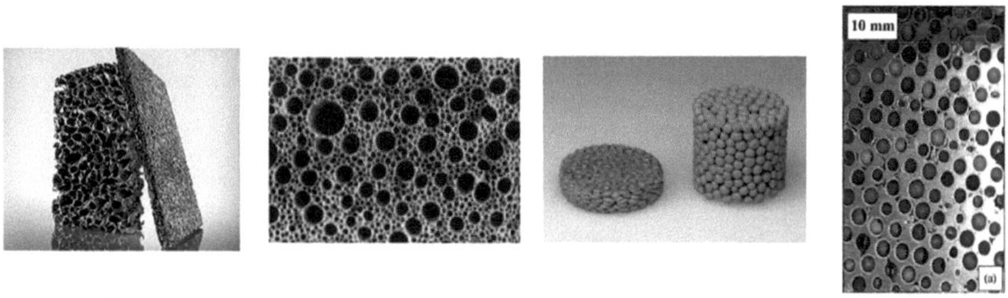

a. Cymat foam by gas melt injection [88]
b. solid-gas eutectic solidification foam [86]
c. Sintered hollow spheres [86]
d. Al-SS casted foam [89]

Fig. 3.37 Some stochastic cellular structures. (**a**) Cymat foam by gas melt injection [86]. (**b**) Solid-gas eutectic solidification foam [84]. (**c**) Sintered hollow spheres [84]. (**d**) Al-SS casted foam [87]

cellular structures has been well developed by various groups [80, 81], which deals with both stochastic cellular structures and non-stochastic cellular structures.

Stochastic cellular structures are often referred to as foams, and they generally exhibit random porosities and pore characteristics including pore size and pore shapes. Figure 3.37 shows some of the commercially available metal stochastic cellular structures, which are mostly made by traditional processes such as casting, sintering, and foaming [82–85]. Due to the randomness of geometries, relative density is employed as the primary design parameter.

Relative density is defined based on the volume ratio between the cellular structure and its geometrical bounding volume, which is the total volume of solid and void combined. Therefore, the relative density ρ_r and the porosity η always meet the following relationship:

$$\rho_r + \eta = 1 \tag{3.1}$$

For cellular components, the geometrical bounding volume could be easily identified as the overall bounding spaces of the components. However, for cellular structures that are constructed by unit cells, the determination of geometrical bounding volume is closely related to the spatial pattern of these unit cell geometries. For example, for the unit cell shown in Fig. 3.38a, the two geometrical bounding volumes shown in Fig. 3.38b and c would correspond to two different cellular structures.

For stochastic cellular structure, based on whether the pores are interconnected, there exist two types of porosities. Open-cell structures have interconnected pores (e.g., Fig. 3.37a), while closed-cell structures have isolated pores (e.g., Fig. 3.37d). In general, open-cell structures exhibit lower mechanical strength compared to closed-cell structures. However, in certain applications such as biomedical implants and catalyst, the pore interconnectivity becomes beneficial.

In general, the mechanical properties of the open-cell cellular structures could be predicted by their relative densities reasonably well by following the Gibson-Ashby theory [80]:

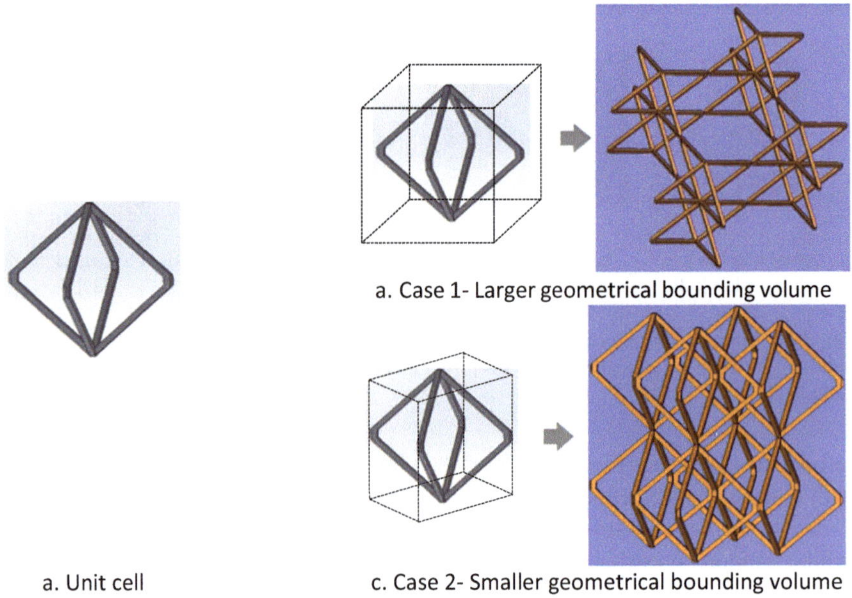

Fig. 3.38 Relationship of geometrical bounding volume and cellular pattern. (**a**) Unit cell. (**b**) Case 1: Larger geometrical bounding volume. (**c**) Case 2: Smaller geometrical bounding volume

$$\frac{E}{E_S} = C_1 \ \rho_r^2 \tag{3.2}$$

$$\frac{\sigma}{\sigma_S} = C_2 \ \rho_r^{1.5} \tag{3.3}$$

$$\frac{G}{G_S} = C_3 \ \rho_r^2 \tag{3.4}$$

where E, E_S, σ, σ_S, G, and G_s are the elastic modulus, yield strength, and shear modulus of the cellular structure and the solid material, respectively, and C_1, C_2, and C_3 are constants that vary between different cellular structures which are usually determined through experimentation. For stochastic cellular structures made by Al alloys, C_1, C_2, and C_3 are approximately 1, 0.3, and 0.4, respectively. Although in some literatures modulus and strength equations with slightly different exponential values were reported, Eqs. (3.2)–(3.4) generally provide reasonable approximation for most cellular structures even including non-stochastic ones.

There are some immediate observations from Eqs. (3.2)–(3.4). Firstly, the modulus of the cellular structure is more sensitive to the relative density variations compared to the yield strength and therefore could be potentially controlled over a larger range. Secondly, as a lot of the cellular structures have relative densities that are smaller than 50%, their elastic modulus, yield strength, and shear modulus would fall below 25%, 35%, and 10% of the solid material properties, respectively. Therefore, compared to solid materials, cellular structures are considerably weaker in absolute properties. Thirdly, the property-to-mass scaling ratios for both the modulus and yield strength of the 3D cellular structure are lower than unity, which means that they do not perform as well as 2D extruded (i.e., 2.5D) structures. To elaborate on this, consider the uniaxial elastic compression of a 2.5D cellular extrusion of perfectly elastic material as shown in Fig. 3.39. The modulus of the structure is proportional to the cross-sectional area, which in turn is proportional to the total relative density of this structure. As a result, the modulus forms a linear relationship with relative density. For 3D cellular structures, since the porosity perpendicular to the compressive loading direction does not contribute to the modulus/stiffness of the structure, the overall modulus-to-weight ratio of the cellular structures will always be lower compared to the 2.5D designs.

It could be readily seen that 2.5D extruded cellular structures are generally more weight efficient under uniaxial compression/tension loading conditions, compared to 3D cellular structures. This seemingly intuitive and simple observation, however, reflects the underlying design mentality for cellular structures. By adopting periodic porosity in the design, a cellular structure is inevitably weaker than the solid materials, and increasing the level of geometry complexity or the dimensionality of porosity does not necessarily correspond to increased performance efficiency either. On the other hand, cellular structures of different designs often exhibit rather unique mechanical and physical characteristics, which can be exploited for applications. In addition, cellular structures also exhibit multidirectional property tailorability, which makes it possible to introduce unique behaviors or more efficient bulk performance via geometry designs.

Cellular structures exhibit unique stress-strain mechanical characteristics. For a stochastic cellular structure made of elastic-perfectly plastic material, the typical

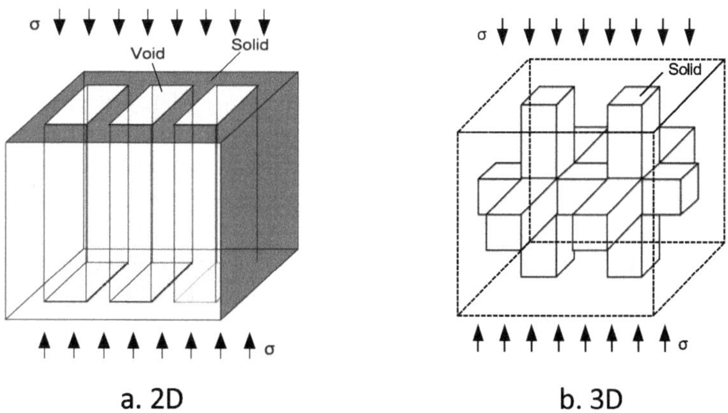

Fig. 3.39 Comparison between 2D and 3D cellular structures under uniaxial compression. (**a**) 2D. (**b**) 3D

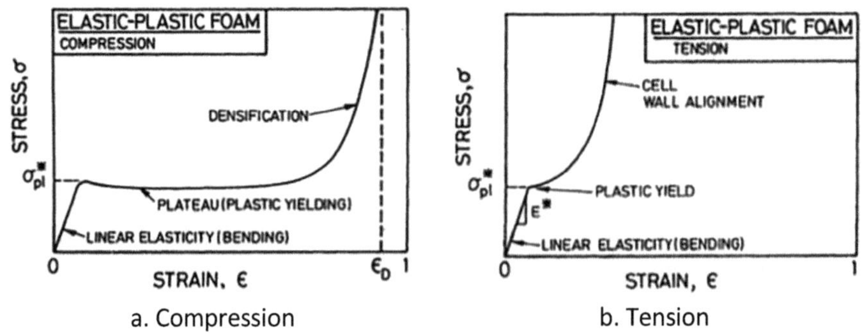

Fig. 3.40 Stress-strain characteristics of stochastic cellular structures under different uniaxial loading [80]. (**a**) Compression. (**b**) Tension

stress-strain curves under compressive or tensile loading are shown in Fig. 3.40 [80]. Under tensile stress, cellular structures exhibit linear elasticity initially, which is dominated by the elastic deformation with the cellular solids. Upon further deformation, some struts/walls in the cellular structures will start to yield/fail, which consequently results in realignment of these solid components along the loading direction. Therefore, upon the initiation of yield, the cellular structure will effectively exhibit "strain-hardening" effect as shown in Fig. 3.40b. Also, when fracture initiates in the cellular structure, it takes place as the fracture of individual struts/walls, with the overall fracture cracking propagating through the cross sections and eventually disintegrating the cellular structures. On the other hand, under compressive stress, although the elastic behaviors of the cellular structures still remain the same, the yield of the structure progresses much differently. Under compression, the failed debris struts/walls will be trapped and compacted in the main structures and therefore will continue to contribute to the overall strength of the structure as obstruction of deformation. As a result, the cellular structures exhibit a plateau stage in which the stress remains largely constant while the strain accumulates as shown in Fig. 3.40a. Toward the end of the plateau stage, most of the cellular struts/walls will have failed, and the compacted structures with debris will begin to behave similar to the solid material in the so-called densification stage. The existence of plateau stage is of great interest to many energy-absorbing applications with cellular structures, since it would allow for energy absorption over a large range of deformation with the amount almost proportional to the amount of deformation introduced into the structures.

Figure 3.41 further demonstrates the progression of compressive failure within a stochastic cellular structure. Due to the random characteristics of the pores, the crack/failure could initiate at a random location with the cellular structure. Once formed, the failure site will cause stress concentration in the neighboring regions, which become the sites for the growing failure front. As a result of the progressive failure, a failure band usually forms for stochastic cellular structures as shown in Fig. 3.41. With further compression, the failure band will expand in width and eventually propagates through the entire structures.

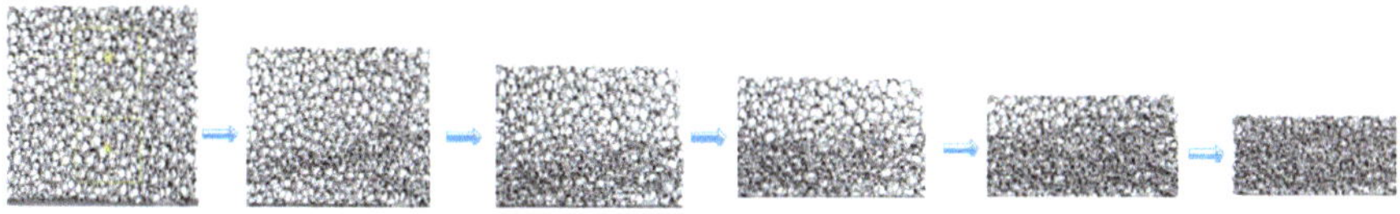

Fig. 3.41 Progression of compressive failure in stochastic cellular structure [88]

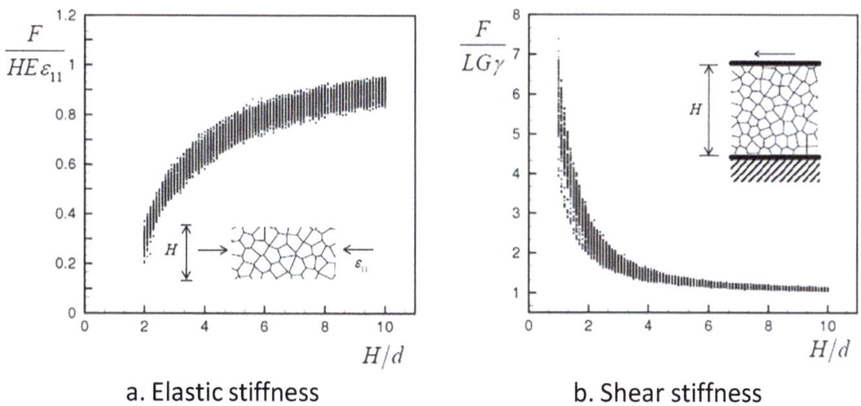

Fig. 3.42 Size effect for 2D Voronoi cellular structure [89]. (**a**) Elastic stiffness. (**b**) Shear stiffness

Despite the somewhat limited designability, one useful characteristic of stochastic cellular structures is that they could potentially exhibit isotropic properties. Such isotropy arises from the size effect of the overall structures, in which the topological characteristics of individual struts/walls along all the orientations are stochastically identical. Correspondingly, the size comparison between the overall structure and its individual strut/wall elements is relevant to the isotropy of the properties. It also turned out that such size effect is also relevant for the mechanical properties of cellular structures, regardless of their geometry stochasticity. Both experimental and theoretical works are available for the characterization of size effects for cellular structures of different designs. For stochastic cellular structures, it is generally suggested that with increasing number of cells in the structure (i.e., larger structure size to unit cell size ratio), the elastic modulus, yield strength, and compressive strength of the cellular structures increase, whereas their shear modulus and shear strength decrease. Figures 3.42 and 3.43 illustrate the size effect of elastic modulus and shear properties with 2D Voronoi cellular structures and 3D stochastic cellular structures [89, 90]. As the structure-to-unit size ratio reaches 8–10 along the loading directions, the size effects for elastic modulus and compressive strength of the structures converge to relatively consistent levels. Also, as the structure-to-unit size ratio reaches 3–4 perpendicular to the shear loading directions, the size effects for shear properties converge to relatively consistent levels. These thresholds are often considered as guidelines for achieving asymptotic homogenization with the cellular

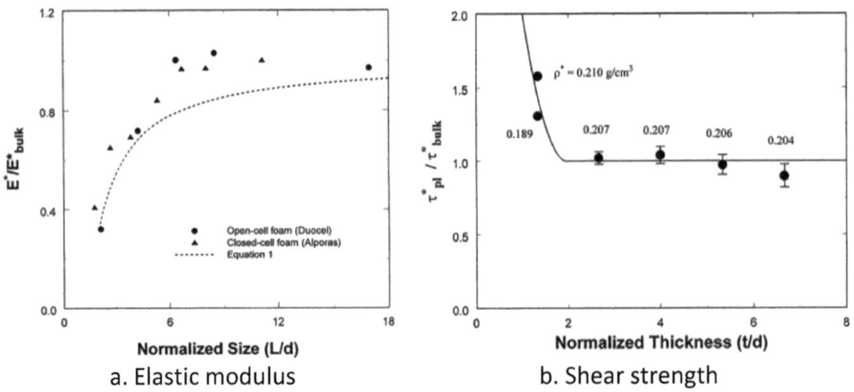

Fig. 3.43 Size effect for 3D stochastic cellular structures [90]. (**a**) Elastic modulus. (**b**) Shear strength

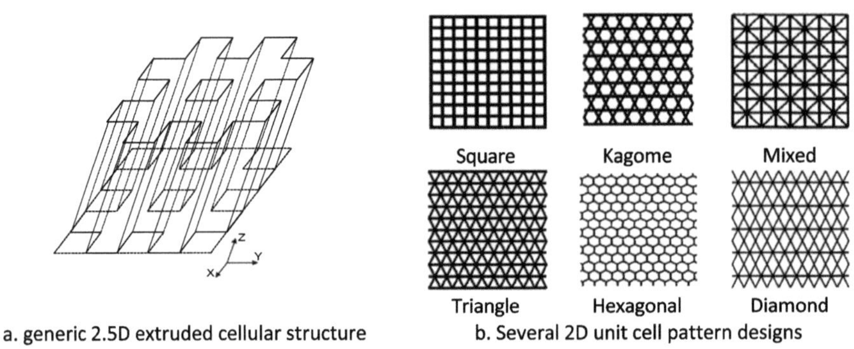

Fig. 3.44 2.5D extruded cellular structure design. (**a**) Generic 2.5D extruded cellular structure. (**b**) Several 2D unit cell pattern designs

structure designs, which effectively treat them as equivalent continuum material voxels in the designs.

Due to their increased designability, non-stochastic cellular structures are often considered advantageous for applications compared to non-stochastic designs. The geometries of the cellular structures are most commonly determined by unit cell geometry designs, but could also be defined by other means such as spatial shape functions.

For 2.5D cellular structure designs, the unit cell geometry could be represented by a 2D profile and a third axis of "extrusion" direction, as illustrated in Fig. 3.44a. The design of the unit cell geometry is required to be 2D space filling and periodic. Such pattern design is well studied in the research of plane symmetry group. In addition, the third axis (z-axis in Fig. 3.44a) is usually designed to be perpendicular to the pattern plane so that Cartesian coordinate system could be established for stress analysis. In general, the in-plane mechanical properties (i.e., mechanical properties along the direction in x–y plane shown in Fig. 3.44a) are considerably lower than the out-of-plane mechanical properties (i.e., mechanical properties along the z-direction). This could be further demonstrated by the following equations for hexagonal honeycomb structures following the axis convention used in Fig. 3.44a:

$$\frac{E_{x(y)}}{E_S} \propto \rho_r^3 \tag{3.5}$$

$$\frac{E_Z}{E_S} \propto \rho_r \tag{3.6}$$

$$\frac{\sigma_{x(y)}}{\sigma_S} \propto \rho_r^2 \tag{3.7}$$

$$\frac{\sigma_Z}{\sigma_S} \propto \rho_r \tag{3.8}$$

Comparing Eqs. (3.5) and (3.7) with Eqs. (3.2)–(3.4), it could be seen that the in-plane properties of the 2.5D extruded cellular structures exhibit higher sensitivity to the relative density, which means that at lower relative densities, the structures exhibit significantly weakened mechanical properties. Therefore, many of the current applications with this type of cellular structure focus on exploiting their out-of-plane properties which scale linearly with relative densities (Eqs. (3.6) and (3.8)).

The selection of unit cell geometry has significant influence on the mechanical performance of the 2.5D extruded cellular structures. Figure 3.44b shows multiple 2D unit cell pattern designs, whose mechanical properties are shown in Fig. 3.45 [91]. Comparing different designs, many structures exhibit rather linear relationship with relative densities except for hexagonal honeycomb, and the diamond design exhibits the highest overall performance. On the other hand, for different mechanical property criteria, different designs perform differently. For example, the square structure exhibits very high elastic modulus but very low shear modulus and yield strengths. Some of these differences are more intuitive and easy to explain, such as the low shear modulus of the square structure, while the others are less obvious, such as the difference between the Kagome structure and the hexagonal honeycomb structure. It must also be noted that the comparison illustrated in Fig. 3.45 is established with only relative density, which means that other geometry parameters (e.g., the angles between the walls, the width/length ratios of the unit cell bounding box) are arbitrarily set. Therefore, the comparison among different geometry designs is only valid under specific geometry setting and does not apply to general cases. Such conclusion also applies to 3D cellular structure designs, which exhibit even higher degrees of freedom and thus more complicated geometry-property characteristics. For designers, other design guidelines should be followed instead.

One of the most significant design guidelines for cellular structures is the structural connectivity. This concept was originally established for mechanical truss

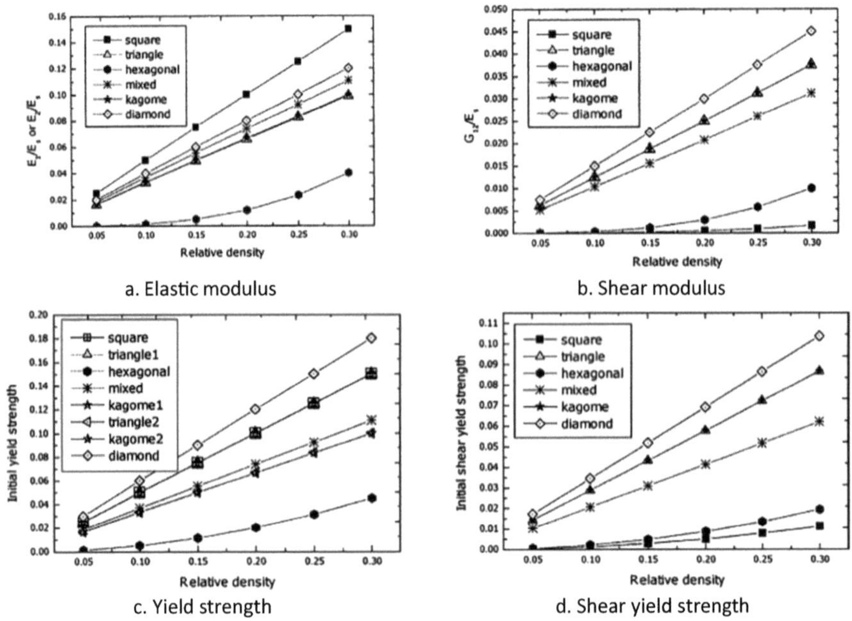

Fig. 3.45 Comparison of different 2D unit cell pattern designs [91]. (**a**) Elastic modulus. (**b**) Shear modulus. (**c**) Yield strength. (**d**) Shear yield strength

structures to indicate structural rigidity and stability. In particular, with strut-based cellular structures, the Maxwell's stability criterion determines the statistical and kinematical determinacy of a pin-jointed frame. For 2D structures, the Maxwell's criterion is expressed as [92]

$$M = b - 2j + 3 \tag{3.9}$$

where M is the Maxwell's stability number, and b and j are the number of struts and joints of the framework, respectively. Figure 3.46 illustrates different 2D framework systems that exhibit different Maxwell's stability numbers. $M < 0$ indicates a kinematically instable framework system (Fig. 3.46a), which upon loading is unable to retain the original shape or construction. $M = 0$ indicates a kinematically determinant system with no redundancy (Fig. 3.46b), while $M > 0$ indicates a kinematically over-determinant system (Fig. 3.46c) that requires additional compatibility equations to fully demonstrate. For AM cellular structures, the joints usually exhibit sufficient rigidity and therefore should not be considered as pin joints. However, the Maxwell's stability criterion could still be applied to determine the dominant deformation mechanism of the structure. For cellular structures with $M < 0$, the dominant deformation mechanism is strut/wall bending, and the structure is termed bending-dominated structure. For cellular structures with $M < 0$, the dominant deformation mechanism is strut/wall stretching/compressing, and therefore the structure is termed stretch-dominated structure. For cellular structures with $M = 0$, a mixed

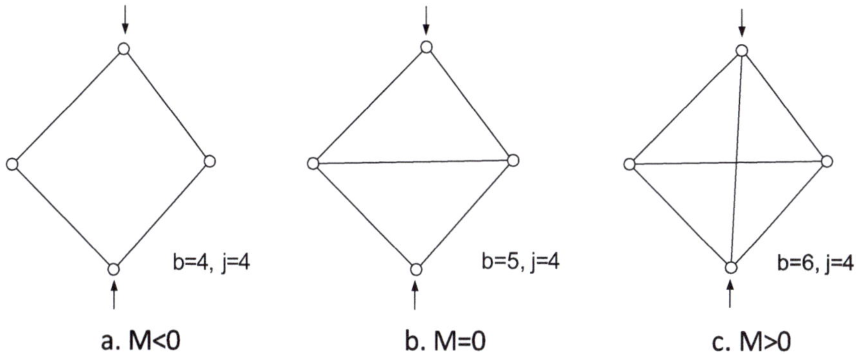

Fig. 3.46 2D illustration of Maxwell's stability of different frameworks/structures [92]. (**a**) $M < 0$. (**b**) $M = 0$. (**c**) $M > 0$

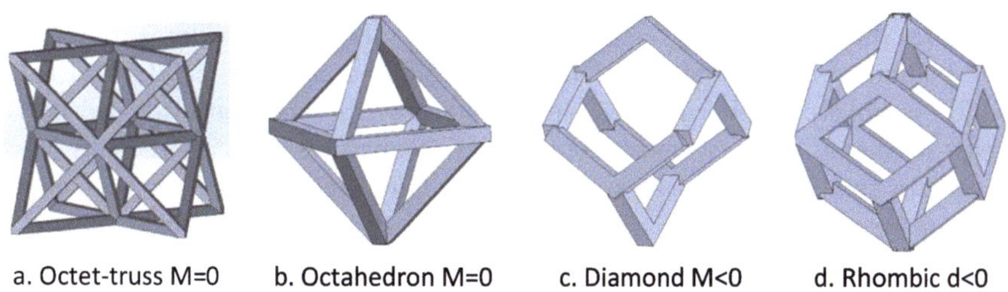

Fig. 3.47 Maxwell stability of different 3D cellular designs. (**a**) Octet truss $M = 0$. (**b**) Octahedron $M = 0$. (**c**) Diamond $M < 0$. (**d**) Rhombic $d < 0$

deformation mode is often observed; however, the structure could still be considered primarily stretch dominated.

Similar to the design of 2.5D extruded cellular structures, the deformation mechanisms of the 3D cellular structures could also be identified with the help of the Maxwell's stability criterion, which is given in Eq. (3.10) for 3D cellular structures:

$$M = b - 3j + 6 \tag{3.10}$$

Some examples of unit cell designs for 3D cellular structures are shown in Fig. 3.47, with indicated Maxwell's stability M value. It should be noted that although the Maxwell's criterion is indicative of a range of mechanical properties of cellular structures, its calculation is not always straightforward due to the existence of shared elements in the unit cells. In general, the analysis of the deformation mechanism for 3D cellular structures is more complex. In addition, the behaviors of the structures are also more sensitive to the boundary conditions (e.g., free boundary or boundary constraints), which will be elaborated further in later discussions.

There exist significant differences of mechanical characteristics between stretch-dominated and bending-dominated cellular structures. Bending-dominated structures usually exhibit lower modulus and strength compared to stretch-dominated

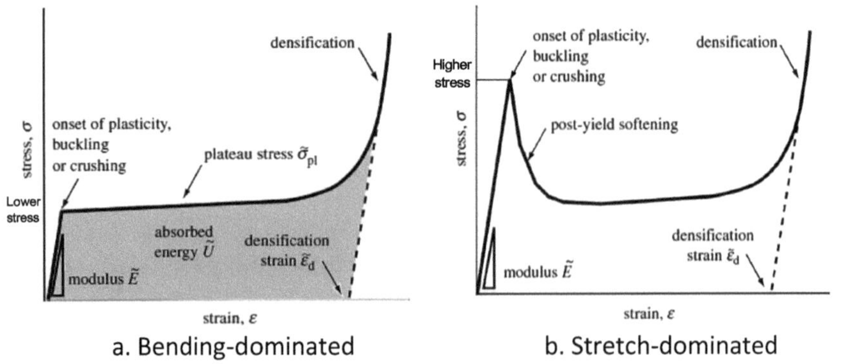

Fig. 3.48 Stress-strain characteristics of different deformation mechanisms [92]. (**a**) Bending dominated. (**b**) Stretch dominated

structures due to the lower resistance of strut and thin-wall structures to bending deformation. As shown in Fig. 3.48a, the bending-dominated structures usually exhibit the classical plateau stress-strain curve, which has been discussed previously in detail. In fact, most commercial stochastic cellular structures have Maxwell's stability $M < 0$ and exhibit bending-dominated deformation mechanism. In comparison, the stretch-dominated structures exhibit a slightly different stress-strain characteristic. As shown in Fig. 3.48b, the initial failure of stretch-dominated structures usually occurs at higher stress levels. However, upon initial failure, the local catastrophic of structure collapse will often cause sudden drop of stress as a result of structural relaxation, which can be termed as post-yield softening. Such structural collapse is also often accompanied by macroscopic geometrical distortion of the structures, which sometimes alters the deformation mechanism of the remaining structures into bending-dominated structures by introducing loading misalignment. The example stress-strain curve shown in Fig. 3.48b demonstrates such case, in which the consequent deformation exhibits plateau stage before densification occurs. However, in some cases, the local collapse does not cause the change of deformation mechanism, and consequently the post-yield softening could happen again when another local collapse takes place.

A second design guideline is related to the relationship between the cellular geometry and the loading direction. Unlike the dominant deformation mechanism guideline, this design guideline is loading case specific. When some of the unit cell elements of the cellular structure are aligned to the primary loading direction, the structure usually exhibits higher stiffness along that particular direction. As illustrated in Fig. 3.35a and b, the reinforced diamond structure has stretch-dominated design; however, the design orientation illustrated in Fig. 3.49a would result in higher stiffness under the same relative density level, due to the alignment of the central reinforcement element along the primary loading direction. Similarly for the square structure illustrated in Fig. 3.49c and d, even though the unit cell design is bending dominated, the orientation shown in Fig. 3.35c corresponds to higher

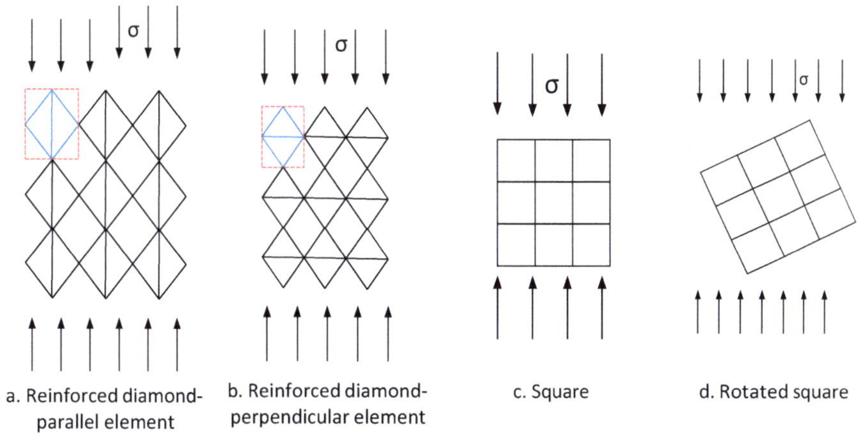

Fig. 3.49 Alignment of cellular element along the loading direction. (**a**) Reinforced diamond: parallel element. (**b**) Reinforced diamond: perpendicular element. (**c**) Square. (**d**) Rotated square

stiffness under the same relative density level due to the alignment of some of the elements along the vertical loading direction.

There exist some difficulties in applying this guideline in real-life applications. Firstly, with many applications, the principal loading directions of the structures might either be compound or undetermined, which makes it difficult to determine the optimal orientations if the unit cell geometry design is already fixed. Secondly, this guideline is not necessarily valid for many other mechanical properties such as ultimate strength. One complicating factor is the potential of elastic failure (i.e., buckling), which leads to sudden change of structural behaviors that is often structurally catastrophic.

A third design guideline is closely related to the previous one, which is the presence of elastic failure/buckling mode within the structure. The presence of buckling mode typically suggests that there exist compression-dominated slender elements, which indicates both high stiffness and catastrophic behavior upon failure. Besides these characteristics, which were already discussed in the previous guideline, the presence of buckling mode could potentially enable unique structural behaviors such as the deformation mode change or shape morphing. There exist various literatures that exploit the bistable and metastable design concepts utilizing elastic buckling [94], although most of them are purely mathematical. Figure 3.50 shows some examples of bistable and metastable cellular designs that have been demonstrated experimentally using AM. These designs have been shown to have promising potentials for energy absorption, due to the large amount of deformation that occurs during the elastic instability-induced strain.

A fourth design guideline for the cellular structure is the auxeticity. Auxeticity refers to the negativity status of the Poisson's ratio of a cellular structure. An auxetic structure exhibits negative Poisson's ratio in at least one of the directions, which means that upon compression along that direction, the lateral dimension would also

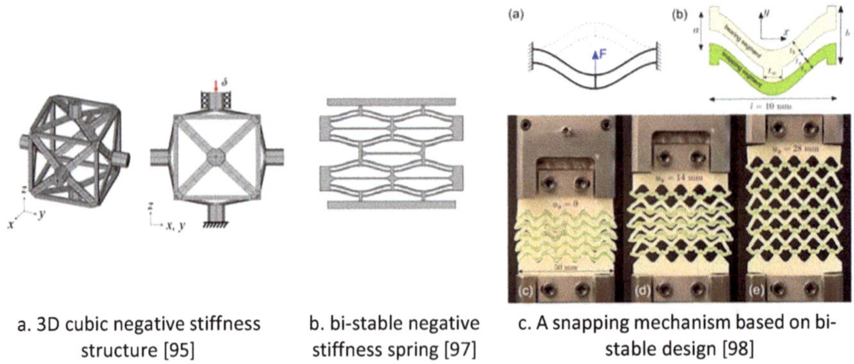

Fig. 3.50 Some examples of cellular designs exploiting elastic instability/buckling for performance. (**a**) 3D cubic negative stiffness structure [93]. (**b**) Bistable negative stiffness spring [95]. (**c**) A snapping mechanism based on bistable design [96]

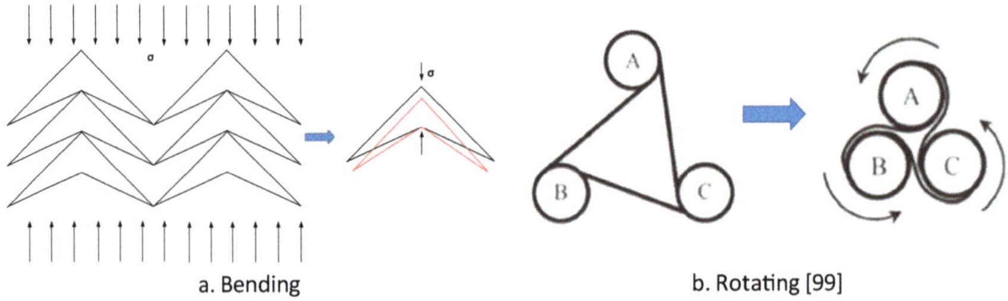

Fig. 3.51 Common auxeticity mechanisms for cellular structures. (**a**) Bending. (**b**) Rotating [97]

reduce. This unusual characteristic of structure arises from the "folding-in" deformation of the cellular network, which could be thought of as analogous to the opening/un-opening of an umbrella. Such deformation mode could be introduced via either bending or rotation of elements/units, as illustrated in Fig. 3.51. Various 2D and 3D auxetic unit cell designs have been proposed in literatures, and some of them have been demonstrated experimentally via AM.

There exist some unusual and appealing characteristics of auxetic structures. Owing to the auxeticity, these structures exhibit very high shear properties and indentation resistance. Figure 3.52 illustrates the comparison of stress concentration and deformation status between two sandwich structures of auxetic and non-auxetic designs. Also, despite having bending-dominated deformation mechanism, auxetic structures also exhibit high energy absorption capability and fracture resistance.

Ultimately, the design guidelines should mainly be employed for preliminary concept design of the cellular structures, and the detailed design that determines the actual properties is still needed. Rigorous analytical modeling of cellular structure has been demonstrated in a large body of works with different cellular designs, and the classic description could be found in [80]. In general, the modeling analysis is carried out with the cellular unit cells, due to the assumed periodicity with the

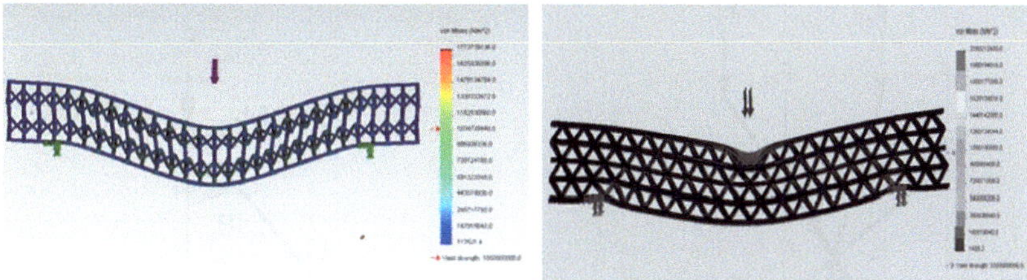

Fig. 3.52 Comparison of indentation stress concentration effects between an auxetic structure and a non-auxetic stretching-dominated structure [98]. (**a**) A reentrant auxetic structure. (**b**) A hexagonal structure

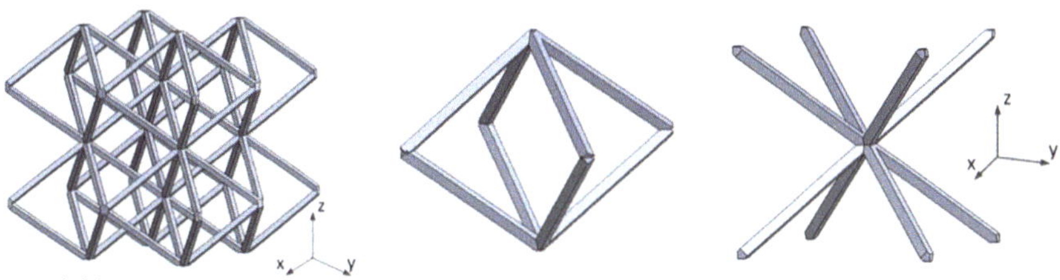

Fig. 3.53 Octahedral cellular structure [99]. (**a**) Lattice structure. (**b**) Unit cell layout 1. (**c**) Unit cell layout 2

structures. An example is given here to demonstrate the typical process flow of the mechanical modeling of unit cell structures.

Figure 3.53 shows a 3D BCC cellular structure design. The unit cell geometry for the cellular structure shown in Fig. 3.53a could be represented by either Fig. 3.53b and c. However, the unit cell layout 2 is easier to model and will therefore be adopted for later modeling analysis.

From the topology of the unit cell, it is obvious that the uniaxial mechanical properties of the BCC lattice could be modeled with identical formulations. Considering a remote compressive stress σ_z applied on the structure along the *z*-direction, the loading of the unit cell is shown in Fig. 3.54a. Since all struts are subjected to identical loading conditions and boundary conditions, an arbitrary strut is taken for modeling, whose loading condition is illustrated in Fig. 3.54b. From force equilibrium, the force components shown in Fig. 3.54b can be determined as

$$F_z = \frac{1}{4}\sigma_Z L_1 L_2 \tag{3.11a}$$

$$M_1 = \frac{1}{2} F_Z L_3 \sin\theta_Z = \frac{1}{8}\sigma_Z L_1 L_2 L_3 \sin\theta_Z \tag{3.11b}$$

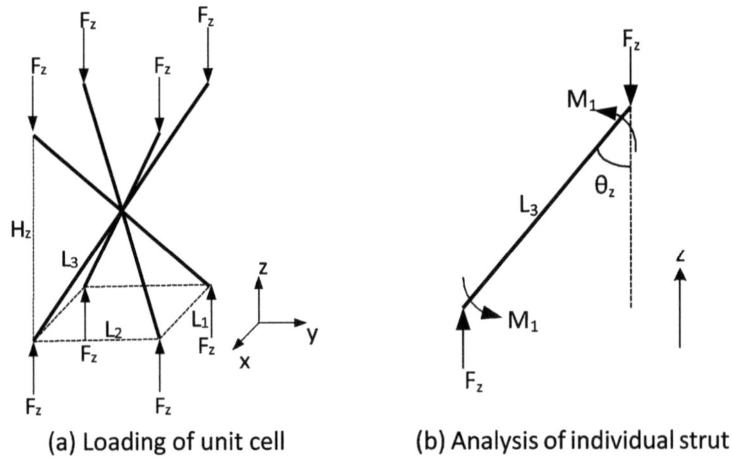

Fig. 3.54 BCC unit cell under compression. (**a**) Loading of unit cell. (**b**) Analysis of individual strut

where L_1 and L_2 are the dimensions of the unit cell in x- and y-directions, L_3 is the length of a half-strut, and θ_z is the slope angle of the strut in relation to the z-axis:

$$\theta_Z = \sin^{-1}\frac{\sqrt{L_1^2 + L_2^2}}{2L_3} \tag{3.11c}$$

Under the loading condition shown in Fig. 3.54b, the strut undergoes a bending/shearing-combined deformation. For metal cellular struts, the axial deformation along the strut axis is relatively insignificant and therefore could be ignored. Employing beam analysis, the deformation of the strut in the z-direction and the direction perpendicular to the z-direction can be obtained as

$$\Delta z = L_3 \sin\theta_z \left(\frac{M_1 L_3}{6EI} + \frac{6F_z \sin\theta_z}{5GA}\right) \tag{3.12}$$

$$\Delta z_\perp = L_3 \cos\theta_z \left(\frac{M_1 L_3}{6EI} + \frac{6F_z \sin\theta_z}{5GA}\right) \tag{3.13}$$

where E and G are the modulus of elasticity and shear modulus of the solid material, I is the second moment of inertia of bending in the plane shown in Fig. 3.40b, and A is the cross-sectional area. If the deflection $\Delta z_\perp$ is further decomposed into deformations in the x- and y-directions, Δx and Δy, then it follows that

$$\Delta x = L_3 \cos\beta_z \cos\theta_z \left(\frac{M_1 L_3}{6EI} + \frac{6F_z \sin\theta_z}{5GA}\right) \tag{3.14}$$

$$\Delta y = L_3 \sin \beta_z \cos \theta_z \left(\frac{M_1 L_3}{6EI} + \frac{6F_z \sin \theta_z}{5GA} \right) \quad (3.15)$$

Therefore, the modulus of the unit cell can be determined as

$$E_z = \frac{L_3 \cos \theta_z \sigma_z}{\Delta_z} = \frac{H_z \sigma_z}{2\Delta_z} = \frac{H_z}{2L_1 L_2 L_3 \sin^2 \theta \left(\frac{L_3^2}{48EI} + \frac{3}{10GA} \right)} \quad (3.16)$$

where

$$H_Z = 2L_3 \cos \theta = 2\sqrt{L_3^2 - \frac{1}{4}\left(L_1^2 + L_2^2\right)}$$

which is also shown in Fig. 3.54a.

The strength of the BCC lattice could be roughly estimated by the initial yield strength of the strut. Such conservative estimation could sometimes be justified by the fact that metal AM cellular structures often exhibit significant quality variability. From Fig. 3.40b, the normal stress and shear stress of the strut could be determined as

$$\sigma_1 = \frac{M_1 u}{I} + \frac{F_z \cos \theta_z}{A} = \frac{\sigma_z L_1 L_2 L_3 \sin \theta_z}{8I} u + \frac{\sigma_z L_1 L_2 \cos \theta_z}{4A} \quad (3.17a)$$

$$\tau_1 = \frac{F_z \sin \theta_z D}{Ib} \quad (3.17b)$$

where u is the distance from the geometrical center of the cross section to the location of interest, D is the moment of area of the cross section, and b is the width of the cross section at the location of interest. Applying von Mises criterion, the maximum allowable stress level σ_m that results in the onset of structure yield is

$$\sigma_m = \frac{4}{L_1 L_2 \sqrt{\frac{L_3^2 \sin^2 \theta_Z}{2I^2} u^2 + \frac{2\cos^2 \theta_Z}{A^2} + \frac{6\sin^2 \theta_Z D^2}{I^2 b^2}}} \sigma_y \quad (3.18)$$

Following Eqs. (3.16) and (3.18), through geometrical parameter design, a range of mechanical properties could be achieved in a particular direction of the BCC lattice as shown in Fig. 3.55. Although the mechanical properties of the BCC lattice along the three principal directions are not independent, due to the broad range of properties, it is still possible to achieve multiaxial optimization.

Despite the mathematical simplicity, analytical modeling was shown to be an effective method for obtaining design insights with the geometry-property relationships, as well as comparing the characteristic behaviors of different cellular unit cell

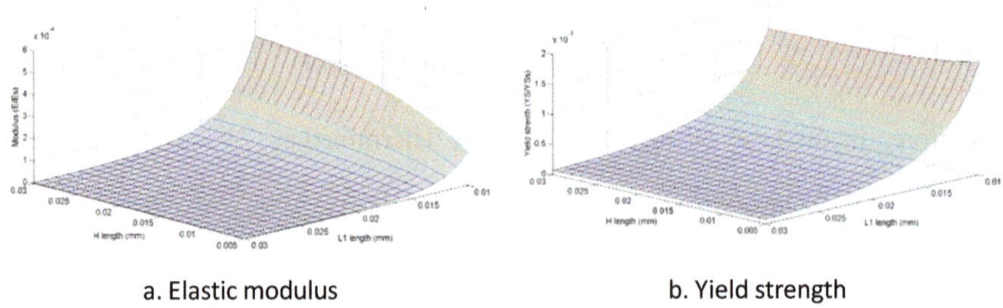

Fig. 3.55 Mechanical properties of BCC lattice as functions of geometrical parameters [99]. (**a**) Elastic modulus. (**b**) Yield strength

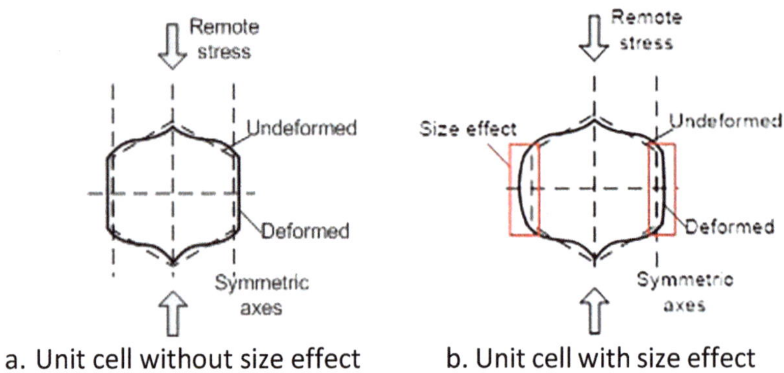

Fig. 3.56 Modeling of size effects for honeycomb structure. (**a**) Unit cell without size effect. (**b**) Unit cell with size effect

topologies. On the other hand, analytical modeling results are usually not quantitatively accurate. There are two major contributing factors for the inaccuracy issue:

1. The modeling conditions for the unit cell structure are typically not satisfied for real-world structures.
2. There typically exist various material property and quality variations within an AM cellular structure.

The first factor can be effectively understood as the size effect of the cellular structures, which was previously discussed for stochastic cellular structures. To further explain about the mechanism of size effect, a honeycomb unit cell under uniaxial compression is illustrated in Fig. 3.56. Under the uniaxial stress applied along the X_2-direction, the unit cell structure exhibits multiple types of symmetry including mirror symmetry and twofold rotational symmetry. When no size effect is considered, the cellular structures are considered to have infinite periodicity, and therefore the unit cell could be considered to also satisfy additional translational symmetry along the X_1-direction, which means that the vertical cell walls (i.e., cell walls along X_2-direction) at two sides of the unit cell could only be subjected to axial

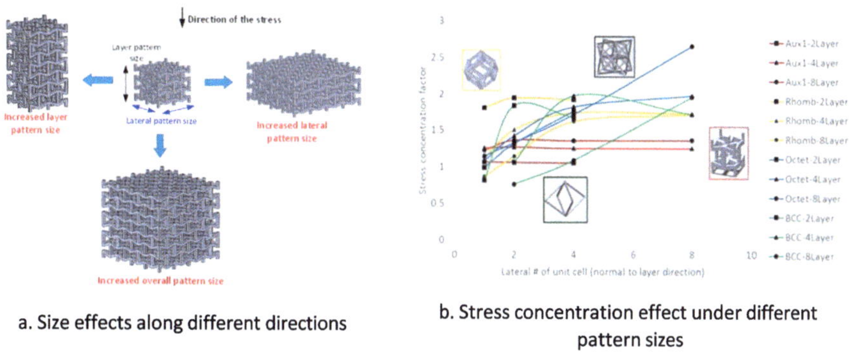

Fig. 3.57 Pattern size effects for some cellular structure designs from numerical study [101]. (**a**) Size effects along different directions. (**b**). Stress concentration effect under different pattern sizes

deformation. This in turn dictates that only compressive stress exists in these cell walls. On the other hand, when the unit cell is located near the boundary of a finite pattern, since the bending moment is not balanced for the vertical boundary cell walls, the cell walls will be subjected to additional bending as shown in Fig. 3.56b. Consequently, the mechanical characteristics of the unit cell would differ from the ideal conditions assumed for the unit cell modeling. It should be noted that although such nonideal boundary conditions could still be included in the analytical model via both force equilibrium and deformation compatibility, the analytical solution becomes increasingly computationally inhibitive due to the large number of variables [100]. Therefore, for the investigation of size effects, numerical and experimental approaches are more commonly employed.

Beside the unbounded free surfaces, fully bounded boundary conditions such as surface solid (e.g., cellular core bonded with skin panels for sandwich structures) also introduce local stress concentration effects near the constrained areas of the cellular structures. The effects of these nonideal boundary conditions may result in either increase or decrease of certain mechanical properties of the cellular structures and therefore should be further analyzed for specific designs. Also, it has been shown that the size effects of cellular structures are manifested not only as the overall structure sizes but also as the relative sizes along different pattern directions. As illustrated in Fig. 3.57, for cellular structures of different unit cell geometry designs, the structures exhibit significantly different size effects and along different pattern directions when subjected to uniaxial compression stress [101]. For example, for the BCC design, the stress concentration effects reach maximum values when the pattern sizes are identical along both the lateral and the layer directions, whereas for the reentrant auxetic structures, the size effect is largely independent to the layer pattern sizes (Fig. 3.57b). As the stress concentration effects are directly related to the failure of the structures, adequate accounting for size effects with cellular structures is critical for structural component designs.

An additional note related to the size effects of cellular structures is the homogenization treatment with many cellular structure design tools. Although the

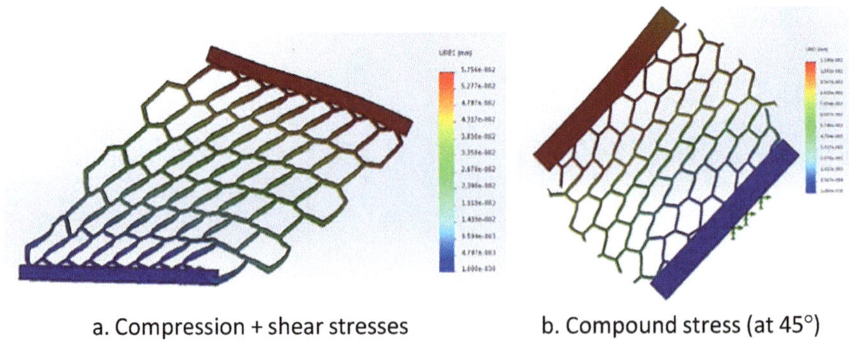

Fig. 3.58 Comparison of a honeycomb structure under two loading cases that are equivalent for a continuum solid. (**a**) Compression + shear stresses. (**b**) Compound stress (at 45°)

homogenization approaches could significantly simplify the computational costs of design by substituting sub-geometries with equivalent continuum, it also imposes the assumption of periodicity with the boundary conditions, effectively neglecting the nonideal and heterogeneous boundary conditions. With finite-size cellular structure designs that are commonly seen with real-world lightweight AM designs, the size effects, or nonideal boundary conditions, could not be ignored. Further, there also exist challenges for treating the unit cells as continuum material voxel. One example of the lack of stress and strain continuities with the cellular structures is illustrated in Fig. 3.58. In Fig. 3.58, the deformation of a hexagonal honeycomb structure under the combined loading of compression and shear stresses of equivalent quantity is significantly different from the case when the same structure is subjected to the equivalent compound stress at a 45° orientation. This can be understood from the fact that the honeycomb structure is constructed by discrete elements, which accommodate deformation compatibility via axial (compression/tensile) and bending deformations. Such characteristics obviously deviate from those of a continuum solid.

Lastly, it is worth noting that the concept of anisotropy and isotropy with cellular structures should be treated with care. Unlike continuum solids, non-stochastic cellular structures exhibit distinct and finite numbers of symmetry axes/planes. Therefore, even with large number of unit cells in a pattern, the structure is still subjected to the same symmetry conditions, which, in combination with the discrete construction as previously discussed, results in angular mechanical properties that likely deviate from continuum anisotropic/isotropic solids. An example is illustrated in Fig. 3.59, in which a 3D BCC cellular structure of relatively large pattern size (=10) is loaded uniaxially along various orientations [102]. The BCC structure exhibits significantly different deformation characteristics and resulting overall deformation that could not be adequately accounted for under a continuum material model framework. Currently, although full-scale structural analysis with FEA can be computationally intense, it might be necessary to achieve accurate results with the pre-damage mechanical analysis of cellular structures.

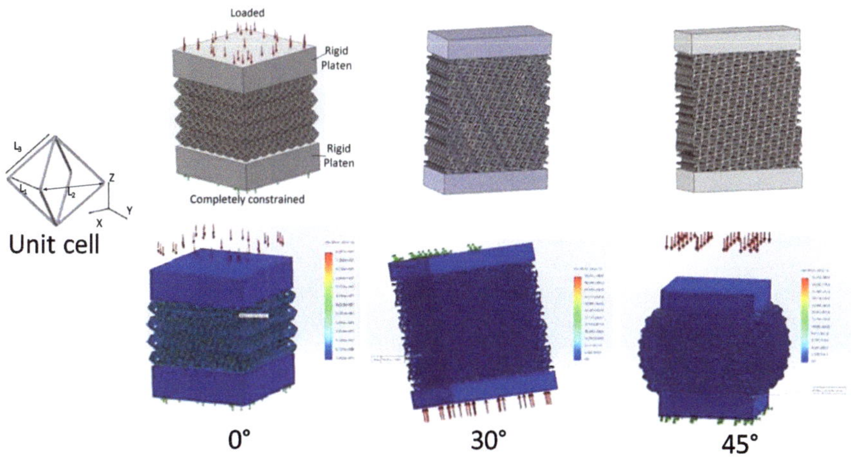

Fig. 3.59 Directional deformation characteristics of a BCC cellular pattern [102]

The second factor that affects the modeling accuracy of cellular structures is closely associated with the intrinsic process characteristics and quality variabilities of different AM processes, as previously discussed. For lightweight features such as thin walls and thin struts, the impact of process-induced stochasticity could be more pronounced due to the comparable dimensional scales of the fabricated features compared to the material feedstock or the processing mechanism, i.e., process discretization resolution.

Although relatively limited literatures are available for the investigation of manufacturing issues with cellular structures, various insights have been gained about the process-related issues with lightweight features, particularly with PBF. With PBF, it has been identified that factors such as energy density, energy beam power, scanning speed, part location, part orientation, scanning strategies, and powder feedstock sizes all have significant effect on the qualify variability of the cellular structures [103–108]. Staircase effect is more significant for cellular structures due to the small dimensions of the geometries. At smaller orientation angles, the staircase effects are influenced by two competing aspects, including the inclined angle effect and the projected cross-sectional area. While the smaller inclined angle determines that the overlap between the layers could be more negatively impacted, it also results in larger projected cross-sectional areas for the same-size feature, as illustrated in Fig. 3.60a. As a result, the minimum achievable orientation angle for the lightweight features can be potentially significantly smaller than 45°. Figure 3.60b and c shows the thin struts fabricated at both 20° and 70° orientations using EB-PBF [109]. The 20° strut was successfully fabricated, although there exist more significant dimensional fluctuation and surface defects.

Powder feedstock influences the quality of the PBF lightweight features in multiple ways. Firstly, powder attaches to the surface of the processed features via sintering, as a result of heat dissipation. As shown in Fig. 3.61a, these surface defects cause dimensional variations on the cellular struts. For the estimation of mechanical

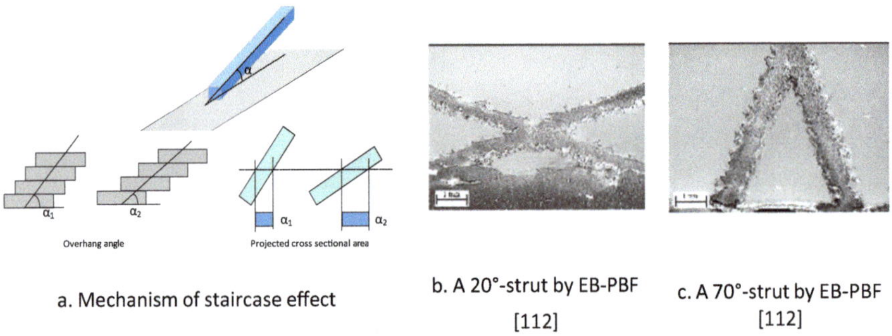

Fig. 3.60 Staircase effects for PBF lightweight structures. (**a**) Mechanism of staircase effect. (**b**) A 20°-strut by EB-PBF [109]. (**c**) A 70°-strut by EB-PBF [109]

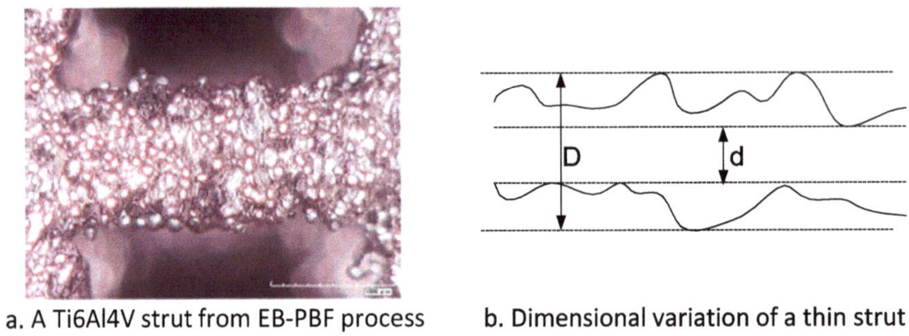

Fig. 3.61 Surface quality issue with thin struts [110]. (**a**) A Ti6Al4V strut from EB-PBF process. (**b**) Dimensional variation of a thin strut

properties, it was suggested that the minimum inscribed feature dimension (d in Fig. 3.61b) is preferred [110]. For bulky structures, these surface defects often serve as crack initiation sites that affect the fatigue performance. In comparison, for cellular structures, these defects could effectively influence the dimensions of the features and introduce local property variabilities. Literatures have shown that the fatigue strength of AM cellular structures is generally lower compared to that of the bulk feature fabricated via the same AM processes [103, 111–113]. Due to the complex geometries and the extensive existence of internal features, cellular structures are generally difficult to perform surface treatment with. As a result, the fatigue performance of the metal AM cellular structures still poses a significant barrier for their structural applications. Secondly, due to the coupling of powder particles with the melt pool during the process, smaller powder particle size could potentially benefit the stability of the melt pool as it advances as single-line scanning typical to thin feature processing [108]. On the other hand, as small-size powder bed often exhibits lower density due to lower spreadability, the ultimate outcome of thin feature quality exhibits nontrivial characteristics and would require additional studies for a more comprehensive understanding.

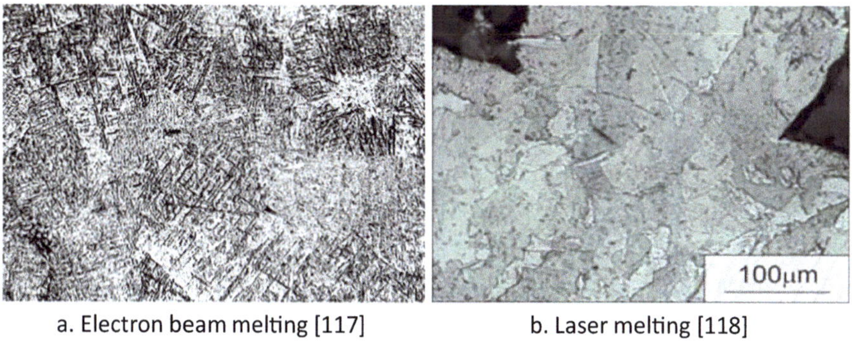

Fig. 3.62 Microstructure of PBF-AM Ti6Al4V cellular struts. (**a**) Electron beam melting [114]. (**b**) Laser melting [104]

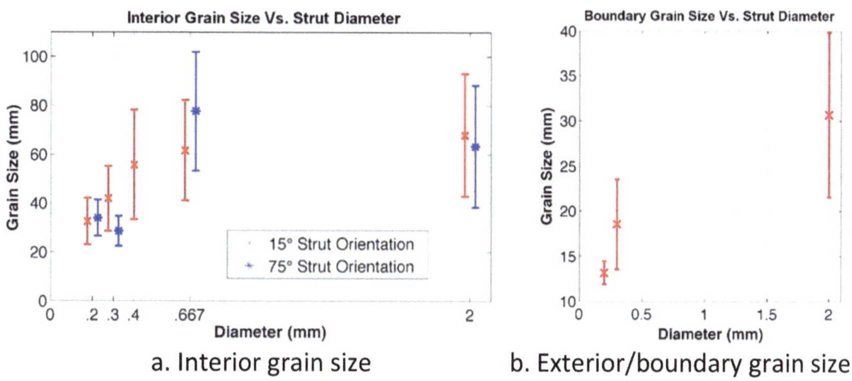

Fig. 3.63 Grain size dependency on orientation and strut diameter for Ti6Al4V fabricated by laser melting [115]. (**a**) Interior grain size. (**b**) Exterior/boundary grain size

The microstructure of the metal cellular structures fabricated via powder-bed fusion AM often exhibits more refined microstructural phases compared to the bulk structures. As shown in Fig. 3.62, for Ti6Al4V cellular structures fabricated via EB-PBF and L-PBF processes, the predominant microstructure is the fine-grained lath α' martensite, which is suggested to be contributed by the rapid cooling effect introduced by large surface area of the cellular geometries [104, 114]. In addition, the size of the prior β grains in the microstructure is also dependent on both strut orientation and strut dimensions [115]. As shown in Fig. 3.63, for Ti6Al4V thin struts fabricated by L-PBF process, with increasing feature dimensions, the microstructural grain size of the Ti6Al4V struts exhibits an increasing trend, while there also exist significant size differences between grains that are close to the exterior surface of the struts versus those that are at the interior of the struts. On the other hand, smaller overhanging angle results in more consistent grain size distribution throughout the strut. Such microstructural variation likely corresponds to

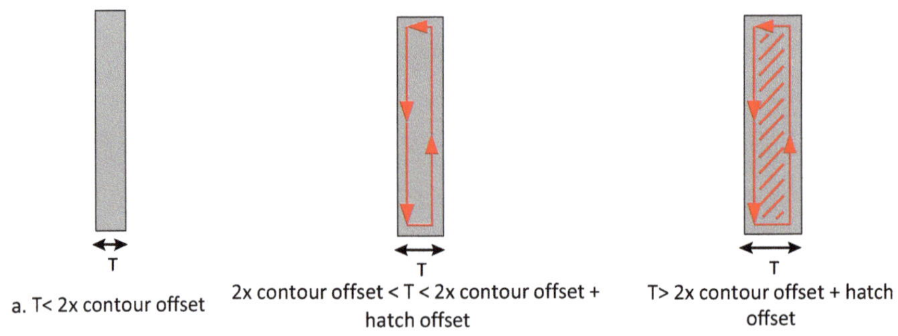

Fig. 3.64 Scanning strategies of lightweight feature with different characteristic dimensions. (**a**) T < 2x contour offset. (**b**) 2x contour offset < T < 2x contour offset + hatch offset. (**c**) T > 2x contour offset + hatch offset

mechanical property changes, indicating strong coupling effects between the geometry and process designs of the cellular structures and their mechanical properties. So far, the comprehensive knowledge for geometry-process-material (GPM) characteristics of PBF thin features is generally lacking. Typically, experimental characterization is carried out to establish empirical cellular feature quality baselines for further design purposes utilizing the same feature designs.

Another important process design variable is the scanning strategies. Although this is not among the commonly adjustable process variables with most of the current PBF systems, its potential impact of the manufacturability should not be overlooked. For example, many L-PBF systems employ the contour + hatch scan strategy. In addition, there also exist beam offset setting parameters. When the feature dimension becomes small as shown in Fig. 3.64, the actual scanning paths become more sensitive to the selection of scanning strategies and the setting of offset [107]. For characteristic dimension (T) of T < 2x contour offset, 2x contour offset <T < 2x contour offset + hatch offset, and T > 2x contour offset + hatch offset, the actual scan paths that are activated under the nominal contour + hatch scan strategy setting are different. When the designed feature dimension is smaller than the minimum threshold value for the contour scan to be activated, then regardless of the design, no actual contour melt would take place. Also, due to both the reduced heat sink volume and reduced hatch vector length, the artificial "overheating" effects due to reduced heat dissipation becomes more pronounced with the lightweight features. As a result, the activation of hatch scan strategy and the percentage of cross-sectional areas covered by hatch scan become prominent factors for determining the thermal evolution of the scanned areas.

In some cases, the choice of specific feature dimensions could result in significant defects with the features. As illustrated in Fig. 3.65, for a cellular strut processed by an EOS M270 L-PBF system, due to the 0.1 mm beam offset setting, the 0.2 mm strut resulted in significant overlap of scan tracks from both the contour scan and a specific "edge" scan from the system [107]. The resulted thin strut

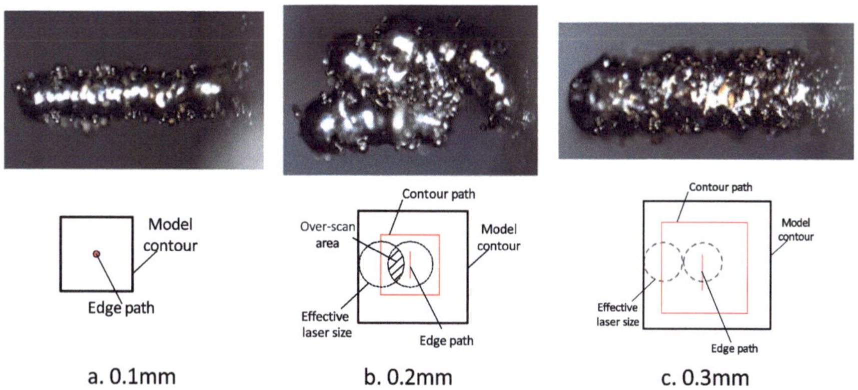

Fig. 3.65 Scanning path overlap for different features under contour + edge scanning (EOS M270) [107]. (**a**) 0.1 mm. (**b**) 0.2 mm. (**c**) 0.3 mm

exhibited severely deformed shape, likely associated with the overheating/overmelting effects from overlapped scan.

For the other AM processes, similar geometry dependency with material characteristics also exists for lightweight features. For example, with material extrusion, due to the contour + infill construction, the small-size lightweight feature processing would be primarily generated by contour toolpaths, resulting in unique shell-type structural characteristics. Also, depending on the size of the extruded tracks, the feasible sizes of the lightweight features might also be somewhat constrained. Ultimately, the accurate design of properties for AM cellular structures requires adequate understanding of the geometry-process-material (GPM) characteristics that necessitates a more integrated design approach that deviates from the traditional design frameworks. For organizations, this might even imply the change of organizational structures and supply chain structures. Despite the current challenges, existing approaches for adopting AM lightweight design in product development still provide feasible pathway toward successful application, as long as the unique design aspects are accounted for.

There exist some research literatures of integrated design for AM cellular structures. An example of simultaneous design for both the geometry and the process for cellular structure is illustrated in Fig. 3.66 [116]. In this example, both the mechanical properties and the thermal residual stress of unit cells at different geometry and process scan strategy conditions are pre-calculated, which are used subsequently used to establish homogenized unit material voxel for the iterative design optimization of both geometry and process. Although the primary design criteria employed in this study were only thermal residual stress, it could be reasonably anticipated that similar frameworks could be utilized toward multifaceted design requirements of AM cellular structures.

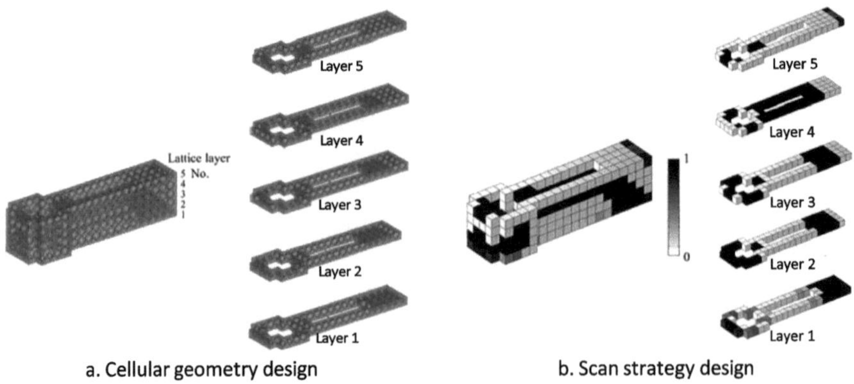

Fig. 3.66 Simultaneous design of both cellular geometry and scan strategy for a generic bracket component for minimization of thermal residual stress [116]. (**a**) Cellular geometry design. (**b**) Scan strategy design

3.4 Material Design Consideration

Commercial AM systems usually come with preset parameter sets for specific types of materials. These are usually the materials that are experimentally tested and verified by the system manufacturers for their properties and qualities. More stringent quality requirements are imposed when the intended applications involve reliability and durability, which corresponds to fewer material options. In general, polymer AM systems offer more material options, including both default ones from the system manufacturers and third-party custom feedstocks, compared to the metal AM systems. When considering about material aspects for AM designs, several key aspects should be closely noted:

(a) Specific AM technology types are often closely associated with specific material classes, which could impose significant restrictions for material selection.
(b) AM design could potentially employ structure/architecture to make up intrinsic material property limitations and therefore ease the material constraint.
(c) Even with known material types, the properties of the materials could be different from the known databases.
(d) There exist compound geometry-process-material (GPM) relationships for the properties of most AM materials.

There exist extensive literatures and databases on the mechanical and physical properties of metal AM materials from experimentation. Many of these reference data were generated from standard testing coupons (e.g., ASTM D536, ASTM E8, ISO 6892), while the process setting employed for the fabrication of the testing coupons is often not consistently defined or not even intentionally controlled/recorded. As a result, much of the existing data should only be used as a general reference for preliminary design or process development purposes.

DED-processed materials exhibit many material characteristics similar to those of welding, due to the considerable similarity of process configurations between the two. Most commonly used materials include steels (e.g., 316 L, 17-4PH), Ni-base superalloys (e.g., IN718, IN625), Al alloys (e.g., Al12Si, AlSi40), and Ti alloys (e.g., Ti6Al4V), which generally exhibit comparable or even superior yield strength and ultimate strength compared to the wrought parts. On the other hand, the ductility tends to be lower [85]. The fatigue properties of the DED materials are significantly influenced by internal material porosity, which is of common occurrence. It has been suggested that due to the larger size and often sharp-shaped porosity characteristic, the DED materials generally exhibit lower fatigue endurance compared to their PBF counterparts [117]. On the other hand, surface and near-surface defects might be of less concern with many current DED applications, due to the post-printing machining finishing operations that are almost always employed to generate the final dimensions.

One distinct material design flexibility with the DED-based AM designs is the capability to consider heterogeneous material compositional control. Compared to PBF, which employs layer-by-layer powder spreading and thus very limited heterogeneous composition control, the DED processes could achieve location-specific material composition control in theory. This is particularly feasible for powder-fed DED systems, in which multiple powder ingredients could be fed through multiple feeding heads into the melt pool with dosage control to realize in situ alloying. Although the multi-material selection is still constrained by their chemical/metallurgical compatibility, additional process adjustment is possible to alleviate such issue. One example is the concept of functionally graded material (FGM) design, in which the compositional ratios of multi-materials could be varied following certain predesigned spatial functions to realize gradual transition of material composition from one location to another [118]. FGM concept has been demonstrated with various multi-material combinations that exhibit low intrinsic compatibility, although the understanding of the influence of particular transition function on the properties and defect occurrence of FGM constructions is still largely lacking. With many real-world applications, heterogeneous material microstructure and properties are considered highly undesirable from the perspective of quality control. With few tools capable of accurate microstructure prediction, such advanced material design concept is still not practical.

With highly directional thermal gradient introduced by the process, the DED microstructure often exhibits characteristic long columnar grain morphology oriented toward the build direction (z-direction). If not carefully controlled, the microstructure could also exhibit extensive grain width. On the other hand, nucleation agents could be added to the material matrix to promote the reduction of grain sizes or even the generation of equiaxed grain morphology when combined with process scan strategy optimization [119]. In general, the materials/structures processed by DED are subjected to additional heat treatment for either thermal residual stress relieving or stabilization of microstructure, although the primary microstructural morphology is usually not significantly altered afterwards.

Compared to the DED, more material options have been demonstrated with PBF, even though the two types of AM processes exhibit sufficient similarities with material compatibility. Also, compared to wire feedstock, powder feedstock introduces additional aspects of attention during the design process.

The powder feedstock employed for both DED and PBF exhibits stochastic size distribution and shape variation. When the energy beam is applied to the powder-bed surface, part of the incipient energy is absorbed by the powder particle, while the remaining energy is carried away with the reflected energy beam. The energy beam could travel in powder bed through multiple reflections and absorptions before being completely absorbed or escaping from powder bed. Due to this phenomenon, the total energy absorption rate (energy incoupling) for the powder bed is generally much higher than that of the solid material [120]. In addition, this also implies that the energy absorption rate for the powder-bed fusion AM is powder batch specific, which sometimes necessitates experimental characterization when accurate energy absorption information is needed. Figure 3.67 shows some of the simulated results for the relationship between energy incoupling and solid material absorption for laser melting process, as well as the energy absorption profile in the depth direction of the powder bed [120]. The highest energy absorption occurs in regions underneath the surface of the powder bed, which is in turn dependent on the powder bed characteristics such as powder morphology and particle sizes.

Although higher powder-bed porosity might promote the overall energy incoupling, it also results in more dispersed distribution of energy and therefore compromises the fabrication quality by reducing geometrical accuracy and introducing internal defects. It should also be noted that the selection of powder size distributions is also closely related to the powder-spreading mechanisms in each individual system. Multiple powder-spreading mechanisms have been employed in various powder-bed fusion AM systems, including roller, comb blade, and sweep blade. It is commonly perceived that the roller mechanism could potentially result in higher powder-bed packing density that is beneficial for quality control of the processes. On the other hand, the additional shear and compressive stress exerted to the powder

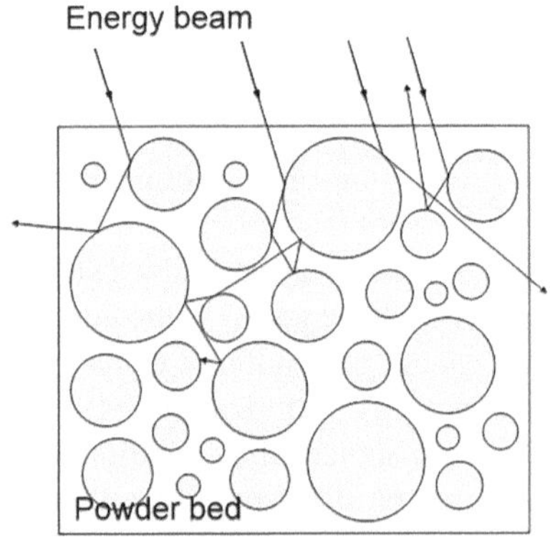

Fig. 3.67 Simulated energy absorption characteristics for laser melting processes [120]. (**a**) Energy incoupling vs. solid material absorption. (**b**) Energy absorption profile vs. powder-bed depth

bed via rolling could potentially lead to loss of accuracies of the parts and even damage to the structures. This is especially significant when the surface is not perfectly flat due to either local defects or part warping, since the roller would interfere with the surfaces. In addition, for powder with relatively low flowability, roller mechanism also tends to be less effective in rendering uniform and complete powder spreading. On the other hand, mechanisms such as comb blade and moving hopper tend to exert less compacting effects on the powder bed; however, they are also more forgiving to surface imperfections. In recent years, noncontact powder-spreading concepts have also been demonstrated, such as vibration nozzle, patterning drum, and gravity-fed hopper [121]. These mechanisms could potentially overcome some limitations associated with the current powder feedstock such as particle size limitation, flowability issue, and material utilization efficiency. On the other hand, without additional compaction effect, the powder bed generated from the noncontact powder-spreading methods might exhibit lower density.

Although currently there is limited consensus on the optimal powder feedstock size distributions for specific part and process designs, it is generally observed that narrower particle size distribution usually corresponds to increased feedstock flowability and spreadability, which is beneficial for the mechanical properties of the materials. Numerous studies have shown that bimodal distribution usually results in higher powder-bed density and consequently better part surface quality, compared to mono-size feedstock [122–127]. A particle size ratio of 1:10 between the two mean particle sizes along with a percentage ratio of 10–15% for the smaller particles was suggested to yield optimal powder-bed density and surface finish [124, 125]. It was suggested that powder with smaller particle sizes is generally more favorable for fabrication as long as it could be successfully spread, since it facilitates more efficient melting that is beneficial for achieving higher part density and surface quality [128]. However, the larger particles in the powder bed also appear to bring about some benefits such as higher elongation [123].

PBF materials typically exhibit anisotropic mechanical properties, due to the directional growth of microstructure under the influence of layer-wise melting/heating, as well as the nonuniform distribution of defects. For both EB-PBF and L-PBF, columnar grains form along the build direction (z-direction) in most cases unless the scanning strategies are intentionally designed to alter the dominant thermal gradient directions. In most studies, it was reported that the horizontally fabricated tensile coupons (i.e., XY or YX orientation per ASTM F2971-13) exhibit higher quasi-static mechanical properties (e.g., yield strength, ultimate strength, elongation) compared to vertically fabricated samples (i.e., z-direction).

Fatigue properties for the as-fabricated PBF metal materials are generally significantly lower compared to the wrought ones, which is largely caused by the combination of internal defects, surface roughness, and thermal residual stress. For example, Figs. 3.68 and 3.69 show some of the compiled results for Ti6Al4V from recent experimental studies for both laser- and electron beam-based systems. As-received AM Ti6Al4V samples exhibit about 0.1–0.15σ_y for endurance level with σ_y representing the quasi-static yield strength of the material, which is only 25–38% of the typical endurance level for solid wrought Ti6Al4V (0.4σ_y). It was

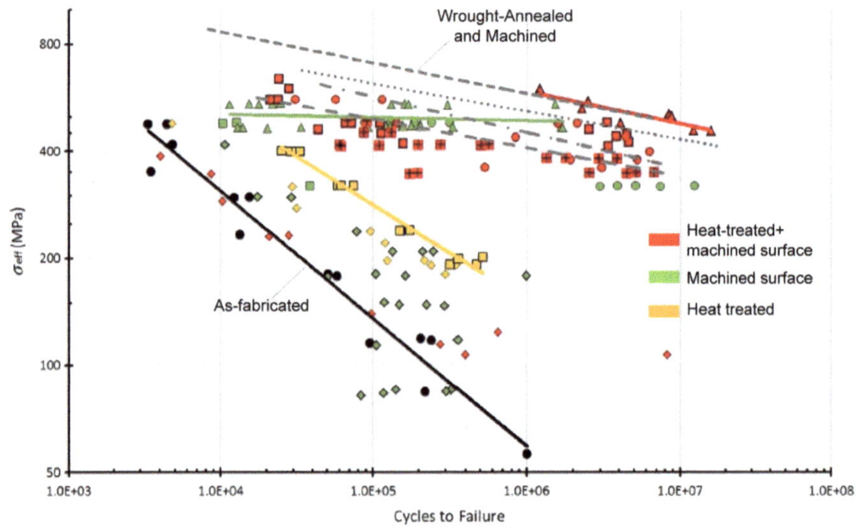

Fig. 3.68 Uniaxial fatigue performance of laser-melted Ti6Al4V parts by laser melting at $R \approx 0$ from various experimental studies [131]

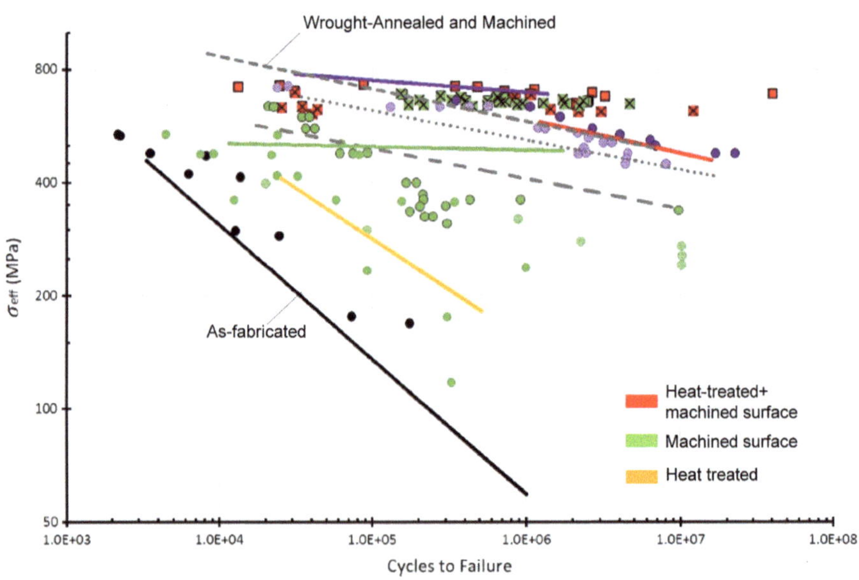

Fig. 3.69 Uniaxial fatigue performance of laser-melted Ti6Al4V parts by electron beam melting at $R \approx 0$ from various experimental studies [131]

suggested that for bulky structures, the internal defects become dominant factors for the deterioration of the fatigue performance of these AM parts, while thermal residual stress issues can be largely alleviated via proper heat treatment. On the other hand, when process parameters and post-processes are properly designed, surface defects caused by surface sintering and process instability become a significant contributor for crack initiation and premature part failure under cyclic loading. Such issue could become more significant when lightweight structures are designed, which generally contain large specific surface areas [111, 129, 130]. Due to the difficulty of surface treatment with AM structures, surface defects appear to be a more difficult challenge for the fatigue performance improvement of AM metal parts. In addition, the columnar grain morphology of both the laser- and electron beam-melted structures also favors low cycle fatigue, which in turn may be directly correlated to the surface crack-dominated failure behavior associated with high cycle fatigue experiments with AM metals [113].

It should be noted that much of the discussions about material properties and material designs with AM are highly coupled with the process conditions. For example, the microstructure and mechanical properties of AM metal materials are inextricably associated with the corresponding process system and process setup. In fact, with most of the AM product designs, the material design is approached via process adjustment.

3.5 Process Design Consideration

Both laser- and electron beam-based AM systems predominantly employ energy beam profiles that could be best described as Gaussian distribution following energy intensity function of Eq. (3.19), as illustrated in Fig. 3.70. In recent years, it was suggested that the Gaussian distribution of energy beam is partially responsible for certain difficulties with the process quality control of the PBF process, such as the process instability, unfavorable thermal gradient, and local evaporation. Novel energy intensity distribution forms such as ring distribution [132, 133], ring-point distribution [134], ring and double-peak distribution [135], and elliptical distribution [136] have been shown to provide various advantages over the traditional Gaussian distribution. The beam profile control can be achieved by multiple

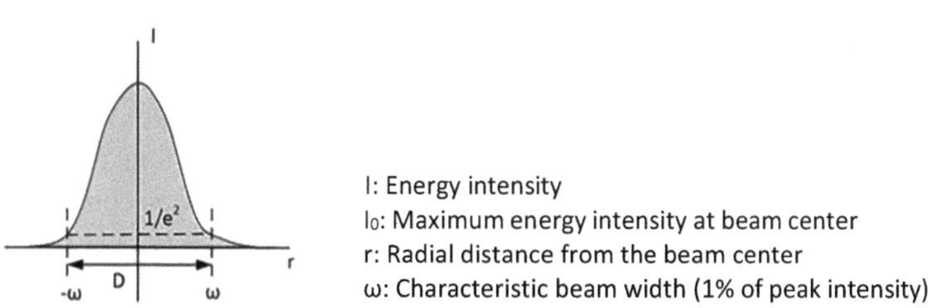

Fig. 3.70 2D representation of Gaussian beam energy intensity distribution

methods, such as adjustable in-line optical modulators (refractive, reflective, or diffractive), zoom lens, beam combiners, and segmented optic fiber [137]. The change of energy intensity profile directly impacts the melt pool characteristics such as shape, size, and temperature distribution, which in turn influences the melt pool dynamics such as Marangoni convection, spatter, as well as solidification process. Ultimately, the microstructure, thermal residual stress distortion, and mechanical properties of the materials are also altered. Due to the nonlinearity of the physics involved in the PBF heating/melting and solidification process, the design of the energy intensity profile for specific material property objectives is a nontrivial task, and currently very little knowledge exists. On the other hand, for specific processing cases, it is possible to employ physics-based simulation to optimize the beam profile for more homogenized temperature distribution [135].

$$I(r,\omega) = I_0 \exp\left(\frac{2r^2}{\omega^2}\right) \tag{3.19}$$

I: Energy intensity

I_0: Maximum energy intensity at beam center

r: Radial distance from the beam center

ω: Characteristic beam width (1% of peak intensity)

For both PBF and DED processes, the process quality is closely associated with the melt pool characteristics, as it defines the minimum process voxel. There exist a number of physics factors that drive the melt pool characteristics and evolution, including heat transfer, phase transformation, dynamic fluid flow and particle flow, and surface forces. Once formed, the melt pool is subjected to forces such as Marangoni convection force, capillary force, surface tension, evaporation recoil pressure, and viscous force. Meanwhile, the melt pool is subjected to continuing heat from the energy source, heat dissipation via surrounding powder bed, solid phase underneath and environmental gas phase, and heat loss due to evaporation and spatter. As a result, the melt pool exhibits highly dynamic characteristics as well as instability that can be caused by mechanisms such as Plateau-Rayleigh instability, keyholing, and wetting with powder bed, which in turn causes various process variabilities and quality fluctuations.

Simulation-based studies have shown that capillary forces exert a significant effect in the formation of continuous melting pool and therefore continuous solid tracks upon solidification, as illustrated in Fig. 3.71 [30, 138–140]. When the dimensions of the melting pool are comparable to the powder particle sizes, the capillary force dominates the behavior of the melting pool, and the solidified track exhibits Plateau-Rayleigh instability as shown in Fig. 3.72, which results in geometrical dimension variations and increased surface roughness. In certain circumstances, under the effect of Plateau-Rayleigh instability, the liquid phases could also break up into spherical droplets and cause the so-called balling effects [141, 142]. Input energy density, which is typically expressed as Eq. (3.20), has been shown to exert significant effect on the balling phenomenon during the melting-solidification of the

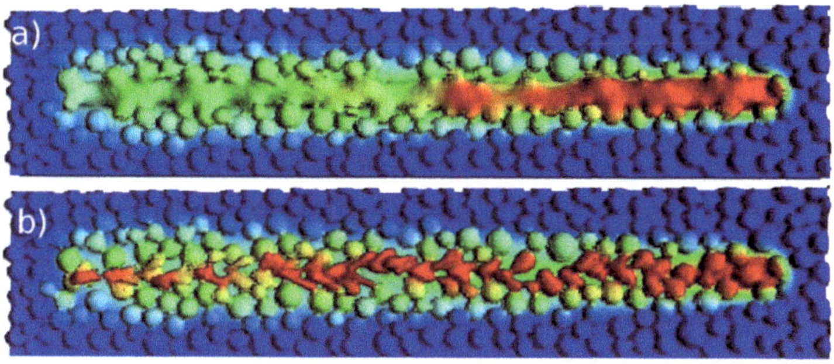

Fig. 3.71 Single-track simulation of laser melting of powder bed. (**a**) Capillary force considered, (**b**) no capillary force considered [30]

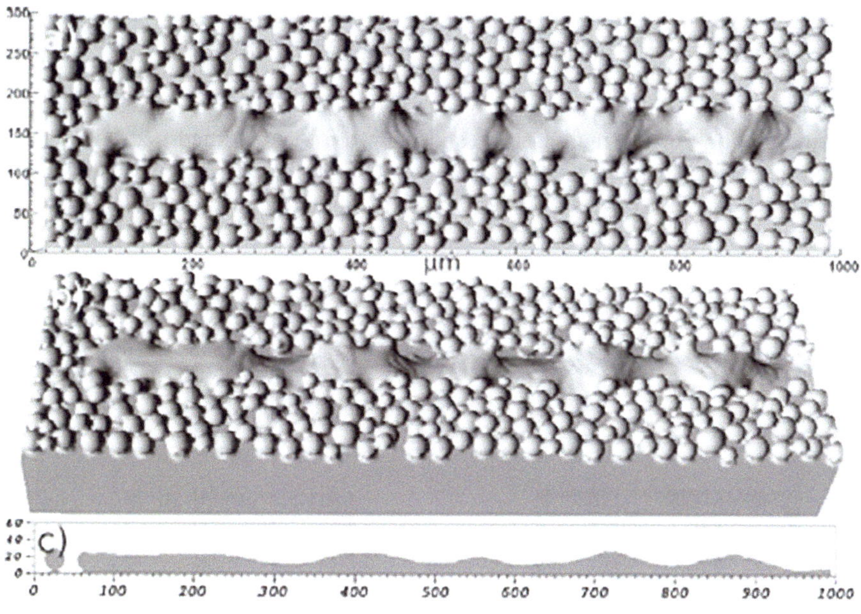

Fig. 3.72 Instability of the single-track melting [30]

powder bed. In Eq. (3.20), E is the energy density in J/mm^3, P is the energy beam power, v is the scanning speed, t is the energy absorption thickness, d is the energy absorption width, and n is the number of repeated scanning. Excessive energy density could potentially cause balling due to the elongated existence time of the liquid phase, which leads to more sufficient coalescence driven by surface energy. However, this process can be complicated by the interaction between the liquid phase and the powder bed. When the process takes place under high energy density and very slow energy beam scanning speed, the balling phenomenon could be avoided, which is suggested to be contributed by the widened melting track width and wetting of liquid phase with the surrounding powder bed [139, 143, 144]. In order to counter the process instability caused by both Plateau-Rayleigh instability and the stochastic powder-bed characteristics, it is suggested that thinner layer thickness and slower beam scanning speed should be adopted when the total input energy density is held constant [139]. Such process setups could potentially result in narrower track width and improved surface finish as shown in Fig. 3.73 due to more sufficient melting and liquid continuity in the vertical directions. The significance of surface capillary effects in the powder-bed melting process also helps understanding the importance of powder-bed density, as higher powder-bed density would facilitate better wetting of liquid and alleviation of balling effects [145].

$$E = \frac{np}{vtd} \tag{3.20}$$

Within the melting pool, due to the energy intensity distribution and rapid motion of the energy sources, the temperature distribution of the melting pools is highly heterogeneous. Due to the temperature gradient, there exists internal convective flow in the melting pool or the Marangoni convection [146]. Marangoni convection affects the morphology of the melting pool and therefore the solidified track. For many metal alloys that have negative temperature-surface tension gradients, the occurrence of Marangoni convection would result in outward flow of the molten liquid as

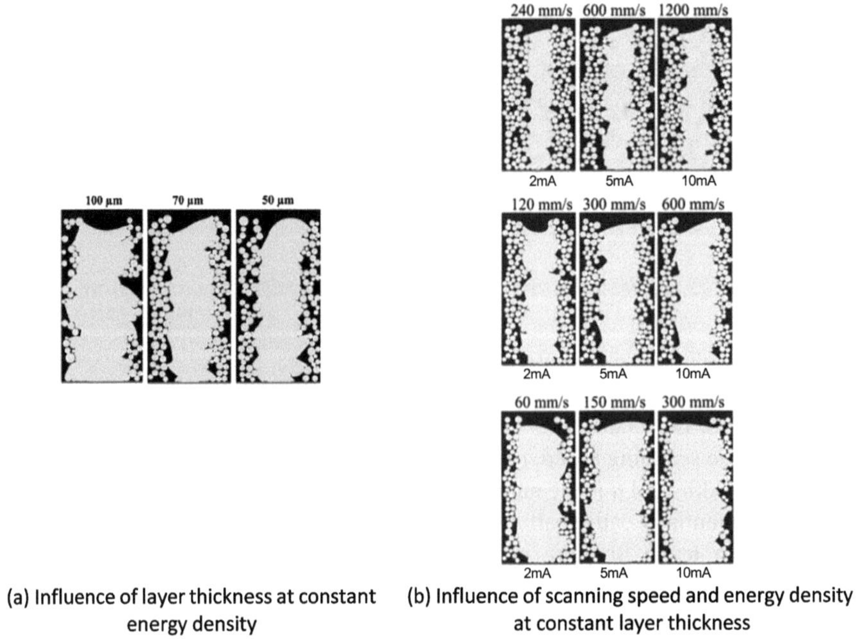

Fig. 3.73 Effect of process parameters on the multilayer single-wall formation [139]. (**a**) Influence of layer thickness at constant energy density. (**b**) Influence of scanning speed and energy density at constant layer thickness

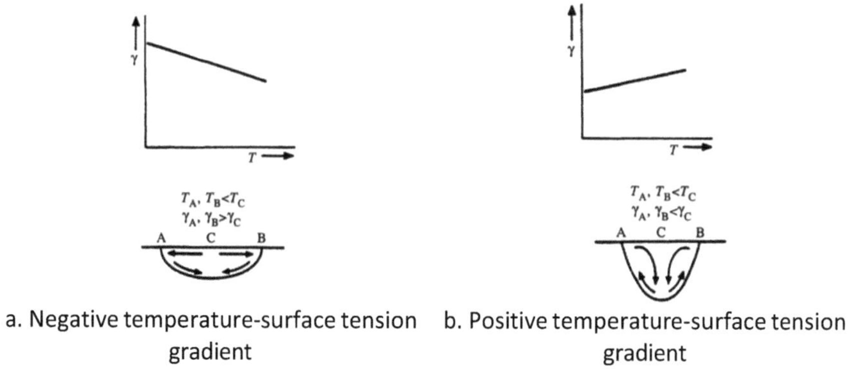

Fig. 3.74 Marangoni effect in powder-bed melting pool [146]. (**a**) Negative temperature-surface tension gradient. (**b**) Positive temperature-surface tension gradient

shown in Fig. 3.74a, since the temperature at the beam center is higher than the surrounding areas. As a result, the melting pool's shape becomes wider and flatter. On the other hand, in case of materials with positive temperature-surface tension gradients, Marangoni convection results in a narrower and deeper melting pool as shown in Fig. 3.74b.

Besides the effect with the melting pool characteristics, Marangoni effect plays an important role in producing continuous smooth track at high input energy densities [147, 148], and it could also be intentionally utilized to create a turbulent effect and to break the surface oxide layers of the substrate for reactive alloys such as aluminum [149].

Evaporation and the consequent gas-phase formation within the melt pool occur when keyhole phenomenon occurs in the PBF process, which is in turn closely associated with the high local input energy density. The generated gas phase could potentially become trapped in the rear side of the melt pool as isolated gas bubbles, which eventually forms small-sized and spherical-shaped pores within the microstructure [150]. Also, the vapor flow interacts with both the melt pool fluid and the surrounding powder particles, which results in denudation along the scanning tracks and particle ejection as spatter [151]. Figure 3.75a illustrates an in situ X-ray image of the melt pool for L-PBF, in which the various phenomena were also marked out, such as the shape of the melt pool and the formation of spatter and denudation [152]. Figure 3.75b illustrates another in situ X-ray image of the formation of keyhole defects [153]. Due to the abrupt increase of energy coupling efficiency of the molten liquid phase to the incident laser, keyhole phenomena and the associated defects are difficult to completely avoid with the existing L-PBF systems. Instead, various efforts are now focusing on the direction of quantifying the stochastic defect occurrence and establishing corresponding quality design factors useful for practical purposes.

Regarding the energy density equation Eq. (3.20), during the process development of a certain material, the layer thickness is usually fixed at small values to

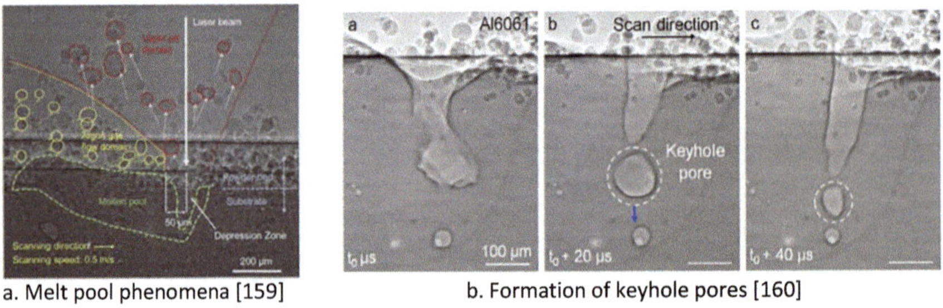

Fig. 3.75 Melt pool dynamics and defect generation in L-PBF, in situ X-ray imaging. (**a**) Melt pool phenomena [152]. (**b**) Formation of keyhole pores [153]

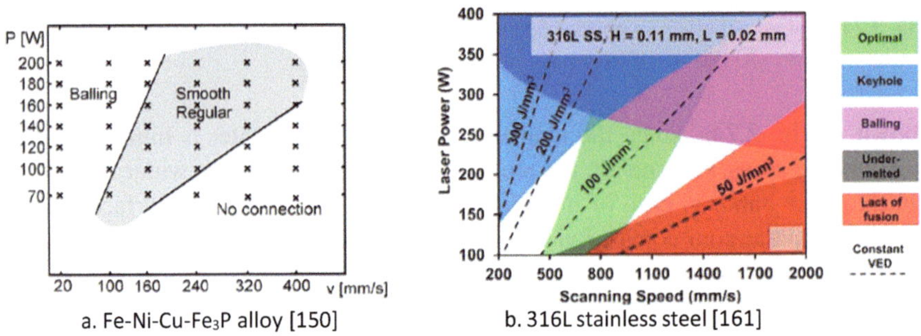

Fig. 3.76 Typical power scan speed process map for L-PBF processes. (**a**) Fe-Ni-Cu-Fe_3P alloy [143]. (**b**) 316 L stainless steel [154]

facilitate the overall quality control as long as the fabrication time is not a paramount priority. Therefore, a common experimental-based approach for process development is to make adjustment with beam power, scanning speed, and scan spacing, while fixing a preselected layer thickness level. When the scan spacing is fixed, various studies have shown that there exist optimal beam power/scanning speed ratio envelopes in which the PBF-AM processes could yield good overall qualities including part densities and strength. As shown in Fig. 3.76, at medium power/velocity ratio ranges, the process generally corresponds to stable melt pool dynamics and low occurrence of porosity defects. On the other hand, at either high or low power/velocity ratios, defects such as balling effects, keyholing, and lack of fusion could occur. It should be noted that the power/velocity-based process mapping only provides a particular tuning strategy for process design that involves only power level and scan speed and therefore may lead to limited optimality. For example, scan strategy often offers a more effective option for alleviating thermal residual stress and microstructural grain orientation, and scan line spacing can often be adjusted together with power and scan velocity to introduce particular cooling rates for the process. Also, the numerical values of these power/velocity process mapping are often specific to the system/material combination used for the experimentation and therefore should be referenced with care. On the other hand, the trends observed from the process mapping are more generalizable.

The limitations of the traditional process mapping for L-PBF as discussed above have been widely acknowledged, due to its negligence of material characteristics that are relevant to the melting process, such as melting temperature, heat capacity, and heat transfer coefficient. Alternatively, normalized enthalpy has been used as a process design indicator, which is the ratio between deposited energy density and the total volumetric melt enthalpy of the material, as defined in Eq. (3.21):

$$\frac{\Delta H}{H_S} = \frac{\alpha P}{\pi \rho C_p T_m \sqrt{D_t v r^3}} \tag{3.21}$$

where ΔH and H_s are deposited energy intensity and total melt enthalpy, α is the absorptivity, ρ is the density of the material, C_p is the material heat capacity, T_m is the material melting temperature, D_t is the material diffusivity, v is the scan speed, and r is the energy beam radius. The normalized enthalpy is shown to be more closely correlated to the shape and size of the melt pool, which are in turn utilized to indicate process stability.

Dimensionless process characteristics are also being increasingly explored as process indicators. Dimensionless characteristics possess several potential advantages over energy density and enthalpy, such as translatability between different process systems and material types and easiness for process mapping analysis. For example, it was suggested that there exist robust correlations between the normalized enthalpy and the melt pool depth normalized by material diffusivity, as illustrated in Fig. 3.77a [155]. Another work suggested that two Peclet-type dimensionless

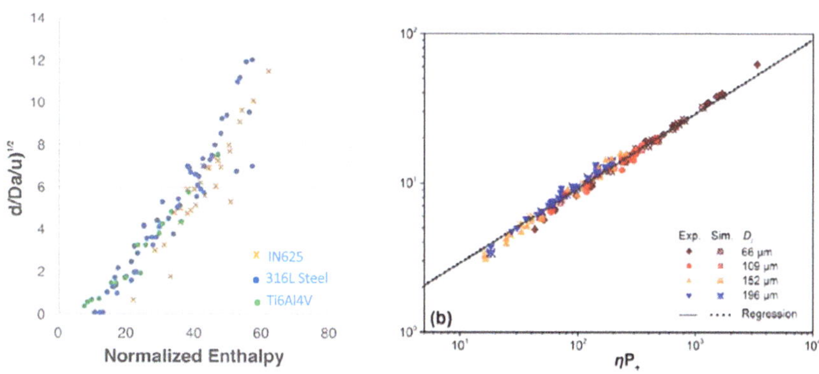

a. Normalized melt pool depth vs. normalized enthalpy. d: melt pool depth, D: thermal diffusivity, a: beam radius, u: scan speed [162]

b. Two Peclet-type parameters, P_e vs. P+, for 316L stainless steel [163]

Fig. 3.77 Correlations between dimensionless parameters for process mapping. (**a**) Normalized melt pool depth vs. normalized enthalpy. d: melt pool depth, D: thermal diffusivity, a: beam radius, u: scan speed [155]. (**b**) Two Peclet-type parameters, P_e vs. $P+$, for 316 L stainless steel [156] parameters, P_e and P_+, can form robust correlations with the scaling factor H_{adi} derived from effective energy input density, as illustrated in Fig. 3.77b, with dimensionless parameters defined in Eq. (3.22) [156].

$$P_e = \frac{bv\rho c}{k} \tag{3.22a}$$

$$P_+ = \frac{Pv\rho c}{k^2 \Delta T} \tag{3.22b}$$

$$H_{\mathrm{adj}} = \frac{b^2 v\rho c \Delta T}{P} \tag{3.22c}$$

In Eq. (3.22), b and v represent melt pool width and laser scan speed; ρ, c, and k represent the density, heat capacity, and thermal conductivity of the material, respectively; $\Delta T = T_m - T$ is the characteristic temperature difference between the melting point (T_m) and processing ambient temperature (T); and P is the laser energy. The establishment of process parameter correlations and relationships is likely dependent on the process principles and physics domains of process window. For example, Eq. (3.22) was suggested to be most effective for conduction-type melt pool, which exhibits minimum keyhole phenomena and less turbulent melt pool dynamics. Also, while there exist similar dimensionless parameters for DED process that establish significant correlations between melt pool dimension and process inputs, the formula might take different forms due to the differences of process setting between the DED and PBF processes [157, 158].

It is important to note that currently the process design knowledge for metal AM is highly restricted to only several common process control parameters, which is highly incomprehensive and sometimes even misleading. There exist various uncontrollable or even inaccessible process parameters between different AM systems that could very likely influence the outcome of material properties and qualities. On the other hand, for establishing adequate understanding of the important aspects for efficient in-house process development, the spectrum of canonical disciplinary knowledge that is required imposes often inhibitive learning curve. There currently exists a gap between the state-of-the-art understanding with the metal AM process physics and the need of real-world process design optimization for products. In the near term, data modeling-based design could potentially provide a solution. As AM community embraces the current wave of technological revolution fueled by advanced data modeling and artificial intelligence, the use of various machine learning tools for AM process and quality modeling will become more widespread. Although the effectiveness of the data-driven modeling approach is yet to be adequately demonstrated for complex process design problems, the prospective appears positive.

3.5.1 Post-printing Processing

Regardless of the optimality of the process, with as-received AM parts, post-processing is usually considered a necessary and integral step of the full process design, even though sometimes this aspect is overlooked during the initial design stage. At the current stage of technology, the as-received AM parts usually do not have all the required microstructural characteristics or surface quality, and correspondingly the post-printing processing could be generally categorized into surface processing and microstructure processing.

Surface processing: Surface finishing is often considered as an integral part of the AM production workflow due to the intrinsic surface texture and roughness with the layer-wise AM process characteristics. There exist a wide range of surface processing techniques for traditional manufacturing-based parts of different materials, and many of them have been evaluated or demonstrated on AM parts. In general, the knowledge for process design for surface finishing AM parts is highly scattered and unstructured, despite the existence of a considerable body of literatures. This is partly due to the complexity of the part and process design conditions, as well as the lack of generality with the specific guidelines and conclusions. From the perspective of process and geometry design, the surface characteristics and qualities are closely related to a larger number of design variables, often in highly coupled manner. For example, the surface textures of material extrusion process are determined by various process parameters such as track diameter, spacing between neighboring tracks, printhead nozzle clearance, printing temperature, printing toolpath, as well as geometry designs [159, 160]. For PBF, the surface textures and qualities of parts exhibit complex characteristics under the combined effect of melt pool and individual powder particles, which are in turn determined by virtually all the process parameters as well as geometry of the parts [161, 162]. Further, the different post-printing surface process methods introduce different resulting surface characteristics and near-surface material microstructural characteristics, potentially resulting in altered material properties [163–165]. On determining the adequate surface conditions for performance, there also exist knowledge gaps that lead to lack of consensus with the processing objectives. Although it is generally recognized that surface textures and defects can act as crack initiation and thus reduce the fatigue endurance of the parts [166–168], recent works have suggested that traditional surface roughness characteristics such as R_a (arithmetic average roughness) might not serve as a reliable metric for determining the criticality of surface defects and thus would fall inadequate as a quality metric for metal PBF parts [168–170]. As such, the effectiveness of different surface processing methods is not well established.

With regard to specific surface process approaches, multiple process technologies have been more commonly employed in "commercial" scales, such as mechanical mass finishing [171, 172], chemical polishing/milling, electropolishing [173, 174], electroplating, shot peening [165, 175], surface coating, machining [163, 164], and abrasive flow polishing [171]. Much of the established knowledge about the advantages and limitations of these technologies from traditional manufacturing still largely applies to the AM cases [176]. For mass finishing processes, accurate and location-specific tolerance control is generally difficult, and the processes usually do not provide high-level dimensional tolerance accuracy. The process development of mechanical mass finishing is also usually based on highly empirical design of experiment (DOE) methods [171, 172]. Chemical solution-based surface treatments such as chemical etching, electropolishing, and electroplating generally possess the capability of accessing surfaces of complex geometries in a more controlled manner, while often introducing nonuniform surface processing rates that are influenced by factors such as the flow of the chemical solutions, the local strength of electrical field, and the presence of surface textures of different dimensional scales. The polishing uniformity is a common issue with most of the solutions, and the effectiveness can be highly dependent on specific part geometries and process histories. Currently, product-specific process development appears to be the most economical and efficient approach for developing surface processing solutions within an enterprise.

Microstructure processing: The distinctive microstructures generated from different AM processes might not be suitable for certain application purposes, and additional microstructural tailoring is needed, such as residual stress relieving, introduction of new material microstructural phases, or material doping. In general, although the microstructures of metal AM materials exhibit unique characteristics, such as directional grain orientations, presence of metastable phases, and particular distribution of dispersive phases, the thermodynamics of heat treatment of these materials still largely follow the canonical knowledge. This is also closely related to the fact that with many application scenarios, the additional microstructural processing such as heat treatments is utilized to introduce more stable microstructural phases. On the other hand, the refined microstructural grain sizes and the generation of metastable phases from the rapid cooling process of metal AM sometimes allow for the introduction of a more desirable microstructural evolution upon heat treatments [177, 178]. Typically, the heat treatment processes for the AM materials require some refinement development based on the known traditional routes; however, the practice is generally consistent with the established methodologies, and there also exist relatively established literatures for a variety of AM metal materials [177, 179, 180]. For relatively new materials (e.g., high-entropy alloys and AM-specific new materials), the heat treatment processes are generally predicted from the combination of computational material thermochemistry tools (e.g., CALPHAD or sintering model) and experimental development.

Another increasingly popular microstructure processing technique for metal PBF parts is hot isostatic pressing (HIP). Although the treatment often involves high-capital-investment equipment, it has been established that HIP treatment could effectively reduce both the internal porosity and surface roughness with PBF parts, which in turn translates to improved fatigue endurance [181, 182]. Besides fatigue strength, HIP process could also potentially improve the other mechanical properties such as ductility, compressive strength, as well as corrosion resistance [183]. On the other hand, HIP tends to be less effective towards near-surface porosities, as well as microstructure grain coarsening that could potentially have negative effects on mechanical properties with certain materials. Due to the overall beneficial effects on the overall reliability of materials, HIP is often considered as a standard procedure for metal AM components in aerospace applications.

Lastly, it should also be noted that there have been increased amounts of interest in the in-process surface and microstructure enhancement treatments, which are considered to be potentially more effective due to the ability to overcome much of the geometry complexity-induced challenges compared to the traditional approaches. For example, for material extrusion process, in situ laser-assisted heating, machining, ironing, and chemical polishing have been demonstrated [184–187]. For both PBF and DED, secondary heating and machining have been employed for both surface and microstructural treatments [175, 188–192].

3.6 Design with AM (DwAM)

The end of this chapter is dedicated to a brief general discussion of the DwAM concept. Although the concept of considering AM technology as designable variables in the overall design process appears to be an abstract and difficult-to-follow one, the differentiation between DfAM and DwAM could be far less significant in practice. To transition from DfAM to DwAM, the designer mainly needs to incorporate AM consideration in early stages of product/structure design process, instead of only during the manufacturing/realization stages. Also, instead of focusing on adapting design to AM capabilities, AM technologies are only considered as manufacturing options. One example of such design concept is the generative design. The concept of generative design is often intertwined with its implementation tools, but in essence, it could be understood as iteratively developing optimized design solutions via either automated algorithms, manual analysis, or a hybrid of both. For single-objective design problems that can be well defined, generative design could readily utilize existing design algorithms such as topology optimization to automatically converge to solutions, which have been demonstrated extensively and sometimes lead to the confusion between these two distinct definitions. An example that better demonstrates generative design concept is illustrated in Fig. 3.78, in which topology optimization algorithm is employed to generate different solutions for a generic bracket based on different design priority settings [193]. As the generated solutions might differ significantly from each other, and as previously discussed, an optimized solution for multi-objective might not be attainable, either due to the difficulty with the algorithm or being conceptually intractable as relative weights of different criteria are difficult to quantify. Further, with many product designs, the criteria can be highly subjective and non-formulatable. In these scenarios, manual design inputs or modifications can be implemented, either to provide additional directions/restrictions to further automatic design optimization or to generate final solutions based on the reference solutions from automated designs. Ultimately, the solution might not converge to the use of AM processes or certain AM materials. Similarly, the solution might not arrive at single process/single

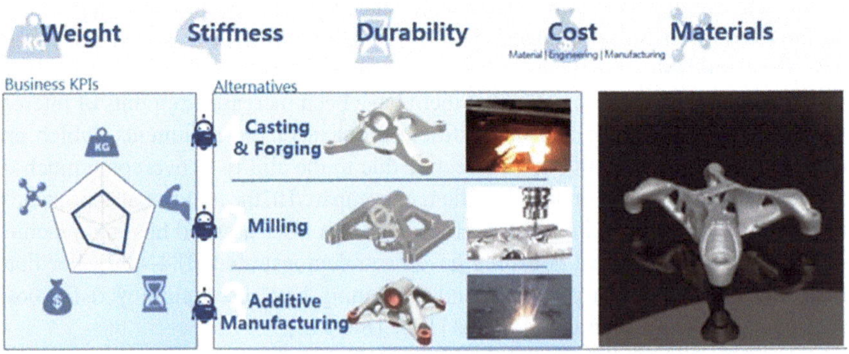

Fig. 3.78 Generative design with multi-objectives and design alternatives generated using topology optimization algorithm [193]

material for the entire structure, but instead suggests an architecture based on composite or assembly concepts, utilizing multiple processes and material types.

Another aspect of DwAM concept is that the digital design tools and AM technologies could potentially enable new solutions and innovations. One good example regarding the novel design concept is metamaterial, which indicates structures that combine intrinsic material characteristics and structural mechanisms to achieve mechanical and physical properties unattainable from traditional materials. There exist research literatures that demonstrate various unique metamaterial design concepts, such as guided light deflection (invisible cloak), tunable acoustic filters, passive-programmable morphing structures, and various mechanical mechanisms. While most of these concepts still remain in early research stage, AM has been extensively employed for the realization of the designs. The maturation of AM metamaterials is currently hindered mainly by material options and system capabilities [194]. Many metamaterial concepts rely on small-size architectural features that are often below 0.1 mm in order to realize the functionality. For the few AM systems that are capable of such resolution level, the volumetric material build speeds are often too slow to be practical. In addition, the material options for these systems are also generally limited. On the other hand, mechanical metamaterials such as auxetic structures, negative stiffness designs, and other mechanical mechanisms are less dependent on feature sizes for achieving designed functionality and thus might possess more immediate potentials for applications.

With DwAM, the process characteristics of AM could be intentionally utilized for achieving innovative design solutions. A good example of utilizing the AM process characteristics for facilitating design objectives is illustrated in Fig. 3.79 [195]. While it is generally undesirable for the material extrusion nozzle to move at a speed too fast for the material feeding mechanism, which might cause inconsistent deposition track dimension or even deposition discontinuity, this process phenomenon could be intentionally employed for the realization of variable-thickness strands or ultrathin strands. Such "out-of-box" design concept requires in-depth

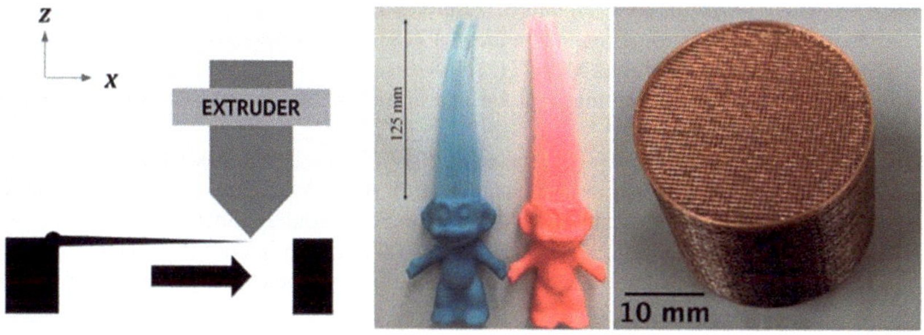

Fig. 3.79 Utilizing rapid and nonsteady material extrusion to print fibrils for sound absorber application [195]

understanding with the AM process and material characteristics, but could lead to more creative and disruptive solutions.

References

1. https://www.ge.com/news/reports/epiphany-disruption-ge-additive-chief-explains-3d-printing-will-upend-manufacturing. Accessed Dec 2023
2. https://www.ge.com/news/reports/transformation-3d-walnut-sized-part-changed-way-ge-aviation-builds-jet-engines. Accessed Dec 2023
3. Lechler T, Fuchs J, Sjarov M, Brossog M, Selmair A, Fatus F, Donhauser T, Franke J (2020) Introduction of a comprehensive Structure Model for the Digital Twin in Manufacturing. In: Proceedings of the 25th International Conference on Emerging Technologies and Factory Automation (ETFA)
4. Mies D, Marsden W, Warde S (2016) Overview of additive manufacturing informatics: "a digital thread". Integr Mater Manuf Innov 5:6
5. Tim DB, Witherell P, Lu Y, Feng S (2017) Towards a digital thread and data package for metals-additive manufacturing. Smart Sustain Manuf Syst 1:75–99
6. Qin Y, Qi Q, Scott PJ, Jiang X (2019) Status, comparison, and future of the representations of additive manufacturing data. Comput Aided Des 111:44–64
7. Prater T (2017) Database development for additive manufacturing. Prog Addit Manuf 2:11–18
8. Coykendall J, Cotteleer M, Holdowsky J, Mahto M (2014) 3D opportunity in aerospace and defense. Deloitte University Press
9. Brlard T, Segonds F, Zamarlola N (2020) G-DfAM: a methodological proposal of generative design for additive manufacturing in the automotive industry. Int J Interact Des Manuf 14:875–886
10. Laverne F, Segonds F, Anwer N, Coq ML (2015) Assembly based methods to support product innovation in design for additive manufacturing: an exploratory case study. ASME J Mech Des 137(12):121701
11. Diegel O, Schutte J, Ferreira A, Chan YL (2020) Design for additive manufacturing process for a lightweight hydraulic manifold. Addit Manuf 36:101446
12. https://www.tctmagazine.com/additive-manufacturing-3d-printing-news/airbus-division-3d-prints- aluminium-components-for-new-eurostar-satellite/. Accessed Dec 2023
13. Horn M, Schmitt M, Schafnitzel M, van Husen A, Wagenblast P, Auernhammer S, Heyder J, Hauck C, Barz J, Ott M, Schlick G, Seidel C (2022) Powder bed fusion of highly filigree copper features using a green laser. Procedia CIRP 111

14. Gokcekaya O, Ishimoto T, Hibino S, Yasutomi J, Narushima T, Nakano T (2021) Unique crystallographic texture formation in Inconel 718 by laser powder bed fusion and its effect on mechanical anisotropy. Acta Mater 212:116876
15. MIL-S-23284A: steel forgings, carbon and alloy, for shafts, sleeves, propeller nuts, couplings, and stocks (1990)
16. Shah J, Connor Z, Robison S, Prucha T (2017) Casting solutions for readiness: metallic materials properties development and standardization (MMPDS), cast high-integrity alloy mechanical property standards (CHAMPS). American Foundry Society
17. Altan T, Nichols DE, Henning HJ, Sabroff AM (1969) Fifth Interim topical report on a study of the mechanics of closed-die forging: calculation of geometrical parameters in designing the forging process for axisymmetric shape
18. Hagen JF, Rasch M, Kohl S, Schmidt M (2018) Geometry dependent microstructures: bug or features? Procedia CIRP. 74:727–727
19. Munk J, Breibarth E, Siemer T, Pirch N, Hafner C (2022) Geometry effect on microstructure and mechanical properties in laser powder bed fusion of Ti-6Al-4V. MDPI Metals. 12:482
20. Plotkowski A, Pries J, List F, Nandwana P, Stump B, Carver K, Dehoff RR (2019) Influence of scan pattern and geometry on the microstructure and soft-magnetic performance of additively manufactured Fe-Si. Addit Manuf 29:100781
21. Li G, Ganeriwala R, Herzog D, Bartsch K, King W (2023) Numerical and experimental investigation of the geometry dependent layer-wise evolution of temperature during laser powder bed fusion of Ti-6Al-4V. Prog Add Manuf 8:961–975
22. Khorasani M, Ghasemi A, Leary M, Sharabian E, Cordova L, Gibson I, Downing D, Bateman S, Brandt M, Rolfe B (2022) The effect of absorption ratio on meltpool features in laser-based powder bed fusion of IN718. Opt Laser Technol 153:108263
23. Cook PS, Ritchie DJ (2023) Determining the laser absorptivity of Ti-6Al-4V during laser powder bed fusion by calibrated melt pool simulation. Opt Laser Technol 162:109247
24. Shen H, Rometsch P, Wu X, Huang A (2020) Influence of gas flow speed on laser plume attenuation and powder bed particle pickup in laser powder bed fusion. JOM 72(3):1039–10521
25. Bitharas O, Burton A, Ross AK, Moore AJ (2021) Visualisation and numerical analysis of laser powder bed fusion under cross-flow. Add Manuf 37:101690
26. Khairallah SA, Anderson AT, Rubenchik A, King WE (2016) Laser powder-bed fusion additive manufacturing: physics of complex melt flow and formation mechanisms of pores, spatter, and denudation zones. Acta Mater 108:36–45
27. Shrestha S, Chou K (2021) A study of transient and steady-state regions from single-track deposition in laser powder bed fusion. J Manuf Process 61:226–235
28. Wu C, Xie P, Zafar MQ, Zhao H (2023) High-fidelity modeling of multilayer building process in electron beam powder bed fusion: build-quality prediction and formation-mechanism investigation. Chin J Mech Eng Add Manuf Front 2:100086
29. Zheng H, Li H, Lang L, Gong S, Ge Y (2018) Effect of scan speed on vapor plume behavior and spatter generation in laser powder bed fusion additive manufacturing. J Manuf Process 36:60–67
30. Korner C, Attar E, Heinl P (2011) Mesoscopic simulation of selective beam melting processes. J Mater Process Technol 211:978–987
31. Zhang XZ, Tang HP, Leary M, Song T, Jia L, Qian M (2018) Toward manufacturing quality Ti-6Al-4V lattice struts by selective electron beam melting (SEBM) for lattice design. JOM 70(9):1870–1876
32. Amani Y, Dancette S, Delroisse P, Simar A, Maire E (2018) Compression behavior of lattice structures produced by selective laser melting: x-ray tomography based experimental and finite element approaches. Acta Mater 159:395–407
33. Reinhart G, Teufelhart S, Riss F (2012) Investigation of the geometry-dependent anisotropic material behavior of filigree struts in ALM-produced lattice structures. Phys Procedia 39:471–479

34. Chen C, Xiao Z, Zhu H, Zeng X (2020) Deformation and control method of thin-walled part during laser powder bed fusion of Ti-6Al-4V alloy. Int J Adv Manuf Technol 110:3467–3478
35. Mostafaei A, Elliott AE, Barnes JE, Li F, Tan W, Cramer CL, Nandwana P, Chmielus M (2021) Binder jet 3D printing—Process parameters, materials, properties, modeling, and challenges. Prog Mater Sci 119:100707
36. Miyanaji H, Orth M, Akbar JM, Yang L (2018) Process development for green part printing using binder jetting additive manufacturing. Front Mech Eng 13(4):504–512
37. Wang YM, Voisin T, McKeown JT, Ye J, Calta NP, Li Z, Zeng Z, Zhang Y, Chen W, Roehling TT, Ott RT, Santala MK, Depond PJ, Matthews MJ, Hamza AV, Zhu T (2018) Additively manufactured hierarchical stainless steels with high strength and ductility. Nat Mater 17:63–73
38. M. M. Kirka, Y. Lee, D. A. Greenley, A. Okello, M. J. Goin, M. T. Pearce, R. R. Dehoff. Strategy for texture management in metals additive manufacturing. JOM. 69(2017), 3: 523–531
39. https://www.digitalengineering247.com/article/materialise-takes-the-pain-out-of-support-generation-for-metal-3d-printing/. Accessed Apr 2024
40. https://www.metal-am.com/articles/design-for-3d-printing-increasing-part-value-through-intelligent-optimisation/. Accessed Apr 2024
41. https://news.europawire.eu/breakthrough-on-laser-powder-bed-fusion-technology-brings-effective-production-of-larger-components-closer-to-mainstream-97185611/eu-press-release/2019/05/29/17/03/49/73574/. Accessed June 2024
42. https://www.additivemanufacturing.media/articles/an-additive-manufacturing-machine-shop-2. Accessed June 2024
43. https://www.mmsonline.com/articles/meeting-the-machining-challenges-of-additive-manufacturing. Accessed June 2024
44. Cloots M, Spierings AB, Wegener K (2013) Assessing new support minimizing strategies for the additive manufacturing technology. Proceedings of Solid Freeform Fabrication (SFF) Symposium, Austin, TX, USA
45. Calignano F (2014) Design optimization of supports for overhanging structures in aluminum and titanium alloys by selective laser melting. Mater Des 64:203–213
46. https://www.digitalengineering247.com/article/simulation-new-driving-force-behind-additive-manufacturing. Accessed July 2024
47. 2024 Industry Report: Additive Manufacturing (2024) SME
48. Krol TA, Zaeh MF, Seidel C (2012) Optimization of supports in metal-based additive manufacturing by means of finite element models. In: Proceedings of Solid Freeform Fabrication (SFF) Symposium, Austin, TX, USA
49. Ameen W, Al-Ahmari A, Mohammed MK, Kaid H (2021) Multi-objective optimization of support structures for metal additive manufacturing. Int J Adv Manuf Technol 116:2613–2632
50. Poyraz O, Yasa E, Akbulut G, Orhangul A, Pilatin S (2015) Investigation of support structures for direct metal laser sintering (DMLS) of IN625 parts. Proceedings of Solid Freeform Fabrication (SFF) Symposium, Austin, TX, USA
51. Subedi SC, Shahba A, Thevamaran M, Thoma DJ, Suresh K (2022) Towards the optimal design of support structures for laser powder bed fusion-based metal additive manufacturing via thermal equivalent static loads. Addit Manuf 57:102956
52. Feng R, Li X, Zhu L, Thakur A, Wei X (2021) An improved two-level support structure for extrusion-based additive manufacturing. Robot Comput Integrated Manuf 67:101972
53. Dumas J, Hergel J, Lefebvre S (2014) Bridging the gap: automated steady scaffolding for 3D printing. ACM Trans Graph 33(4):98
54. Strano G, Hao L, Everson RM, Evans KE (2013) A new approach to the design and optimisation of support structures in additive manufacturing. Int J Adv Manuf Technol 66:1247–1254
55. Hu J, Molavi-Zarandi M, Pasini D (2023) Topology optimization of lattice support structures for cantilever beams fabricated via laser powder bed fusion. Adv Eng Mater 2300976:2300976
56. Cooper K, Steele P, Cheng B, Chou K (2018) Contact-free support structures for part overhangs in powder-bed metal additive manufacturing. MDPI-Invent. 3:2

57. Tan C, Wang D, Ma W, Chen Y, Chen S, Yang Y, Zhou K (2020) Design and additive manufacturing of novel conformal cooling molds. Mater Des 196:109147
58. https://facfox.com/docs/kb/3d-printing-supports-the-ultimate-guide/. Accessed June 2024
59. https://bigrep.com/posts/design-for-additive-manufacturing/. Accessed June 2024
60. Ameen W, Al-Ahmari A, Mohammed MK, Mian SH (2018) Manufacturability of overhanging holes using electron beam melting. MDPI Metals 8:397
61. https://www.3dnatives.com/en/support-less-metal-additive-manufacturing-260220204/. Accessed June 2024
62. Schmelzle J, Kline EV, Dickman CJ, Reutzel EW, Jones G, Thompson TW (2015) (Re) Designing for part consolidation: understanding the challenges of metal additive manufacturing. J Mech Des 137:111404
63. Chahine G, Smith P, Kovacevic R (2010) Application of topology optimization in modern additive manufacturing. In: Proceedings of Solid Freeform Fabrication (SFF) Symposium, Austin, TX
64. Bendsoe MP, Sigmund O (2003) Topology optimization: theory, methods and applications. Springer-Verlag, Berlin
65. Bechtold IT (2013) Structural topology optimization for MEMS design. University of Freiburg
66. Aremu A, Ashcroft I, Hague R, Wildman R, Tuck C (2010) Suitability of SIMP and BESO topology optimization algorithms for additive manufacture. In: Proceedings of Solid Freeform Fabrication (SFF) Symposium, Austin, TX
67. Olason A, Tidman D (2010) Methodology for topology and shape optimization in the design process. Master's Thesis, Chalmers University of Technology, Goteborg, Sweden
68. Meng L, Zhang W, Quan D, Shi G, Tang L, Hou Y, Brietkopf P, Zhu J, Gao T (2020) From topology optimization design to additive manufacturing: today's success and tomorrow's roadmap. Arch Comput Method Eng 27:805–830
69. El Khadiri I, Zemzami M, Nguyen N-Q, Abouelmajd M, Hmina N, Belhouideg S (2023) Topology optimization method for additive manufacturing: a review. Int J Simul Multidiscip Des Optim 14:12
70. Tang Y, Kurtz A, Zhao YF (2015) Bidirectional evolutionary structural optimization (BESO) based design method for lattice structure to be fabricated by additive manufacturing. Comput Aided Des 69:91–101
71. Panesar A, Abdi M, Hickman D, Ashcroft I (2018) Strategies for functionally graded lattice structures derived using topology optimisation for additive manufacturing. Addit Manuf 19:81–94
72. Cheng L, Zhang P, Biyikli E, Bai J, Robbins J, To A (2017) Efficient design optimization of variable-density cellular structures for additive manufacturing: theory and experimental validation. Rapid Prototyp J 23(4):660–677
73. Gervasi VR, Stahl DC (2004) Design and fabrication of components with optimized lattice microstructures. In: Proceedings of Solid Freeform Fabrication (SFF) Symposium. Austin, TX
74. Zhang P, Toman J, Yu Y, Biyikli E, Kirca M, Chmielus M, A. C. To (2015) Efficient design-optimization of variable-density hexagonal cellular structure by additive manufacturing: theory and validation. ASME J Manuf Sci Eng 137(2):021004
75. Ibhadobe O, Zhang Z, Sixt J, Nsiempba KM, Orakwe J, Martinez-Marchese A, Ero O, Shahabad SI, Bonakdar A, Toyserkani E (2023) Topology optimization for metal additive manufacturing: current trends, challenges, and future outlook. Virt Phys Prototyp 18(1):e2181192
76. Zhang K, Cheng G (2020) Three-dimensional high resolution topology optimization considering additive manufacturing constraints. Addit Manuf 35:101224
77. Liu J, Gaynor AT, Chen S, Kang Z, Suresh K, Takezawa A, Li L, Kato J, Tang J, Wang CCL, Cheng L, Liang X, To AC (2018) Current and future trends in topology optimization for additive manufacturing. Struct Multidiscip Optim 57:2457–2483

78. Sabiston G, Kim IY (2020) 3D topology optimization for cost and time minimization in additive manufacturing. Struct Multidiscip Optim 61:731–748
79. Lazarov BS, Wang F, Sigmund O (2016) Length scale and manufacturability in density-based topology optimization. Arch Appl Mech 86:189–218
80. Gibson LJ, Ashby MF (1997) Cellular solids: structure and properties, 2nd edn. Cambridge University Press, New York, NY
81. Ashby MF, Fleck NA, Gibson LJ, Hutchinson JW, Wadley HNG (2000) Metal foams: a design guide. Butterworth Heinemann, Woburn, MA
82. Banhart J, Weaire D (2002) On the road again: metal foams find favor. Phys Today 55:37–42
83. Banhart J (2006) Metal Foams: Production and Stability. Adv Eng Mater 8:781–794
84. Banhart J (2001) Manufacture, characterization and application of cellular metals and metal foams. Prog Mater Sci 46:559–632
85. L.-P. Lefebvre, J. Banhart, D. C. Dunand. Porous metals and metallic foams: current status and recent developments. Advanced Engineering Materials. 10(2008), 9: 775–787
86. http://www.interiorsandsources.com/article-details/articleid/15624/title/alusion-from-cymat-corporation.aspx. Accessed June 2016
87. Vendra LJ, Rabiei A (2007) A study on aluminum-steel composite metal foam processed by casting. Mater Sci Eng A 465:59–67
88. Jang W-Y, Kyriakides S (2009) On the crushing of aluminum open-cell foams: Part I. Experiments. Int J Solids Struct 469:617–634
89. Tekoglu C (2007) Size effects in cellular solids. PhD Dissertation, University of Groningen, the Netherland
90. Andrews EW, Gioux G, Onck P, Gibson LJ (2001) Size effects in ductile cellular solids. Part II: experimental results. Int J Mech Sci 43:701–713
91. Wang A-J, Kumar RS, McDowell DL (2005) Mechanical behavior of extruded prismatic cellular metals. Mech Adv Mater Struct 12:185–200
92. Ashby MF (2006) The properties of foams and lattices. Phil Trans R Soc A 364:15–30
93. Ha CS, Lakes RS, Plesha ME (2019) Cubic negative stiffness lattice structure for energy absorption: numerical and experimental studies. Int J Solids Struct 178-179:127–135
94. Kochmann DM, Bertoldi K (2017) Exploiting microstructural instabilities in solids and structures: from metamaterials to structural transitions. Appl Mech Rev 69:050801
95. Correa DM, Seepersad CC, Haberman MR (2015) Mechanical design of negative stiffness honeycomb materials. Integr Mater Manuf Innov 4:10
96. Rafsanjani A, Akbarzaeh A, Pasini D (2015) Snapping mechanical metamaterials under tension. Adv Mater 27(39):5931–5935
97. Rossiter J, Takashima K, Scarpa F, Walters P, Shape T (2014) memory polymer hexachiral auxetic structures with tunable stiffness. Smart. Mater Struct 23:045007
98. Yang L, Harrysson O, West H, Cormier D (2013) A comparison of bending properties for cellular core sandwich structures. Mater Sci Appl 4:471–477
99. Yang L (2015) Experimental-assisted design development for an octahedral cellular structure using additive manufacturing. Rapid Prototyp J 21(2):168–176
100. Onck PR, Andrews EW, Gibson LJ (2001) Size effects in ductile cellular solids. Part I: modeling. Int J Mech Sci 43:681–699
101. Yang L (2016) A study about size effects of 3D periodic cellular structures. In: Proceedings of the Solid Freeform Fabrication (SFF) Symposium, Austin, TX
102. Yang L (2017) An investigation of property isotropy of 3D periodic cellular structure designs. In: Proceedings of the Solid Freeform Fabrication (SFF) Symposium, Austin, TX
103. Hollander DA, von Walter M, Wirtz T, Sellei R, Schmidt-Rohlfing B, Paar O, Erli H-J (2006) Structural, mechanical and in vitro characterization of individually structured Ti-6Al-4V produced by direct laser forming. Biomaterials 27:955–963
104. Stamp R, Fox P, O'Neill W, Jones E, Sutcliffe C (2009) The development of a scanning strategy for the manufacture of porous biomaterials by selective laser melting. J Mater Sci Mater Med 20:1839–1848

105. Brooks W, Sutcliffe C, Cantwell W, Fox P, Todd J, Mines R (2005) Rapid design and manufacture of ultralight cellular materials. In: Proceedings of Solid Freeform Fabrication (SFF) Symposium. Austin, Texas
106. Tsopanos S, Mines RAW, McKown S, Shen Y, Cantwell WJ, Brooks W, Sutcliffe CJ (2010) The influence of processing parameters on the mechanical properties of selectively laser melted stainless steel microlattice structures. J Manuf Sci Eng 132:041011
107. Yang L, Gong H, Dilip S, Stucker B (2014) An investigation of thin feature generation in direct metal laser sintering systems. In: Proceedings of Solid Freeform Fabrication (SFF) Symposium. Austin, Texas
108. Koju N, Hermes J, Saghaian SE, Yang L (2023) Laser powder bed fusion additively manufactured thin lightweight Ti6Al4V features: an experimental investigation on the influence of powder feedstock, geometry, and process parameters on property/quality. Int J Adv Manuf Technol 130:1541–1561
109. Cansizoglu O (2008) Mesh structures with tailored properties and applications in hips stems. PhD Dissertation, North Carolina State University, Raleigh, NC
110. Yang L, Harrysson O, West II H, Cormier D (2011) Design and characterization of orthotropic re-entrant auxetic structures made via EBM using Ti6Al4V and pure copper. In: Proceedings Solid Freeform Fabrication (SFF) Symposium. Austin, Texas,
111. Li SJ, Murr LE, Cheng XY, Zhang ZB, Hao YL, Yang R, Medina F, Wicker RB (2012) Compression fatigue behavior of Ti-6Al-4V mesh arrays fabricated by electron beam melting. Acta Mater 60:793–802
112. Edwards P, Ramulu M (2014) Fatigue performance evaluation of selective laser melted Ti-6Al-4V. Mater Sci Eng A 598:327–337
113. Gong H, Rafi K, Starr T, Stucker B (2012) Effect of defects on fatigue tests of as-built Ti-6Al-4V parts fabricated by selective laser melting. In: Proceedings of Solid Freeform Fabrication (SFF) Symposium. Austin, Texas
114. Murr LE, Gaytan SM, Medina F, Martinez E, Martinez JL, Hernandez DH, Machado BI, Ramirez DA, Wicker RB (2010) Characterization of Ti-6Al-4V open cellular foams fabricated by additive manufacturing using electron beam melting. Mater Sci Eng A 527:1861–1868
115. Zhang S, Dilip S, Yang L, Miyanaji H, Stucker B (2015) Property evaluation of metal cellular strut structures via powder bed fusion AM. In: Proceedings of Solid Freeform Fabrication (SFF) Symposium. Austin, Texas
116. Takezawa A, Guo H, Kabayashi R, Chen Q, To AC (2022) Simultaneous optimization of hatching orientations and lattice density distribution for residual warpage reduction in laser powder bed fusion considering layerwise residual stress stacking. Addit Manuf 60A:103194
117. S.-H. Li, P. Kumar, S. Chandra, U. Ramamurty. Directed energy deposition of metals: processing, microstructures, and mechanical properties. Int Mater Rev. 68(2023), 6: 605–647
118. Feenstra DR, Banerjee R, Fraser HL, Huang A, Molotnikov A, Birbilis N (2021) Critical review of the state of the art of multi-material fabrication via directed energy deposition. Curr Opin Solid State Mater Sci 25:100924
119. Zhang G, Lu X, Li J, Chen J, Lin X, Wang M, Tan H, Huang W (2022) In-situ grain structure control in directed energy deposition of Ti6Al4V. Addit Manuf 55:102865
120. Kruth JP, Wang X, Laoui T, Froyen L (2003) Lasers and materials in selective laser sintering. Assem Automat 23(4):357–371
121. Mussatto A (2022) Research progress in multi-material laser-powder bed fusion additive manufacturing: A review of the state-of-the-art techniques for depositing multiple powders with spatial selectivity in a single layer. Results Eng 16:100769
122. Boivie K (2001) Limits of loose metal powder density in the interstation. In: Proceedings of Solid Freeform Fabrication (SFF) Symposium, Austin, TX, USA
123. Spierings AB, Herres N, Levy G (2010) Influence of the particle size distribution on surface quality and mechanical properties in additive manufactured stainless steel parts. In: Proceedings of Solid Freeform Fabrication (SFF) Symposium, Austin, TX, USA

124. Spierings AB, Levy G (2009) Comparison of density of stainless steel 316L parts produced with selective laser melting using different powder grades. In: Proceedings of Solid Freeform Fabrication (SFF) Symposium, Austin, TX, USA
125. Karapatis NP, Egger G, Gygax P-E, Glardon R (1999) Optimization of powder laser density in selective laser sintering. In: Proceedings of Solid Freeform Fabrication (SFF) Symposium, Austin, TX, USA
126. Liu B, Wildman R, Tuck C, Ashcroft I, Hague R (2011) Investigation the effect of particle size distribution on processing parameters optimisation in selective laser melting. In: Proceedings of Solid Freeform Fabrication (SFF) Symposium, Austin, TX, USA
127. Shi Y, Zhang Y (2006) Simulation of random packing of spherical particles with different size distributions. ASME Int Mech Eng Congr, Heat Transfer, Chicago, IL
128. Konrad C, Zhang Y, Shi Y (2007) Melting and solidification of a subcooled metal powder particle subjected to nanosecond laser heating. Int J Heat Mass Transf 50:2236–2245
129. Hrabe NW, Heinl P, Flinn B, Korner C, Bordia RK (2011) Compression-compression fatigue of selective electron beam melted cellular titanium (Ti-6Al-4V). J Biomed Mater Res B Appl Biomater 99:313–320
130. Yavari SA, Wauthle R, van der Stok J, Riemslag AC, Janssen M, Mulier M, Kruth JP, Schrooten J, Weinans H, Zadpoor AA (2013) Fatigue behavior of porous biomaterials manufactured using selective laser melting. Mater Sci Eng C 33:4849–4858
131. Li P, Warner DH, Fatemi A, Phan N (2016) Critical assessment of the fatigue performance of additively manufactured Ti-6Al-4V and perspective for future research. Int J Fatigue 85:130–143
132. Perez-Ruiz JD, Galbusera F, Caprio L, Previtali B, de Lacalle LNL, Lamikiz A, Demir AG (2024) Laser beam shaping facilitates tailoring the mechanical properties of IN718 during powder bed fusion. J Mater Process Technol 328:118393
133. Kliner DAV, O'Dea B, Lugo J, Farrow RL, Hawke R, Hodges A, Stephens R, Foley B, Almonte K, Kehoe B, Hamilton M, Luetjen C, Small J, Balsley D, Karlsen S, Archer C, Martinsen R (2022) High- productivity laser powder-bed fusion tools enabled by AFX fiber lasers with rapid tunable beam quality. Procedia CIRP 111
134. Chechik L, Schwarzkopf K, Rothfelder R, Grunewald J, Schmidt M (2024) Materials dependent influence of ring/spot beam profiles in laser powder bed fusion. Addit Manuf Lett 9:100211
135. Holla V, Kopp P, Grunewald J, Wudy K, Kollmannsberger S (2023) Laser beam shape optimization in powder bed fusion of metals. Addit Manuf 72:103609
136. Shi R, Khairallah SA, Roehling TT, Heo TW, McKeown JT, Matthews MJ (2020) Microstructural control of in metal laser powder bed fusion additive manufacturing using laser beam shaping strategy. Acta Mater 184:284–305
137. Liner DA, Farrow RL, Lugo J, O'Dea B, Hawke R, Victor B, Gross K, Hodges A, Brown A, Pruyn J, Stephens R, Stephens R, Foley B, Almonte K, Kehoe B, Hamilton M, Reynolds M, Luetjen C, Lowder T, Karlsen S, Balsley D, Antoina M, Kokki T, Swenson D, Archer C, Martinsen R, Hepp M (2022) Advanced metal processing enabled by fiber lasers with tunable beam properties. In: Proceedings of SPIE 11981, Fiber Lasers XIX: Technolgoy and Systems, 119810C
138. Khairallah SA, Anderson A (2014) Mesoscopic simulation model of selective laser melting of stainless steel powder. J Mater Process Technol 214:2627–2636
139. Korner C, Bauereiβ A, Attar E (2013) Fundamental consolidation mechanisms during selective beam melting of powders. Model Simulat Mater Sci Eng 21(8):085001
140. King W, Anderson AT, Ferencz RM, Hodge NE, Kamath C, Khairallah SA (2015) Overview of modelling and simulation of metal powder bed fusion process at Lawrence Livermore National Laboratory. Materials Science and Technology 31(8):957–968
141. Kruth J-P, Levy G, Klocke F, Childs THC (2007) Consolidation phenomena in laser and powder-bed based layered manufacturing. CIRP Ann Manuf Technol 56(2):730–759

142. Tolochko NK, Mozzharov SE, Yadroitsev IA, Laoui T, Froyen L, Titov VI, Ignatiev MB (2004) Balling process during selective laser treatment of powders. Rapid Prototyp J 10(2):78–87
143. Kruth JP, Froyen L, Van Vaerenbergh J, Mercelis P, Rombouts M, Lauwers B (2004) Selective laser melting of iron-based powder. J Mater Process Technol 149:616–622
144. Akhtar S, Wright CS, Youseffi M, Hauser C, Childs THC, Taylor CM, Baddrossamay M, Xie J, Fox P, O'Neill W (2003) Direct selective laser sintering of tool steel powders to high density: part B- the effect on microstructural evolution. In: Proceedings of Solid Freeform Fabrication (SFF) Symposium, Austin, TX, USA
145. Attar E (2011) Simulation of selective electron beam melting processes. PhD dissertation. University of Erlangen-Nurnberg
146. Mills KC, Keene BJ, Brooks RF, Shirali A (1998) Marangoni effects in welding. Philos Trans Math Physi Eng Sci 356:911–925
147. Xiao B, Zhang Y (2007) Marangoni and buoyancy effects on direct metal laser sintering with a moving laser beam. Numer Heat Transfer Part A 51:715–733
148. Antony K, Arivazhagan N (2015) Studies on energy penetration and Marangoni effect during laser melting process. J Eng Sci Technol 10(4):509–525
149. Louvis E, Fox P, Sutcliffe CJ (2011) Selective laser melting of aluminium components. J Mater Process Technol 211:275–284
150. Li E, Zhou Z, Wang L, Zou R, Yu A (2022) Modelling of keyhole dynamics and melt pool flow in laser powder bed fusion process. Powder Technol 400:117262
151. Chen H, Tao W (2020) Spattering and denudation in laser powder bed fusion process: multiphase flow modelling. Acta Mater 196:154–167
152. Guo Q, Zhao C, Escano LI, Young Z, Xiong L, Kezzaa K, Everhart W, Brown B, Sun T, Chen L (2018) Transient dynamics of powder spattering in laser powder bed fusion additive manufacturing process revealed by in-situ high-speed highenergy x-ray imaging. Acta Mater 151:169–180
153. Qu M, Guo Q, Escano LI, Clark SJ, Fezzaa K, Chen L (2022) Mitigating keyhole pore formation by nanoparticles during laser powder bed fusion additive manufacturing. Addit Manuf Lett. 3:100068
154. Agrawal AK, Rankouhi B, Thoma DJ (2022) Predictive process mapping for laser powder bed fusion: A review of existing analytical solutions. Curr Opin Solid State Mater Sci 26:101024
155. Rubenchik AM, King WE, Wu SS (2018) Scaling laws for the additive manufacturing. J Mater Process Technol. 257:234–243
156. Yang Y, Grossmann A, Kuhn P, Molleney J, Kropholler L, Mittelstedt C, Xu B-X (2022) Validated dimensionless scaling law for melt pool width in laser powder bed fusion. J Mater Process Technol 299:117316
157. Lim CWJ, Zhang Y, Huang S, Chan WL (2023) A dimensionless analysis to select directed energy deposition process parameters for proper clad formation. Int J Adv Manuf Technol 125:947–963
158. Gordon JV, Narra SP, Cunningham RW, Liu H, Chen H, Suter RM, Beuth JL, Rollett AD (2020) Defect structure process maps for laser powder bed fusion additive manufacturing. Addit Manuf 36:101552
159. Kechagias JD (2024) Effects of thermomechanical parameters on surface texture in filament materials extrusion: outlook and trends. F1000Research 13:1039
160. Garcia-Martinez F, Carou D, de Arriba-Perez F, Garcia-Mendez S (2023) Toward data-driven research: preliminary study to predict surface roughness in material extrusion using previously published data with machine learning. Rapid Prototyp J 29(8):1640–1652
161. Townsend A, Senin N, Blunt L, Leach RK, Taylor JS (2016) Surface texture metrology for metal additive manufacturing: a review. Precis Eng 46:34–47
162. Obilanade D, Dordlofva C, Torlind P (2021) Surface roughness considerations in design for additive manufacturing- a literature review. In: International Conference on Engineering Design, Gothenburg, Sweden

163. Tascioglu E, Kaynak Y, Poyraz O, Orhangul A, Oren S (2023) The effect of finishing-milling operation on surface quality and wear resistance of Inconel 625 produced by selective laser melting additive manufacturing. In: International Conference on Advanced Surface Enhancement. Singapore
164. Kaynak Y, Tascioglu E (2018) Finish machining-induced surface roughness, microhardness and XRD analysis of selective laser melted Inconel 718 alloy. Procedia CIRP 71:500–504
165. Zhang H, Cao S, Li C, Li B, Qian G (2023) Laser powder bed fused 304L steel shot-peened with various ceramic shot sizes: Surface/subsurface characteristics, tensile behavior, and fatigue behavior. Int J Plasticity 171:103784
166. Elambasseril J, Rogers J, Wallbrink C, Munk D, Leary M, Qian M (2023) Laser powder bed fusion additive manufacturing (LPBFAM): the influence of design features and LPBF variables on surface topography and effect on fatigue properties. Crit Rev Solid State Mater Sci 48(1):132–148
167. Sanaei N, Fatemi A (2020) Analysis of the effect of surface roughness on fatigue performance of powder bed fusion additive manufactured metals. Theor Appl Fract Mech 108:102638
168. Nasab MH, Romano S, Gastaldi D, Beretta S, Vedani M (2020) Combined effect of surface and volumetric defects on fatigue assessment of AiSi7Mg fabricated via laser powder bed fusion. Addit Manuf 34:100918
169. du Plessis A, Beretta S (2020) Killer notches: The effect of as-built surface roughness on fatigue failure in AlSi10Mg produced by laser powder bed fusion. Addit Manuf 35:101424
170. Gockel J, Sheridan L, Koerper B, Whip B (2019) The influence of additive manufacturing processing parameters on surface roughness and fatigue life. Int J Fatigue 124:380–388
171. Jamal M, Morgan MN (2017) Design process control for improved surface finish of metal additive manufactured parts of complex build geometry. MDPI Invent 2:36
172. Nsengimana J, Van der Walt J, Pei E, Miah M (2019) Effect of post-processing on the dimensional accuracy of small plastic additive manufactured parts. Rapid Prototyp J 25(1):1–12
173. Kim US, Park JW (2019) High-quality surface finishing of industrial three-dimensional metal additive manufacturing using electrochemical polishing. Int J Precis Eng Manuf-Green Technol 6:11–21
174. Yang L, O'Neil C, Wu Y (2017) The use of electropolishing surface treatment on IN718 parts fabricated by laser powder bed fusion process. In: Proceedings of Solid Freeform Fabrication Symposium (SFF), Austin, TX
175. Sunny S, Yu H, Mathews R, Malik A (2021) A predictive model for in situ distortion correction in laser powder bed fusion using laser shock peen forming. Int J Adv Manuf Technol 112:1319–1337
176. Maleik E, Bagherifard S, Bandini M, Guagliano M (2021) Surface post-treatments for metal additive manufacturing: progress, challenge, and opportunities. Addit Manuf 37:101619
177. Fiocchi J, Tuissi A, Biffi CA (2021) Heat treatment of aluminium alloys produced by laser powder bed fusion: A review. Mater Des 204:109651
178. Schrodera J, Mishurovaa T, Fritscha T, Serrano-Munoza I, Evansa A, Sprengela M, Klausb M, Genzelb C, Schneiderc J, Bruno G (2021) On the influence of heat treatment on microstructure and mechanical behavior of laser powder bed fused Inconel 718. Mater Sci Eng A 805:140555
179. Teixeira Ó, Silva FJG, Atzeni E (2021) Residual stresses and heat treatments of Inconel 718 parts manufactured via metal laser beam powder bed fusion: an overview. Int J Adv Manuf Technol 113:3139–3162
180. Murr LE (2018) A metallographic review of 3D printing/additive manufacturing of metal and alloy products and components. Metall Microstruct Anal 7:103–132
181. Li P, Warner DH, Pegues JW, Roach MD, Shamsaei N, Phan N (2019) Investigation of the mechanisms by which hot isostatic pressing improves the fatigue performance of powder bed fused Ti- 6Al-4V. Int J Fatigue 120:342–352
182. Krakhmalev P, Fredriksson G, Thuvander M, Åsberg M, Vilardell AM, Oikonomou C, Maistro G, Medvedeva A, Kazantseva N (2020) Influence of heat treatment under hot iso-

static pressing (HIP) on microstructure of intermetallic-reinforced tool steel manufactured by laser powder bed fusion. Mater Sci Eng A 772:138699
183. Jamhari FI, Foudzi FM, Buhairi MA, Muhamad N, Mohamed IF, Sulong AB, Radzuan NAM (2023) Impact of hot isostatic pressing on surface quality, porosity and performance of Ti6Al4V manufactured by laser powder bed fusion: a brief review. J Tribol 36:1–15
184. Nigam A, Tai BL (2024) Effects of in-process surface finishing on part strength in polymer material extrusion additive manufacturing. Addit Manuf 80:103960
185. Han P, Zhang S, Yang Z, Riyad MF, Popa DO, Hsu K (2023) In-process orbiting laser-assisted technique for the surface finish in material extrusion-based 3D printing. MDPFI Polymers 15:2221
186. Caputo M, Rashwan O, Waryoba D, McDade K (2022) Surface texture and thermos-mechanical properties of material extruded and ironed polylactic acid. Addit Manuf 59:103084
187. Montalti A, Galie G, Pignatelli E, Liverani A (2024) Enhancing surface roughness of material extrusion additive manufacturing components via an innovative ironing process. Virtual Phys Prototyp 19(1):e2401929
188. Gisario A, Barletta M, Veniali F (2022) Laser polishing: a review of a constantly growing technology in the surface finish of components made by additive manufacturing. Int J Adv Manuf Technol 120:1433–1472
189. Wang Y, Chen Y, Wen C, Huang K, Chen Z, Han B, Zhang Q (2023) The process planning for additive and subtractive hybrid manufacturing of powder bed fusion (PBF) process. Mater Des 237:111732
190. Hasanabadi M, Keshavarzkermai A, Asgari H, Azizi N, Gerlich A, Toyserkani E (2023) In-situ microstructure control by laser post-exposure treatment during laser powder-bed fusion. Additi Manufact Lett 4:100110
191. Zhou J, Zhou X, Li H, Hu J, Han X, Liu S (2022) In-situ laser shock peening for improved surface quality and mechanical properties of laser-directed energy-deposited AlSi10Mg alloy. Addit Manuf 60:103177
192. Yoon H, Liu P, Park Y, Choi G, Choi P-P, Sohn H (2022) Pulsed laser-assisted additive manufacturing of Ti-6Al-4V for in-situ grain refinement. Sci Rep 12:22247
193. https://edstechnologies.com/blog/generative-design-a-new-era-in-product-design/
194. Askari M, Hutchins DA, Thomas PJ, Astolfi L, Watson RL, Abdi M, Ricci M, Laureti S, Nie L, Freear S, Wildman R, Tuck C, Clarke M, Woods E, Clare AT (2020) Additive manufacturing of metamaterials: a review. Addit Manuf 36:101562
195. Johnston W, Sharma B (2021) Additive manufacturing of fibrous sounds absorbers. Addit Manuf 41:101984

Chapter 4
Laser Powder-Bed Fusion for Metals

4.1 Introduction to LPBF

Laser powder-bed fusion (LPBF) is a metal additive manufacturing process in which a high-power laser fuses layers of metallic powder in a controlled environment to produce components with complex geometries. This chapter examines the entire LPBF workflow and details the structure and operation of LPBF machines, covering laser sources, beam steering systems, environmental controls, and powder handling. A comparison of LPBF with traditional manufacturing approaches is made, emphasizing the influence of thermal history in shaping the final part quality. The chapter concludes by reviewing qualification and certification standards and business considerations. It offers an overview of how LPBF is applied across industries and how cost drivers factor in its broader adoption.

4.1.1 Theoretical LPBF Workflow

In principle, the LPBF workflow begins with a design concept converted into a digital model, considering critical design-for-additive factors such as overhang angles, powder escape holes, and internal channels. Machine preparation follows, ensuring that each subsystem (laser source, beam steering, gas flow, powder handling, etc.) is in proper working order [1]. Process parameters such as layer thickness, scan strategy, and laser power are set during digital file preparation, and support structures may be generated. Fabrication begins once the machine and the file are ready; the powder is spread and then selectively fused by the laser, layer by layer, until the part is complete. After fabrication, the powder is removed and handled for reuse, stress relief is conducted to reduce residual stress, and additional heat treatments (e.g., hot isostatic pressing or HIP) may be applied to enhance mechanical properties. Finally,

K. Hsu et al., *Additive Manufacturing of Metals*, Springer Series in Advanced Manufacturing, https://doi.org/10.1007/978-3-032-04961-2_4

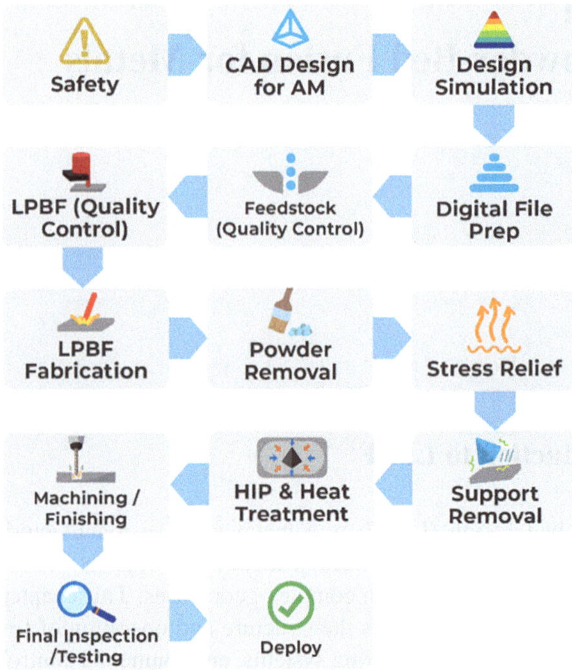

Fig. 4.1 Theoretical LPBF workflow

any supports are removed, finishing operations are performed as needed, and the part undergoes inspection. A theoretical overview of the LPBF workflow is illustrated in Fig. 4.1. In practice, the process often deviates from this theoretical workflow. Furthermore, builds regularly result in design flaws, suboptimal build orientations, or inadequate scan parameters that require iterating through the presented workflow.

4.2 LPBF Machine Overview

LPBF machines are complex, highly integrated systems. A typical LPBF machine consists of several interdependent subsystems, including the laser source, beam steering optics, build chamber, inert gas flow and filtration unit, powder delivery/spreading mechanism, and computer control system. Figure 4.2 illustrates a schematic of the subsystems involved in a typical LPBF machine. These subsystems must work together to selectively melt metal powder layer by layer and produce high-quality parts. Suppose any one subsystem deviates from its optimal performance (for example, an unstable laser output or improper gas flow), the process can become less controlled, and part quality may suffer [2]. Modern LPBF machine designs emphasize tight integration and coordination of all subsystems to ensure

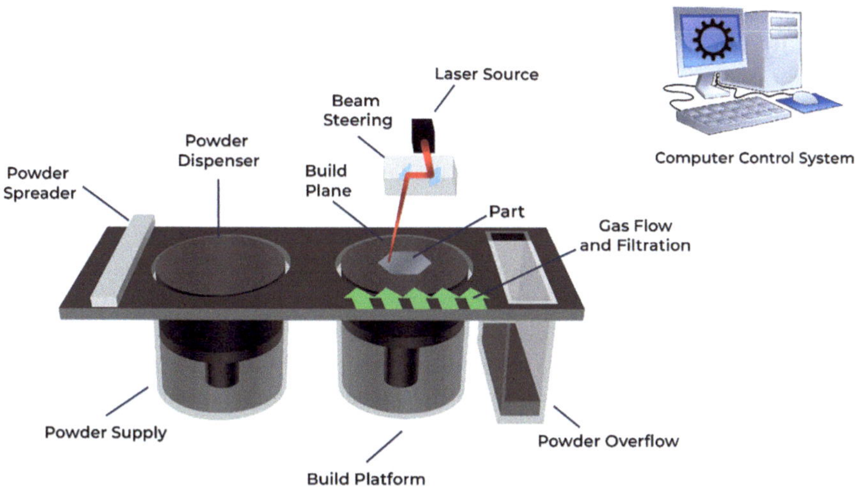

Fig. 4.2 Schematic of a typical LPBF machine

process stability and repeatability. For instance, recent technological improvements have focused on faster and more precise lasers, multi-laser configurations, improved recoaters, uniform inert gas flow, better environmental controls, and advanced sensors, which are all aimed at increasing build speed and enhancing part quality [3]. Industry standards are also emerging to guide machine performance, safety, and process consistency (e.g., ASTM F3303-18, ISO/ASTM 52931:2023). In the following sections, each of the six essential subsystems of a typical LPBF machine is defined and described in detail, with attention to their functions, key components, operating principles, quality control considerations, and references to recent technical literature.

4.2.1 Laser Source

The laser source in an LPBF machine provides the concentrated energy needed to melt metallic powders. Modern LPBF systems almost exclusively use high-power continuous-wave ytterbium fiber lasers operating at near-infrared wavelengths (~1060–1070 nm) [4]. Fiber lasers are favored for their robust beam quality, high electrical-to-optical efficiency, and reliability. A typical commercial machine's laser is a fiber-coupled module delivering a collimated beam into the optical scanner; standard power ratings range from about 200–500 W in standard machines up to 1 kW in the latest high-productivity systems [3]. For example, a midsized LPBF printer might use a single 400 W fiber laser. In contrast, newer large-frame machines employ multiple fiber lasers (e.g., four lasers of 500 W or even 1000 W each) to simultaneously scan different regions of the powder bed [3]. The laser output is typically Gaussian or quasi-Gaussian in profile, focused to a small spot (on the order of 50–100 μm diameter) on the powder bed to achieve the power densities (~10^6–10^7 W/cm^2) required for melting metal powders [5].

The laser's output characteristics are critical, including power, beam mode quality, and stability. Key parameters such as focus spot size, divergence, and beam symmetry are often characterized following laser beam measurement standards (e.g., ISO 11146) to ensure that the laser meets the required specifications [5]. High beam quality is desired, as it produces a tight focal spot and consistent energy density, directly affecting melt pool shape and track stability. The laser is steered dynamically across the powder bed (via the beam steering subsystem) and selectively melts the powder particles where it is focused. Because galvanometer scanners require a finite acceleration time, many systems employ a technique called skywriting, in which the laser power is modulated in coordination with mirror motion so that the laser is only at full power when the scanner is moving at constant velocity [6]. This prevents overexposure at turnaround points and improves the fidelity of scan vectors. Advanced research has shown that controlling laser power at microsecond timescales (for example, gradually ramping the laser power over a few hundred microseconds at scan endpoints) can further reduce spatter and pore defects associated with abrupt laser on/off events [6]. In most commercial machines, however, such fine laser modulation and its timing are preprogrammed by the machine's control software and not directly adjustable by end users.

Even small laser power or focus fluctuations can introduce variability in melt pool dimensions and porosity. Thus, machines often include routines to verify laser performance before or during a build. Standard checks involve measuring the laser power at the build plane using a power sensor to ensure that it matches the commanded value [2] and inspecting the beam profile (e.g., a burn-paper test or camera-based beam diagnostics) to detect any degradation in beam shape. Hilton and Gray [2] reported that most lasers in LPBF have some inherent power variation, and additional energy losses can occur as the beam passes through optics. Regular calibration can compensate for these effects. Another concern is focal spot alignment and focus drift. If the build plane moves out of focus (due to powder-bed height error or thermal expansion), the spot size increases and energy density drops, risking a lack of fusion. Many machines allow the operator to set a focus offset or have an autofocus adjustment in the optical head to maintain the correct focal distance. Any contamination on optical elements (like the protective window in the laser path) can also attenuate or distort the beam. Over time, smoke and spatter can deposit on the internal optics; therefore, machines employ protective cover glasses and sometimes a purge of clean gas across the laser window to keep it clear. These protective optics are replaced or cleaned periodically as part of machine maintenance. In multi-laser systems, an important quality concern is the alignment between lasers. Multiple beams must be co-registered on the build plate within tight tolerances. Misalignment between multiple laser optical systems can cause stitching defects where their melt tracks meet. Thus, multi-laser machines undergo alignment calibration tests (e.g., scanning known patterns and measuring offsets) to ensure that all lasers produce uniform results across the build area [2]. When properly controlled, a high-quality laser source enables stable melt pools and is essential to achieving good part density and mechanical properties.

4.2.2 Beam Steering System

According to the digital part design, the beam steering system directs the laser beam across the powder bed in precise patterns. In many LPBF machines, this subsystem consists of a two-axis galvanometer scanner and an f-theta scan lens (also called a flat-field lens). The galvanometer mirrors (typically one for the *X*-direction and one for *Y*) are lightweight reflective mirrors mounted on fast servo motors that can rapidly pivot to scan the laser beam across the build plane. The f-theta lens then focuses the deflected beam onto a flat plane, maintaining the focal spot size across the entire build area. The beam steering optics translate the digital toolpaths (vector scan patterns for each layer) into physical laser motion on the powder bed. In this standard configuration, as shown in Fig. 4.3, the collimated laser beam enters the scan head, reflects off the *X* and *Y* galvo mirrors, passes through the f-theta lens, and exits through a protective glass window into the sealed build chamber, striking the powder layer at the desired location.

Key components of the beam steering system include the following: (1) Galvanometer mirrors consisting of two small mirrors with mounted galvanometer motors and optical encoders for feedback: These provide high-speed angular positioning (often with response times of a few hundred microseconds) to trace the laser

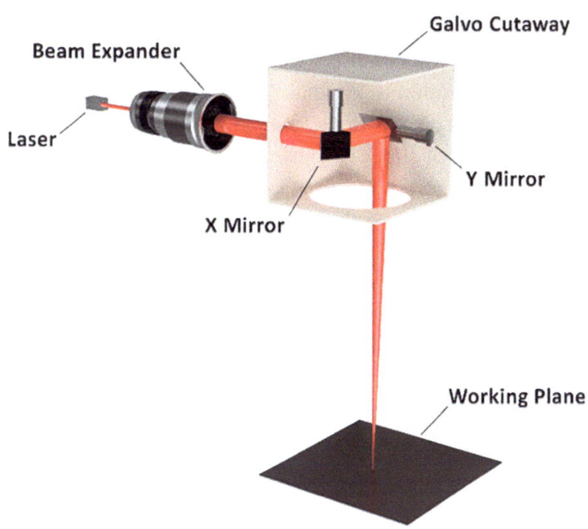

Fig. 4.3 Schematic of a typical beam steering system in LPBF

along the prescribed scan paths. (2) Galvo control electronics with a servo control unit that receives digital scanning commands (vectors or raster points) from the computer controller and drives the mirror angles precisely. (3) F-theta lens, which is a specialized multielement lens that focuses the beam and corrects for the spherical field, producing a flat focal field (the term "f-theta" refers to the linear relationship between the focal spot position and the scan angle, helping achieve uniform spot size and shape across the scan area). (4) Beam expander or dynamic focus (optional), where some systems incorporate a dynamic focus unit (e.g., moving lens or deformable mirror) to adjust focus on the fly, especially in large-area systems. However, in many designs, the f-theta lens has a fixed focal length chosen to cover the entire build plate. (5) Protective window typically consists of a disposable transparent plate (often fused silica) that sits above the build area to shield the costly scan lens from debris or spatter. A purge flow of inert gas may be directed across this window to prevent smoke deposition. All optical components are typically antireflection coated for the laser wavelength to maximize transmission and avoid heating. The optical train is usually enclosed to keep it clean and maintain calibration. Calibration targets or photodiodes may also be present at the periphery of the build area to allow the machine to periodically self-check the scanner's alignment and accuracy.

During a build, the beam steering system follows the scan strategy for each layer, which is defined by the machine's control software (based on the sliced CAD model). The galvanometer mirrors steer the laser along vectors typically at scanning speeds of about 0.5–2 m/s for most builds, though capable of much higher speeds (5–7 m/s in older machines and up to ~10 m/s in some newer systems) when using high-power lasers and scanning large hatch areas [3, 5]. The f-theta lens ensures that the focal spot remains small and consistent in shape, whether the beam is at the center or edge of the build plate. However, slight optical distortions and field curvature can occur, which manufacturers compensate for via calibration maps (software correction of scan positions) so that the actual laser spot position aligns with the commanded position within a small error (often ±30–50 μm across the field). The scanning mirrors are controlled in an analog fashion. For each vector segment, the controller sends a time-varying voltage to each galvo corresponding to the desired mirror angle as a function of time. Because the mirrors have inertia, the controller either slows down at tight corners or uses skywriting delays (as mentioned earlier) to ensure that the laser intensity on the powder is uniform. Scan path planning is an integral operational aspect in which vectors are ordered and sometimes divided into sub-exposures to minimize abrupt direction changes that the galvos must execute. When multiple lasers are used, the machine's control coordinates the scanners to avoid interference, and typically, each laser scans a different zone of the build, but in overlap regions, they are timed not to cross paths simultaneously or one waits briefly to prevent their laser beams from interacting. High-end multi-scanner systems also perform stitching calibration by scanning a test pattern where the two lasers' fields meet and adjusting mirror offsets so that the two different lasers can make a continuous weld track with no mismatch. Overall, the beam steering executes the raster (hatch fills), contours, and any other scan patterns (such as skywriting lead-ins) with precision and repeatability, typically achieving positioning accuracy on the order of tens of micrometers and timing accuracy in the microsecond range for laser on/off events.

The performance of the scanning optics directly impacts the geometric accuracy and surface quality of printed parts. One concern is positional accuracy and drift. Galvanometer scanners can drift out of calibration over time (due to thermal effects or mechanical creep), causing small offsets where the laser melts each layer. This can lead to dimensional inaccuracies or even shift contours between layers. To mitigate this, many machines run a scanner calibration check, for example, marking reference points on a build plate or using a built-in camera system, to realign the coordinate system if needed before critical builds. Another issue is beam distortion at field edges. Even with f-theta correction, the spot may elongate slightly or the laser incidence angle at the powder bed may become non-normal at the extreme edges of the build area, which can affect melt pool dynamics. Some large-area systems incorporate dynamic focusing to maintain optimal focus across the field or restrict the highest precision builds to the central area of the plate. Mirror vibrations or overshoot can introduce glitches in scan lines (manifesting as wobble or uneven hatch spacing). Hence, scanner tuning and maintenance are important; the galvo motors have feedback controllers that must be tuned for a critically damped response without oscillation. Manufacturers often specify a settling time for the mirrors (e.g., they might not exceed a certain acceleration to avoid ringing). Quality issues also arise from optical contamination. Spatter or condensate deposits on the scan lens or protective window can partially block or scatter the laser, which reduces delivered power and can distort the beam shape, leading to erratic melt pools.

Synchronization of the laser and the scanner is another important aspect. Timing errors between laser on/off commands and mirror motion can create start-point defects (e.g., slight puddles or "dots" at the beginning of scan vectors if the laser fires a moment too early before the mirror is up to speed). Techniques like skywriting and laser power ramping (as noted earlier) are intended to alleviate these defects [6]. Finally, in multi-laser machines, cross talk between lasers must be managed. If two laser beams come too close, the plume or spatter from one can interfere with the other, causing local defects. Recent studies confirm that byproducts blown by the gas flow from one laser can attenuate another laser's beam if scanning in parallel nearby [7]. The control software may enforce a minimum separation distance or delay when two melt pools are adjacent to maintain quality. Regular verification of the scanner (using test patterns or reference artifacts) is recommended in production environments [8], and some aerospace standards require documented scanner accuracy and stability as part of machine qualification (ISO/ASTM 52941:2020).

4.2.3 Environmental Control Chamber

LPBF is conducted inside a sealed environmental control chamber with a tightly regulated atmosphere and safety enclosure. The primary function of the chamber is to maintain an inert, low-oxygen environment around the powder bed during fabrication. Metallic powders (especially reactive alloys like titanium or aluminum) can oxidize or ignite in the presence of oxygen at the high temperatures reached during laser melting. Thus, the build chamber is purged with an inert gas (commonly argon or nitrogen for less reactive materials) to typically achieve oxygen levels below about 0.1% (1000 ppm) before and during the build. The chamber also physically contains the powder and spatter ejected, protecting the operators and the external environment. It often includes interlocks to prevent the system from opening when unsafe (e.g., filled with inert gas or when the powder bed is hot).

The environmental subsystem includes the chamber structure itself, usually a robust metal enclosure that houses the build plate, recoater, and powder reservoirs, as shown in the example given in Fig. 4.4. It has a door or lid that can be sealed gastight (with gaskets and clamps). There are ports or glove-box-style gloves for handling powder or removing parts, though many machines now have automated powder handling to minimize manual interaction. An inert gas supply system connects to the chamber, often via solenoid valves, to displace the air out of the chamber. The gas (argon or nitrogen) can come from compressed gas cylinders or on-site generators; argon is predominant for most metal LPBF because it is inert to all metals [4]. Nitrogen is sometimes used for steels or nickel alloys, but is generally avoided for titanium, aluminum, and other metals that form nitrides or are sensitive to oxygen pickup [4]. The chamber will have an oxygen sensor (and sometimes a moisture sensor) to monitor the atmosphere's composition. A pressure relief or vent system is also part of the chamber design to prevent overpressure and allow controlled reintroduction of air after a build. Additionally, heating elements or

Fig. 4.4 An example of an LPBF chamber is one containing a build area, powder reservoir, and powder spreading mechanism

temperature control may be integrated into the chamber walls or build plate. While LPBF generally does not heat the entire chamber to high temperatures (unlike electron beam PBF, which operates hot), some machines allow preheating the substrate or maintaining a moderate ambient temperature (e.g., 80–200 °C) to reduce thermal stresses. If present, this is controlled and monitored as part of the chamber environment. The chamber interior also includes internal lighting and cameras in many systems, to observe the powder bed under inert conditions. All electrical feedthroughs, laser entry (through a sealed window), and motion feedthroughs for the recoater and build piston are carefully sealed to maintain atmosphere integrity. Lastly, safety interlocks tie into the chamber. For example, the laser is prevented from firing if the chamber is not sealed and inerted, and the door is locked whenever the laser or powder motion is active.

During the build, a continuous flow of inert gas is maintained (this relates to the gas flow subsystem, discussed separately below), and the chamber atmosphere is generally kept slightly above ambient pressure to prevent air ingress. Argon, being heavier than air, also helps blanket the powder bed. As the process goes on, metal vapor and reaction products are carried away by the flowing gas and filtered, so the bulk atmosphere remains mostly argon/nitrogen. The oxygen level is monitored throughout; a slow rise in oxygen could indicate a leak or consumption of argon, and the system would alarm or auto-purge if a set point is exceeded. Modern chambers also control humidity with incoming gas typically dry (dew point well below 0 °C) to avoid moisture, which the powder could otherwise absorb and lead to porosity (from hydrogen outgassing) or oxidation. After the build is complete and the part/powder has cooled sufficiently, the chamber is carefully vented back to air (often through a filter to avoid blowing powder dust out). Only then can the operator open the chamber to retrieve the part. In exceptional cases or research settings, chambers have been used with controlled atmosphere compositions or reduced pressures. For instance, experiments have shown that reducing the pressure or adding specific gases can influence the melt pool and spatter dynamics [4, 9]. However, standard practice is to use an inert gas at near-atmospheric pressure to prevent oxidation.

Maintaining a pure inert atmosphere is crucial for part quality and repeatability, with oxygen contamination being a primary concern. Even low levels of O_2 can be absorbed into the melt pool. In titanium alloys, for example, oxygen pickup will increase the strength and hardness but significantly reduce ductility, potentially causing brittleness in the final part [10]. Dietrich et al. [10] demonstrated that Ti-6Al-4V specimens with oxygen content around 1000 ppm in the chamber had higher tensile strength but lower fatigue life than those built in ultralow oxygen conditions. This indicates that controlling the oxygen level avoids oxidation defects and keeps material properties within specification limits. For materials like Ti-6Al-4V, standards often require the build atmosphere to be below some maximum O_2 concentration (e.g., 0.2%) to ensure that the final oxygen in the alloy stays within grade limits [11]. Oxide formation on the powder or partially melted surfaces can also hinder layer bonding. If the chamber has a leak or the purge was insufficient, one might observe discoloration of the powder or parts (bluish or dark oxides) and increased porosity from oxide entrapment. Thus, a best practice is to log the oxygen level during every build and verify that it remains low; any excursion might necessitate scrapping or at least extra heat treatment of the part to ameliorate oxygen effects. Thermal control in the chamber also matters (perhaps not as extreme as in electron beam processes), as the chamber temperature can rise due to the hot powder bed and laser energy. Excessive heat accumulation might affect electronic components or cause powder caking. High-end machines may have water-cooled walls or heat exchangers to keep the chamber at a stable temperature. If the chamber temperature is too low or fluctuating, water vapor from small leaks could condense or powder might pick up moisture from ambient air during loading; hence, many facilities pre-dry powders and ensure that the chamber has low humidity. The cleanliness of the chamber between the builds is another quality factor since spatter and condensate can be deposited on the chamber walls and ceiling. If these deposits flake off, they could contaminate subsequent builds. Regular cleaning or using easy-to-clean surfaces (e.g., removable liners) is often recommended. When maintained correctly, the inert chamber enables LPBF to process even highly reactive metals (like Ti, Al) with minimal oxide inclusion, resulting in parts that meet stringent aerospace and medical quality requirements [12, 13]. The chamber is thus a foundation for process quality and operational safety in LPBF.

4.2.4 Gas Flow and Filtration

The gas flow and filtration system in an LPBF machine manages inert gas circulation through the build chamber and the removal of process emissions. While the chamber provides an inert atmosphere, a constant gas flow (typically argon) is directed across the powder bed during the build. This cover gas flow serves two critical purposes: First, it clears the plume of metal vapor and ejecta produced by the laser-material interaction. Second, it carries spatter particles and fumes out of the build area into filters. In effect, the gas flow ensures an "optically clear" path for the laser by sweeping away smoke, and it helps prevent redeposition of spatter onto the freshly laid powder layer [14]. Figure 4.5 illustrates a typical gas flow configuration in an LPBF system. Although not shown in the figure, the filtration part of the subsystem then captures these byproducts (soot, condensate, and spattered particles) to prevent them from recirculating or escaping into the environment. Together, the gas flow and filtration maintain both the quality of the build and the safety/cleanliness of operation.

The gas flow system comprises a gas inlet nozzle or manifold, a blower or pressure regulator, and a return outlet leading to filters. In many designs, a laminar flow of argon enters at one side of the build area (often via a long slit nozzle spanning the width of the build plate) and exits on the opposite side into a plenum. The flow is generally horizontal across the powder bed, just above the build surface. A recirculation fan or pump drives the flow; some machines use a closed-loop recirculating design where the argon is continuously reused (passing through filters and cooled,

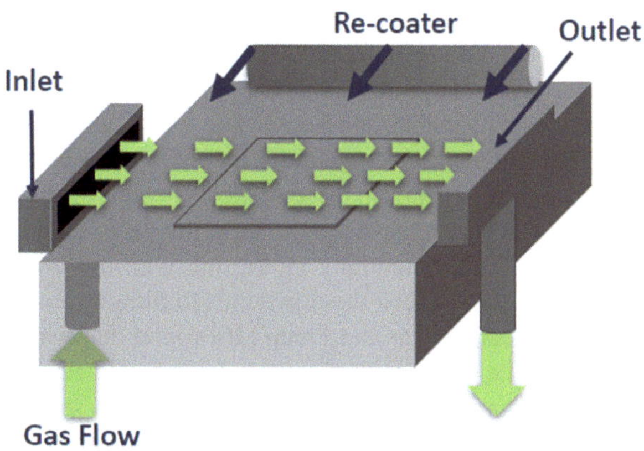

Fig. 4.5 Illustration of gas flow within the chamber

then back to the inlet), while others might use a once-through flow for smaller chambers. The user adjusts the flow rate, typically on the order of a few meters per second in the build chamber, though it must be carefully tuned (too high can disturb the powder bed, and too low allows plume stagnation). The filtration system usually includes a primary filter (often a sintered metal or ceramic filter element that can trap submicron particles) and sometimes a secondary HEPA filter for fine particulates. Metal condensate from the vapor plume can be extremely fine (nanoparticles), so high-efficiency filters are needed. These filters are housed in sealed canisters and are part of the inert loop (to avoid exposure of hot reactive particles to oxygen until they cool). Many LPBF machines incorporate sensors like differential pressure gauges across the filter to monitor clogging. As the filter loads up with soot, the pressure drop increases, and the machine may prompt filter replacement after a certain threshold. Some advanced systems have auto-cleaning filters (e.g., using pulses of gas to knock off particles into a collection bin) or burnout ovens where the filter can be heated in a safe environment to oxidize and remove the accumulated combustibles after a build (this is done outside the machine, in a controlled way). The gas flow path also includes flow straighteners or diffusers to promote uniform flow. For example, honeycomb screens might be used in the inlet to ensure a laminar flow profile. Gas flow adjusters (like dampers or valves) may also allow the user to optimize flow distribution if needed for different materials. Finally, there are safety components such as flashback arrestors or oxygen sensors in the vent line that support in case of any combustion in the filter; these prevent fire from propagating back to the chamber.

The inert gas flow is activated once the chamber is inerted and the build starts. A continuous stream of argon sweeps across the build plane just above the powder bed, typically flowing from the side of the recoater arm toward the opposite side. The direction is often chosen to push spatter away from critical areas. For example, if the recoater moves from left to right, the gas might flow right to left, carrying spatter away from the freshly coated side. The flowing gas captures the bright plume of metal vapor that erupts above the melt pool (especially during keyhole mode melting) and draws it out. Without sufficient flow, this plume can linger and attenuate the incoming laser beam (Bidare et al. [15] observed up to 40% laser energy attenuation in worst-case stagnant conditions). The gas flow also entrains ejected particles; therefore, during melting, some molten droplets (spatter) are expelled by the vapor jet; the cross-flow helps push these particles toward the outlet. It is important to be aware that if the laser scans with the flow, the vapor cloud and spatter are blown before the laser's path, potentially interfering with the next portion of the scan. Contrastingly, scanning against the flow tends to blow byproducts behind the laser, keeping the path clear. Anwar and Pham [16] found that when the laser scan direction was parallel to the gas flow, the resulting parts had lower strength, attributing this to more laser attenuation by byproducts; conversely, scanning against the flow improved part density [16]. Accordingly, many scan strategies are designed to avoid aligning with the gas flow for long distances. As the gas carries fumes and particles into the outlet, it passes through the filter system. The hot plume cools rapidly upon mixing with the larger chamber volume and contacting the cooler surfaces in the filter. Solidified soot and particles are trapped in the filter media. The cleaned gas (argon) then recirculates back (if a closed loop) or is vented and replaced with fresh argon. In recirculating systems, an impurity buildup (like accumulating oxygen or moisture) is prevented by bleeding out a small fraction of gas and adding fresh argon as make-up. Throughout the build, the gas flow rate is usually kept constant. However, if a sudden pressure spike or filter clog occurs, the system may adjust the fan speed to compensate or pause the build if flow falls outside an acceptable range. Some machines allow tuning the flow velocity or direction with interchangeable nozzle geometries for different materials (since optimal flow for titanium may differ from steel, due to different spatter characteristics). As builds proceed layer by layer, the chamber may gradually get a haze if the filter is not efficient or very fine nanoparticles pass through. That is why multistage filters are common (coarse prefilter and fine primary filter). At the end of a build, the gas flow is turned off when the powder has cooled before venting the chamber. Filters loaded with reactive metal powder must be handled carefully and typically remain under inert gas until they can be removed and sealed. Some facilities have fire-safe cabinets for storing used filters, as they can be a combustion risk if exposed to air while still hot.

Proper gas flow is intimately linked to build quality in LPBF as studied by Elkins et al. [17] on a commercial LPBF system, an EOS M290. Insufficient or uneven flow can cause process instability. Ladewig et al. [14] showed that reducing the cross-flow velocity led to more spatter redeposited on the powder bed and parts. These spatter particles can fuse into subsequent layers, causing porosity or surface defects. They also found areas near the gas outlet, where flow might decelerate, and saw increased surface roughness and porosity [14]. Similarly, if the filter begins to clog and the flow rate drops mid-build, the later layers might suffer higher defect rates due to the lingering plume. It has been observed that a lack of adequate gas flow can transition the melt mode toward keyhole with more vapor and porosity [18]. Machine builders often adjust the nozzle position or angle to avoid direct drafts on the powder; the flow is meant to skim just above it. There is an optimal window of flow rate that clears smoke but does not erode the bed (Shen et al. [19] found an upper limit beyond which defects increase again). Gas flow directionality effects can lead to anisotropic properties across the build plate. As noted, scanning relative to flow can change outcomes; also, parts located downstream (near the outlet) sometimes show different microstructure or defect content than those upstream. Berez et al. [20] observed that specimens built near the outlet of a particular machine had lower fatigue life and more porosity, likely due to spatter accumulation in that region.

Reactive dust in filters is a safety and quality issue. A clogged filter covered with titanium powder in the air could ignite easily if exposed to an ignition source. To prevent this, filters in LPBF are usually inert during operation, and the system (as well as user handling) is properly grounded. Some standards (SAE AMS 7003) recommend safe filter designs and handling procedures. Quality-wise, a well-maintained filtration means that the argon loop stays clean, with no back-contamination of the part. If one notices soot deposits on the build plate or part, it is a sign that the filtration may not be effective enough. Monitoring the gas flow is therefore part of quality control. NIST, for example, has developed diagnostic methods to measure gas speed in situ with hot-wire anemometers to verify uniform flow in LPBF machines [21]. In production, one might include a simple flow verification (ensuring that the flow meter reads the expected value and the differential pressure across the filter is normal) as a checklist item before a build. Finally, as the build progresses, the captured particulate in filters or chamber surfaces is essentially material that did not end up in the part. Excessive spatter generation can indicate suboptimal parameters. Research and industrial practice strongly indicate that a well-designed gas flow can significantly reduce porosity and improve mechanical properties of LPBF parts [14, 22]. Although not directly "melting" material, this subsystem profoundly influences melt pool dynamics and final part integrity.

4.2.5 Powder Management and Spreading System

The powder management and spreading system handles the delivery of metal powder to the build area and the creation of each thin powder layer required for the LPBF process. Essentially, this subsystem stores the feedstock powder, doses each layer's worth of powder, spreads it evenly across the build plate, and often recycles or collects the excess powder. It is sometimes called the recoater or powder-bed handling system. A consistent, uniform powder layer (typically 20–50 μm thick for most metals) is crucial each time, as it sets the stage for the laser to melt the intended cross section. Any irregularity in powder deposition can translate into defects in the built part. Thus, the recoating mechanism, powder supply design, and operation are vital for building reliability.

Most LPBF machines use either a blade recoater or a roller recoater to spread powder. A blade recoater is a straight edge (resembling a squeegee or knife) that traverses the build plate, pushing powder from a supply onto the build area. Blades

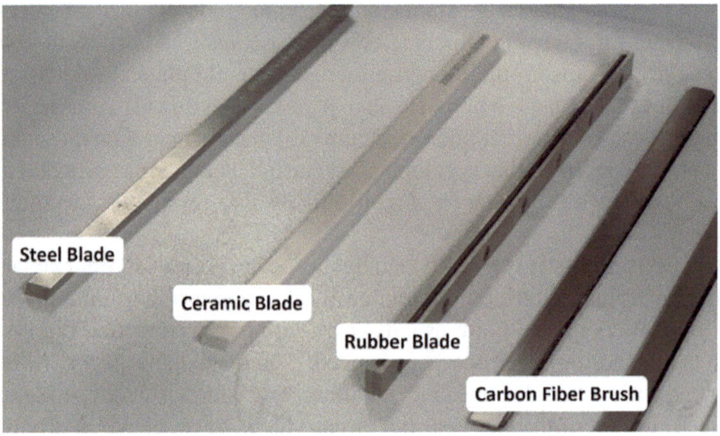

Fig. 4.6 Example of various blade materials used for recoating. Note that some systems may also utilize noncontact spreading mechanisms

can be made of hard materials like stainless steel or ceramic (a hard recoater) or softer, elastic materials like rubber or polyurethane (a soft recoater). Examples of a variety of different blade materials are shown in Fig. 4.6. Hard blades are durable and can help pack the powder, but they can damage a part if a collision occurs; soft blades are more forgiving in a crash and can conform slightly to the powder, though they may wear faster. A roller recoater is a rotating cylinder (metal or elastomer-coated) that dispenses and smooths powder as it rolls over the bed. Rollers can also compact the powder layer gently via their rotation. Most commercial machines currently use fixed blades or wiper-style recoaters, while a few use rollers [3, 23]. In addition to the recoater, the subsystem includes one or more powder hoppers or reservoirs. A common layout is the twin-hopper configuration in which a feed hopper (or powder supply platform) on one side of the build area holds fresh powder and lifts incrementally to provide powder, and a catch hopper on the opposite side collects the overflow powder that is pushed off the build plate with each layer. The build platform (where the part is being made) sits between them. In each layer cycle, the feed hopper piston rises a set distance to present powder, the recoater spreads it across the build plate (which has lowered after the previous layer), and excess powder is skimmed into the catch hopper. Some machines have a single powder supply source that dispenses a measured amount of powder in front of the recoater for each layer (this is more common in small machines). There are designs with multiple feed cartridges for quick material changeover or systems with multiple powder types (for functionally graded material research, e.g., Aerosint's multi-material recoater concept). Supporting equipment includes powder sieves and conveyors, often externally but integrated in an automated setup. After a build, unused powder is sieved to remove agglomerates or debris and can be reused in subsequent builds (with proper quality checks). The machine may have an internal sieve connected by tubes, so overflow powder is automatically sieved and returned to the feed hopper during a long build. Another component is the build platform elevator. This motorized platform precisely lowers (typically by the layer thickness, e.g., 30 μm) after each layer is processed, making room for the next layer of powder. This *Z*-axis mechanism's positioning accuracy and repeatability are important to ensure layer-thickness consistency (Hilton and Gray [2] noted tests for elevator accuracy as part of machine health). The recoater motion itself is driven by a linear actuator or motorized carriage, moving at a set speed (commonly 50–200 mm/s, though some machines can recoat faster for thin powders). Finally, the subsystem includes sensors like level sensors (to detect if enough powder is in the feed hopper) and sometimes collision sensors on the recoater (to detect if it strikes something, often by monitoring motor torque or a deflection of the blade). All these components work together to supply and spread powder layers throughout the build reliably.

At the start of a build, the build plate is at its top position (the powder bed starts essentially at zero height), and a layer of powder is spread to form the initial layer on the substrate plate. The feed hopper is filled with powder (either manually or via automated feeding from a larger container). The machine might initially spread a slightly thicker layer to cover the plate fully. Once the laser melts the first layer, the build plate drops by one layer thickness (e.g., 50 μm). The recoater then traverses where the feed piston pushes up a certain volume of powder above the plane of the build. As the recoater moves across, it drags that powder, filling in the area and leaving a thin, even coat on top of the previously solidified layer. Any surplus powder that the part or build plate does not accommodate is pushed into the overflow hopper. The recoater typically travels in one direction per layer (some machines alternate directions each layer to reduce bias, while others always move the same way). Often, the recoating speed is chosen based on powder flowability. If powder has poor flowability and is spread too fast, it may not spread uniformly, leading to streaks or bare spots, and if it is spread too slowly, it adds unnecessary time and could cause powder piling. Empirical and modeling studies have shown that recoating speed influences the packing density of the powder bed [24, 25]. For instance, Meier et al. [24] found that powder layer uniformity is affected by particle size and recoater dynamics, in which higher speeds can cause more heterogeneity unless the powder is very free-flowing. In practice, speeds around 100 mm/s are common for 30 μm layers with standard 15–45 μm powder. The thickness of each layer is set by the downward movement of the build platform; the recoater ideally glides at a fixed height above the previously solidified surface (the blade might have a slight compliance or the roller a known gap). After recoating, the next laser scan occurs and fuses the new layer. This cycle repeats thousands of times. Throughout, the powder management system must ensure a continuous supply of powder. For long builds, machines may require refilling the feed hopper midway. Many industrial systems have a powder feeder that can add powder from an external reservoir through valves without stopping the process. Unused powder from the overflow is recycled either after the build or continuously. In some designs, after each layer or a set of layers, the overflow hopper can dump powder back to the supply (potentially via a sieve). Often, overflow powder is just collected and then manually reintroduced after sieving. A critical aspect of operation is avoiding recoater crashes. If a part being built distorts or warps upward (which can happen due to thermal stresses), it may protrude above the layer height, and the recoater can hit it. Soft recoaters or spring-loaded recoaters are used to mitigate damage. A crash can cause a layer of powder to be uneven (if the recoater was obstructed) or even shift the part or machine calibration. Some machines monitor the recoater motor current, where a spike indicates a collision, upon which the machine can pause and alert the operator. Another strategy is height mapping, where, before each layer, the machine can do a quick scan (with a laser distance sensor or by probing with the recoater lightly) to detect if anything on the bed is too high. If a tall spot is found, the machine might adjust (by raising the recoater or, in some cases, terminating the build to prevent damage). After the build, the remaining loose powder is brushed or blown off the parts and collected. The powder handling system then facilitates unloading. Here, the build plate with parts is removed, the powder can be vacuumed, or the whole powder bed can be lifted and sieved. Automated powder handling systems often connect directly to the machine. Once a build is done, the powder is conveyed for recycling, and the machine is automatically refilled for the next job (important in high-volume production to reduce turnaround time). Maintaining a stable powder environment (no humidity, minimal contamination) is important throughout the operation. Machines often circulate the inert gas for the process and to keep the powder fluidized and prevent clumping. Any powder that spills or escapes the intended path is contained within the chamber and will be cleaned up later to avoid operator exposure.

The quality of each powder layer has a direct impact on the final part. Layer uniformity is paramount since thin or thick spots in the powder layer can translate to uneven melting. A thinner layer than expected means that the laser will remelt more of the previous layer and possibly cause local overheating or variation in microstructure. In contrast, a thicker local layer may not fully melt through, leaving unmolten islands (lack of fusion defects). Ensuring uniform layer thickness depends on the recoater and powder properties. If the powder has poor flow (e.g., irregular particles or has absorbed moisture), it may clump or form piles instead of a smooth carpet. That is why powder quality (flow rate, particle size distribution) is monitored and often improved by sieving and drying before use (ISO/ASTM 52907:2019 provides methods to characterize powder flowability for AM). Particle size distribution (PSD) affects how densely a powder layer can pack; a very broad PSD might lead to higher packing density but could increase the risk of segregation (fine particle drifting). Meier et al. [24] emphasized that particle size and interparticle adhesion (due to oxides or moisture) influence layer uniformity. Thus, virgin or properly recycled powder within spec is a quality concern. Many manufacturers specify that after a certain number of reuse cycles, powder should be refreshed with a percentage of new powder to maintain consistent PSD and chemistry (ASTM F3456-22 provides guidelines for powder reuse schemes in LPBF, particularly for medical applications). Recoater type and condition also matter. A damaged or worn recoater blade (with nicks or bends) can leave streaks of uneven powder. Operators inspect recoater edges and replace them after a certain amount of use. A ceramic blade that chips can scratch the base plate or create powder lines. Roller recoaters, if not properly calibrated to the correct gap, might either scrape too hard (disturbing the previous layer) or leave too much powder (forming a ridge). Khorasani et al. [3] reviewed that different recoater geometries (blade vs. roller) and motions affect spreadability, but most commercial systems still favor blades due to their simplicity. Some studies indicate that rollers can achieve slightly higher packing for very fine layers [26], but maintaining rollers (keeping them clean and smooth) is another challenge. Recoating speed must be optimized. Too fast recoating can cause a "bow wave" of powder, leaving a wake of lower density behind the blade. Mitterlehner et al. [27] found that recoating above 200 mm/s led to irregular surface defects for a given powder, recommending staying below that speed. On the other hand, Snow et al. [28] observed that higher recoater velocities in their setup increased powder spreadability by preventing the settling of particles during the pass. These differences imply that the optimal recoating parameters may depend on the specific powder and recoater type. Thus, quality control involves validating the recoating process, for example, measuring the actual layer thickness achieved and the density of the powder bed (sometimes test layers can be spread and then vacuumed off to measure how much powder was laid or using optical inspection to ensure no bald spots). Some machines include a camera that images each new powder layer; if it detects an anomaly (like a stripe of missing powder or an object on the bed), it can pause the build. This is increasingly used as an in situ quality check. Recoater crashes or part swelling can be catastrophic to part quality. Even a minor collision can shift powder or the part. If a blade hits a part, it might knock the part slightly, which could lead to layer misalignment and geometric error. In worse cases, it can strip away some of the part's previously built layers. Soft recoaters mitigate damage since they may ride over the part and continue, but that area might get less powder coverage (resulting in a local thin layer and subsequent porosity).

Quality control procedures often involve inspecting the part mid-build using the recoater camera or a height sensor to detect if a feature is growing too tall. Powder contamination is another concern. Cross-contamination can occur if different materials are used in the same machine (e.g., stainless steel powder accidentally mixed with Inconel). This can drastically affect mechanical properties if not caught. Thus, machines often dedicate powder modules to specific materials, and a complete cleanout is performed when changing powders. Standards like ASTM F3049 provide guidelines for characterizing metal powders and detecting contaminants. Also, foreign objects (like a screw or previous build debris in the powder) can cause recoater jams or end up inside a part. Good housekeeping and sieving practices prevent this. Finally, powder aging (gradual change in powder properties with reuse) can affect layer quality, where reused powder may have more fines (from broken particles or spatter), which can change flow behavior. If the powder gets too fine, it may need to be refreshed or entirely replaced to maintain consistent layer formation. Monitoring powder chemistry (oxygen content increasing with reuse, for instance) is also important since oxidized powder might not spread or melt properly. Overall, the powder management and recoating system must reliably produce uniform layers to ensure that each part's cross section has the material to melt. Recent research underscores the importance of recoater design and powder properties in achieving dense, homogeneous parts [29, 30]. By controlling recoating speed, using appropriate recoater materials, and keeping powder within spec (dry, correct particle size, contamination-free), manufacturers can significantly reduce build failures and defects originating from the powder layer. Many modern LPBF machines incorporate features like automated powder feeding, closed-loop recoater motion control, and real-time layer monitoring to enhance the robustness of this subsystem [3].

4.2.6 Computer Control System

The computer control system is the "brain" of the LPBF machine, executing the digital instructions (the sliced CAD model) and coordinating the precise timing of events in the physical machine. The control system typically includes the following: (1) Central processing unit (CPU)/controller that often consists of an industrial PC or a dedicated microcontroller/PLC that runs the main control software: Modern machines might use a PC for high-level coordination and specialized FPGA/DSP boards for real-time laser/galvo control. (2) Motion controllers/drivers or electronics that control motors (for recoater and build platform axes) and galvanometer scan mirrors: For galvanometers, dedicated scan controllers receive vector data and output analog signals to the galvo amplifiers with high timing precision. For linear motions (Z-stage, powder feeders), stepper or servo motor drivers are used, often integrated with the PLC. (3) Laser control unit that interfaces with the laser source to turn it on/off and modulate power: This might be a digital analog interface or TTL control for the laser's acousto-optic modulator or direct diode current control. It must synchronize with scanner position data to implement things like skywriting. (4) Sensor inputs such as thermocouples (chamber or baseplate temperature), oxygen-level meter, pressure sensors, motor encoders (to verify recoater and Z positions), optical sensors (for powder-bed imaging or melt pool monitoring if available), and interlocks (doors, etc.): There is typically an interface board gathering these signals. (5) User interface and software involving a computer program (with GUI) that allows an operator to set up a build (import STL or slice file, arrange parts, choose parameters) and then supervises the build progress. This software often includes the slicing and scan path planning engine, which may be done offline with separate software. It also logs data from the build for quality records. (6) Firmware/real-time software to achieve microsecond-level control: For instance, the galvo control might run at kHz rates to update positions; laser gating is controlled with microsecond resolution. These tasks are handled by firmware on a motion control card or real-time operating system (RTOS) routines. (7) Data storage and communication that includes memory to store the layer data (each layer can be a set of scan vectors or a raster image in some cases) and communication links between subsystems (for example, CAN bus or EtherCAT links for motion control, and Ethernet/USB for high-level PC to microcontroller communication): In essence, the control system spans from high-level job management to low-level execution, and it must be reliable against disturbances (like power fluctuations or sensor noise) to avoid any interruption that could ruin a build.

Software errors or crashes are a latent threat, and any crash could instantly ruin a build and possibly the machine (if, say, the laser stayed on or a motor did not stop). Industrial LPBF controllers, therefore, typically run on proven, stripped-down codebases where possible and sometimes on a real-time OS to guarantee execution timing. The user interface might be separate from the real-time logic to prevent a UI freeze from affecting the process. Also, power outages are considered. Today, many systems have an uninterruptible power supply (UPS) or at least a procedure to handle power loss, such as remembering the last completed layer, so the build might be continued (though in practice, continuing a build after a power loss is challenging due to layer oxidation and alignment issues; most often, the part is lost). Another quality aspect is traceability and logging. For quality assurance, the control system should log critical parameters throughout the build (laser power, scan speed, temperatures, etc.). If a part is found defective, these logs can help identify if any anomaly occurred (say, O_2 rose at layer 200 or laser power dropped). The controller's ability to provide a build report is important in regulated industries.

While traditional LPBF control is largely open loop (following predetermined scan patterns and parameters), there is a push toward closed-loop control where sensor feedback (like melt pool monitoring) is used to adjust parameters in real time [31]. For example, a pyrometer could detect a too-hot melt pool, and the controller might then reduce laser power on the fly. Research by Scime and Beuth [32] and others on machine learning-driven real-time quality control suggests that future controllers might incorporate algorithms to detect defects (e.g., via imaging) and then adapt subsequent layer parameters or trigger recoater adjustments. Current machines, however, typically use the control system more for interlayer adjustments (for instance, if a part is overheating, the user might manually pause and change parameters). As monitoring technology integrates, the control system will manage those inputs and possibly perform iterative learning (e.g., adjusting scan strategy in later layers based on early layer results). From a user perspective, the control software should be user-friendly, but not allow damaging mistakes. There are usually built-in safe ranges for parameters; the software might prevent a user from setting a layer thickness or laser power outside the proven range for that material/machine configuration, as those could cause failures.

4.3 Traditional Manufacturing Versus LPBF

LPBF is a process in which material and shape are formed simultaneously. This approach contrasts sharply with traditional methods like casting, forging, and machining, where the material is first created or preformed and then shaped. In subtractive processes such as CNC machining, a solid billet (often cast or wrought beforehand) is machined away to reveal the final geometry. In forging, a material is deformed into shape under heat and pressure, and in casting, metal is melted and poured into a mold to solidify into the desired shape. The material's initial state is established before the final shaping step in all these conventional techniques. LPBF, by contrast, joins and solidifies raw material powder in situ as it builds the part, effectively combining material creation and geometric shaping into one process. In other words, LPBF builds a structure from the ground up ("bottom-up"), directly from digital data. In contrast, a forged or machined part relies on preexisting material stock that is sculpted or formed into shape [1]. This fundamental difference means that LPBF can produce parts with as-built microstructures that differ significantly from those of wrought or cast materials [33]. Traditional formative and subtractive methods benefit from well-established procedures and typically yield predictable material properties due to decades of experience controlling their processes. For example, forging processes are well understood and governed by models of deformation and recrystallization that allow engineers to predict the resulting grain structure and properties [34]. LPBF's integrated approach, on the other hand, introduces new complexities but also offers unique capabilities.

4.3.1 Thermal Cycles and Cooling Rates

One of the most striking differences between LPBF and conventional metalworking is the thermal profile experienced by the material. LPBF uses a high-power laser to melt fine powder layers, leading to rapid, localized heating and cooling cycles. Each point in the material may be melted and then solidified in milliseconds, and this process is repeated layer upon layer. The cooling rates in LPBF are incredibly high, typically on the order of 10^4–10^6 °C/s (i.e., 10^5–10^7 K/s) in the melt pool. Such rates are several orders of magnitude higher than those in traditional processes like casting, where cooling might occur at only around 10–10^2 °C/s (tens of degrees per second). These rapid solidification conditions in LPBF produce far-from-equilibrium microstructures, for example, very fine cellular dendrites or even metastable phases that would not form under the slower cooling of casting or welding [35, 36]. Furthermore, the layer-wise building means that each new layer can reheat the underlying material (through conduction of laser heat from subsequent passes), causing multiple thermal cycles. The result is a complex thermal history with many reheating and cooling events for any volume in an LPBF part. By contrast, the entire volume typically cools monotonically from the melt in a cast part. Figure 4.7 illustrates such thermal cycling differences between casting and LPBF, which better illustrates this difference. In a forged part, the thermal cycle is dominated by the furnace heating and controlled cooling or post-forging heat treatment. The peak temperatures and cooling rates in forging or machining are much lower. Forging, for example, may involve slow cooling in the die or air, on the order of only a few °C/s, allowing ample time for diffusion and phase transformations. While LPBF's extremely high cooling rates lead to ultrafine microstructures and high residual stresses, the slower cooling of a casting promotes coarse dendritic structures, and a forged piece often develops a refined but equilibrium microstructure after subsequent heat treatments [37].

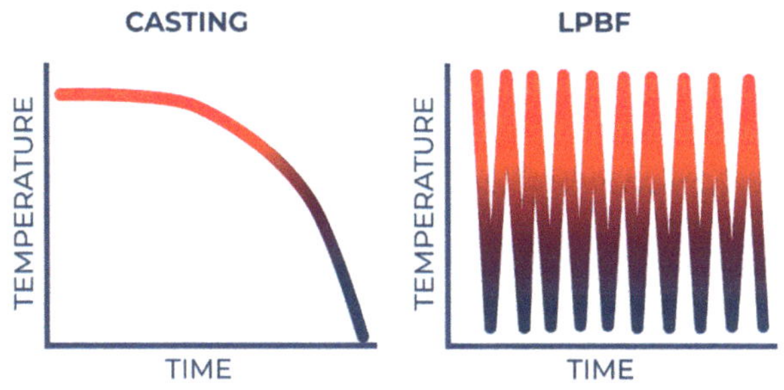

Fig. 4.7 Comparison of thermal cycling of castings versus LPBF

4.3.2 Microstructure and Part Qualification Challenges

The consequences of LPBF's complex thermal cycling are evident in the microstructure and the challenges of part qualification. Because LPBF builds geometry layer by layer, the thermal history varies with the part's geometry and scan strategy. Sections of different thickness, orientation, or scan length can cool at different rates, leading to heterogeneous microstructures within the same part. Traditional manufacturing methods generally produce more homogeneous microstructures or at least well-characterized gradients. In forging, the material undergoes plastic deformation and controlled cooling that are better understood and can be modeled with established metallurgy principles. Wrought (forged) alloys typically have predictable microstructural outcomes (such as refined equiaxed grains or aligned fiber texture) because engineers can rely on decades of empirical data and models relating forging parameters to material properties [34]. Moreover, standards and specifications exist for critical forged components (e.g., required forge ratios, heat treatments, and inspections) to ensure consistent quality. By contrast, the microstructure in an LPBF build is sensitive to numerous process parameters (laser power, scan speed, hatch spacing, etc.) and even the part design itself. The result is a coupling between design and material properties that is much stronger than in conventional processes. In LPBF, a change in geometry can alter the cooling rate and hence the microstructure and mechanical performance in that region of the part.

This complexity directly impacts part qualification and certification. Industries such as aerospace and biomedical engineering demand reliable, repeatable material behavior, but the intricate thermal history of LPBF parts complicates their qualification. Each new LPBF design may require extensive testing and process tuning to ensure that it meets requirements, because one cannot assume that the microstructure will be the same as a previously qualified design, even if the same machine and parameters are used. Conventional processes have a qualification advantage since their process-structure relationships are well established, so a forging produced to a given specification (e.g., a standard alloy, a defined temperature and deformation schedule, followed by heat treatment) can be expected to exhibit known properties. For example, a forged aerospace bracket in Ti-6Al-4V will have a lamellar $\alpha + \beta$ microstructure and specific fatigue properties, with decades of experience correlated with the processing route. In LPBF Ti-6Al-4V, however, the as-built microstructure might be acicular α′ martensite due to rapid cooling, with possible columnar grain structures aligned to the build direction. Depending on part design, parts may require additional analysis and possibly post-processing (like hot isostatic pressing or other heat treatment) to achieve properties equivalent to the forged part. Researchers have highlighted that the process variability in metal AM is a significant challenge for qualification efforts. Frazier [38] noted that additive processes show "drastic variability" at different stages of fabrication, posing significant challenges in developing qualification standards. In practice, qualifying an LPBF part often involves comprehensive testing (material coupons, nondestructive evaluation, process monitoring data) to verify that, despite the complex thermal history, the part meets all performance criteria [39]. By comparison, conventional cast or forged parts benefit from mature quality control methodologies (such as well-defined acceptance criteria for microstructure and defects).

4.3.3 Geometrical Complexity and Design Freedom

Along with these challenges, LPBF offers design freedom unattainable with traditional manufacturing. Since material is added selectively where needed, LPBF can create complex internal geometries and intricate features that would be difficult or impossible to produce by casting, forging, or machining. For example, LPBF can fabricate internal channels, lattice structures, conformal cooling passages, and hollow sections without requiring assembly or sacrificial cores. A conventional casting can form some internal cavities using sand cores. However, there are limits to the complexity and minimum feature size, and removing internal sand or supporting material can be problematic. Machining internal voids is often impossible unless the design is split into multiple parts that are then welded or bolted together. Forging, a solid-state deformation process, cannot produce internal hollow features; any cavities or complex internal profiles must be machined afterward or created by assembling multiple pieces. In LPBF, internal features are built layer by layer along with the external geometry, so one can design free-form passages, cellular infill structures, and other innovations with minimal additional processing [33]. After printing, usually the only post-processing needed for such features is removing loose powder from channels and possibly some surface finishing. This capability means that engineers can consolidate what would traditionally be assemblies of many parts into single, monolithic designs, and they can optimize shapes purely for performance (e.g., lightweight lattices or optimized fluid flow paths) without the usual manufacturing constraints.

The design freedom of LPBF is frequently cited as one of its most significant advantages. Because LPBF does not require special tooling for each new shape (unlike casting, which needs a mold, or forging, which needs a die), complexity comes at little extra cost. Intricate lattice mesh structures, organic free-form shapes, and topology-optimized forms are all feasible in one step. LPBF enables the direct production of geometries that could only be imagined or approximated previously. This includes, for instance, conformal cooling channels in injection molds (which improve cooling efficiency but cannot be drilled with traditional methods) and integral hollow fuel nozzles with internal mixing chambers (which would otherwise require laborious brazing of sub-components). By eliminating the need for joints or assembly, LPBF expands design possibilities and can improve reliability (fewer parts and interfaces) and reduce weight. Minimal post-processing is another benefit. While LPBF parts may require heat treatment to relieve residual stress and some machining for critical tolerances or surface finish, they generally do not need the extensive subtractive shaping that castings or forgings often undergo (e.g., finish machining all mating surfaces).

4.4 Process, Structure, Properties, and Performance in LPBF

As discussed, LPBF distinguishes itself from many traditional manufacturing methods because it merges material processing and part production in a single step. The process, structure, properties, performance (PSPP) relationship provides a framework for understanding how each stage of LPBF influences final part functionality. In essence, the interplay of process conditions, resulting structures (both macro- and microscale), and measurable properties converges to determine performance. The process encompasses how the laser fuses powder material under specific machine parameters, such as layer thickness, scan pattern, and energy density. Each fused layer experiences localized high-temperature gradients followed by rapid solidification. These conditions and feedstock characteristics collectively define the structural outcomes [1]. The structure includes macro-level features like dimensional accuracy, surface roughness, and microstructural attributes (e.g., grain size, phase composition, precipitates). LPBF's rapid and localized heating enables unique microstructural formations that can be difficult to replicate in conventional processes [40]. While this layer-by-layer control opens pathways for tailoring novel alloys or functionally graded materials, it can also introduce quality control challenges if the process is not closely managed. Properties refer to the physical and mechanical characteristics of the fabricated part, such as density, tensile strength, hardness, and thermal conductivity [34]. By fine-tuning process parameters, users can alter microstructure and, consequently, part properties. The culmination of process, structure, and properties manifests as the overall performance of the LPBF component. How well the part endures operational loads, resists fatigue, and maintains functionality throughout its service life defines performance [41].

4.4.1 Flaws and Defects in LPBF

LPBF-produced parts can suffer from flaws and defects from the process's rapid solidification and complex thermal cycles. These defects are generally categorized by scale into macroscale defects (affecting the part geometry or integrity on a larger scale) and microscale defects (internal or surface imperfections on the order of the melt pool or powder particles). Each type of defect has distinct formation mechanisms, visual indicators, effects on mechanical performance, and strategies for mitigation. This section discusses the major LPBF defects, which include warpage, cracking, lack of fusion, balling, keyhole porosity, and gas porosity. We then provide further details on their causes, characteristics, impact on properties, and how they relate to process variables and controls.

4.4.1.1 Macroscale Defects

Macroscale defects in LPBF are those readily observed at the part or layer level (often visible to the naked eye or low-power microscopy). They include warpage/distortion, cracking (which can be further classified as delamination or solidification cracks), lack of fusion (LOF) voids, and balling phenomenon.

Warpage (Distortion)

Warpage refers to the built part's unwanted bending, curling, or distortion. It is primarily caused by accumulating residual thermal stresses during the cyclic heating and cooling in LPBF. As each layer is rapidly melted and solidified, the material contracts; if this contraction is constrained (for example, by the build plate or previously solidified material), tensile stresses develop. When the residual stress exceeds a certain threshold, it can lead to part deformation (e.g., corners of the part lifting) or even delamination of layers. Warpage often manifests as edge curl or lifted sections at the base or corners of the part and, in severe cases, can cause the part to lose contact with the build platform. Visually, a warped LPBF part may deviate from design dimensions or show a bent shape; it may also be indicated in situ by the recoater blade hitting a warped layer or by an observed gap under a curling layer.

The presence of warpage is detrimental to dimensional accuracy and can compromise part functionality. Moreover, warpage is intimately linked with residual stress, where excessive warpage is a clear sign of high internal stresses, which can later contribute to cracking or reduced fatigue life in service. Warped parts might not meet tolerance specifications, require extensive post-processing, or be rejected entirely. Ensuring geometric fidelity is critical in engineering applications, so warpage must be minimized. Figure 4.8 shows a picture of warpage occurring on thin-wall features, likely due to the high aspect ratio of the design, which later resulted in process delamination.

Fig. 4.8 Warpage defect of a thin wall

Thermal gradients and part geometry strongly influence warpage. Key approaches to avoid warping include preheating the build area, optimizing scan strategies, applying support structures, adjusting interlayer cooling time, or applying a post-build stress relief. Elevating the base-plate temperature (e.g., 100–300 °C for some alloys) before and during the build reduces thermal gradients and residual stresses, thereby lessening warpage. Preheating has proven especially effective for high-strength or crack-prone alloys (e.g., tool steels) by avoiding large temperature swings. However, extremely high preheats may alter the microstructure or even increase certain defects if not optimized. Using stress-relief scan patterns (such as island scanning or rotating stripe patterns) distributes heat more uniformly and avoids the accumulation of stress in one direction. Shorter scan lengths and alternating scan directions between layers help reduce net distortion. Similarly, avoiding over-constraining certain regions (e.g., scanning large cross sections in one continuous path) can mitigate distortion. Supports that anchor the part to the build plate and support overhangs can counteract deformation by holding down areas prone to lift. Support structures also increase heat conduction away from the part. Allowing a slight dwell between layers can permit stresses to relax incrementally. Some LPBF machines let the part cool after each layer or after a certain number of layers to mitigate distortion. Finally, a stress-relief heat treatment can alleviate residual stresses and thus reduce the driving force for warping and cracking. This does not prevent warpage during fabrication, but it is necessary to relieve built-in stress before removing the part from the plate (usually, the part is heat-treated while still attached to the plate).

It is worth noting that in-process monitoring can aid in warpage control. Modern systems use sensors to detect if a layer has shifted or raised, triggering corrective actions or aborting the build to prevent a recoater crash. Through careful control of the above factors, warpage can be minimized even in high-stress materials. For example, one study noted that an optimized support layout and scan pattern eliminated visible distortion in a high-temperature alloy build that would otherwise warp significantly.

Cracking

Cracking manifests fracture paths in the material during or after the LPBF process. Cracks in LPBF parts generally fall into two categories: delamination (cold) cracks and solidification (hot) cracks. Delamination cracks occur when previously deposited layers separate or crack due to accumulated tensile stresses at lower temperatures. In contrast, solidification cracks form at high temperatures during the rapid solidification of the melt pool.

Delamination cracking (cold cracking) is a type of cracking that occurs when layers separate from each other or the build plate under excessive residual stress. As noted above, the thermal contraction of each layer can create tensile stress; in materials that cannot plastically accommodate this stress (often high-strength or low-toughness alloys), the part may crack between layers or at the interface with supports. Delamination cracks often appear parallel to the layers (horizontal cracks) or at the base (where the part may tear away from supports or the plate). Visually, they may be seen as a part-height crack or a portion of the build that has lifted and cracked off. In situ, a delamination event may be accompanied by an audible "ping" (as the standard ASTM E3353 notes, acoustic signatures have been observed during cracking events). Poor part design (e.g., sharp internal corners or large cross sections that exacerbate stress) and inadequate support can contribute to this type of cracking.

Solidification cracking (hot cracking) forms during the cooling of the melt pool, typically along grain boundaries, when the last portions of liquid fail to bridge shrinking solidified material adequately. In alloys with susceptible chemistry (for example, those with a wide solidification range or trace impurities that cause low-melting films, such as sulfur/phosphorus in steels or hot cracking-prone nickel superalloys), the rapid cooling and high thermal gradients in LPBF can lead to intergranular cracks as the material cannot accommodate the shrinkage. These cracks are usually oriented along the build or thermal direction (often vertical through layers). They may be microscopic in width (hard to see without microscopy), though they manifest as visible cracks in severe cases. Solidification cracks (hot tears) often occur in materials like high-carbon tool steels or some Ni-based alloys if processed without countermeasures. They tend to follow prior melt pool or columnar grain boundaries and can be detected via metallographic inspection.

Cracks are critical defects because they act as fracture initiators and drastically reduce load-bearing capability. Even small microcracks can propagate under stress or during post-processing (e.g., heat treatment as seen in strain-age cracking of some superalloys). The presence of cracks in an LPBF part usually renders it unacceptable for high-performance applications, as cracks can lead to early failure under tensile or fatigue loading. For instance, studies on LPBF tool steel have shown that even a small percentage of the crack area can reduce elongation to near zero and significantly lower fatigue life. An example of cracking can be seen in Fig. 4.9.

To avoid cracking, the LPBF process must be tailored to reduce the magnitude of residual stresses and metallurgical crack susceptibility. As with warpage, increasing the build temperature is an often-used strategy. Preheating the substrate (often 200 °C or higher for crack-prone alloys) allows the part to cool more slowly, reducing thermal gradients and preventing the formation of brittle microstructures, thus mitigating delamination and solidification cracking. Controlled cooling or active cooling rate control during the build (if available) can ensure that the material has more time to relieve stress or fully solidify without cracking. However, it must be noted that overly high preheats can sometimes promote hot cracking in certain alloys by keeping them in the vulnerable semisolid state longer. Hence, an optimal preheat exists for each material. Alternating with preheating during the process, post-process heat treatment can also manage cracking. While this does not prevent cracks that form during the build, an immediate stress-relief anneal on the build plate can prevent further crack propagation and is generally recommended. Crack-susceptible alloys are sometimes intentionally kept on the build plate for a furnace heat treatment (or hot isostatic pressing) to heal any microcracks and relieve stress.

Some alloys (e.g., Al-Si alloys, austenitic steels like 316L) are inherently weldable and resist cracking, while others (like Al 7075 or some tool steels) crack readily. For crack-susceptible alloys, using modified powders or new compositions designed for AM (for example, adding grain refiners or changing composition to

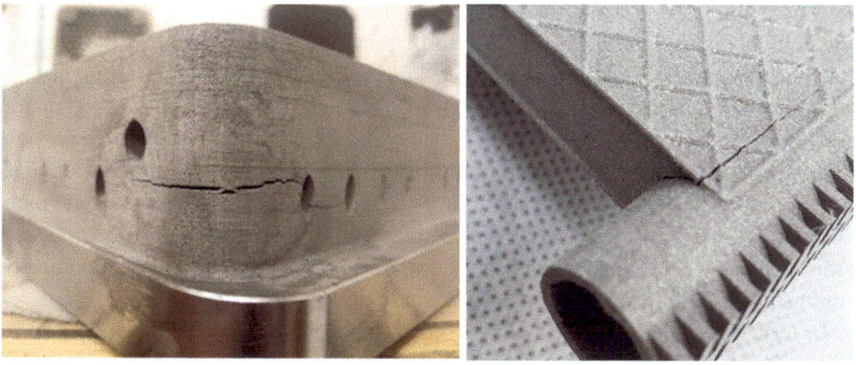

Fig. 4.9 Cracking defects on LPBF parts

narrow the solidification range) can be effective. Researchers have developed modified titanium and nickel alloys that print crack free by adjusting composition to avoid deleterious phases.

Avoiding large contiguous areas that solidify all at once (which build up stress) is important. Segmentation of scans (islands) and alternating scan directions between layers help disrupt the formation of a continuous high-stress path. Also, using contour scans (outlines) that delay the cooling of edges or custom scan order can help. Thin walls and sharp corners are high-stress concentrators; designers often fillet corners or use intermediate supports to reduce stress concentration. A shorter layer time (or pauses to allow stress relaxation) can also mitigate crack formation.

Quality control for cracks typically involves visual inspection of the part (looking for visible surface cracks or layer separations) and NDT methods like dye penetrant or X-ray CT scanning for internal cracks. In-process, acoustic sensors or imaging can sometimes detect a crack as it forms, allowing for layer-wise intervention (this area is under active research). Through a combination of material, process, and design optimizations, LPBF can produce crack-free parts even in traditionally crack-prone materials; for example, with a 200 °C preheated platform and an island scan strategy, AISI H13 tool steel samples have been produced with no visible cracks.

Lack of Fusion (LOF) Defects

Lack of fusion defects are volumetric porosity flaws caused by incomplete melting or bonding between adjacent scan tracks or layers. Essentially, LOF defects are unfused regions where the powder or previous layer did not reach the melting temperature, leaving an internal cavity (often irregularly shaped). These defects are considered macro because they tend to be larger voids (typically tens to hundreds of microns) with irregular geometry, often spanning multiple particles or layer thicknesses, and they significantly disrupt the continuity of the material.

Lack of fusion occurs when the energy density is too low to fully melt the powder or properly remelt the surface of the previous layer. This can happen due to insufficient laser power, overly high scan speed, too large hatch spacing, or excessive layer thickness, leading to a shallow or narrow melt pool that fails to overlap with neighboring tracks. In practice, LOF porosity often appears at the overlap between scan tracks or at layer interfaces, especially if there is misalignment or if the scan pattern leaves gaps. Visually, on a polished cross section, lack-of-fusion porosity has an irregular, often angular shape (since it conforms to unmelted particle shapes or the contour of melt pools) and may be found clustered in planes or lines corresponding to the hatch pattern. Figure 4.10 (cross-section image) shows typical LOF defects appearing as dark cavities between melt tracks. These defects can sometimes be visible on the part surface as well, for example, a poor top surface might show small linear grooves or pits where underlying LOF occurred.

In terms of scale, individual LOF voids are usually larger than gas pores (e.g., 100–300 μm long, and often flattened). Lack-of-fusion defects are extremely detrimental to mechanical performance, as they introduce notches and reduce

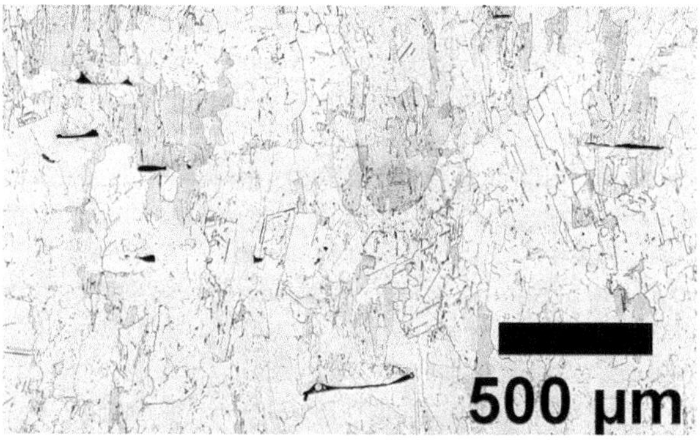

Fig. 4.10 Lack of fusion caused by insufficient melt pool penetration

load-bearing cross-sectional area. It is well documented that LOF porosity can reduce both tensile strength and fatigue life more severely than spherical gas porosity. For instance, in one study, samples with LOF defects had 10% lower tensile strength and 30% lower elongation than fully dense samples [42]. The sharp corners of LOF voids act as stress concentrators where cracks can initiate under stress. In fatigue, lack-of-fusion flaws (especially those near surfaces) dramatically reduce life, as cracks will nucleate at these defects even under relatively low cyclic stresses. Thus, minimizing LOF is critical for high-performance parts.

Lack of fusion is fundamentally a result of insufficient energy input or coverage, so the remedies involve ensuring adequate melting and overlap. Raising the laser power, reducing the scan speed, decreasing the hatch spacing, or using a smaller layer thickness are all measures that deliver more energy to fully melt the powder and remelt the edges of adjacent tracks. Generally, a higher energy supply and/or a tighter overlap between scan lines significantly reduces LOF defects. For example, if a lack of fusion is observed, one might slow the scanning speed or use a second laser pass to fuse any remaining lightly sintered regions. Improper scan strategies (such as all tracks in one layer oriented the same without sufficient rotation between layers) can stack LOF defects. By rotating the scan pattern between layers (e.g., 67° rotation each layer in a chessboard strategy), any LOF gap left in one layer gets remelted by tracks in the next layer. Additionally, scans of contours and infill must be tuned so that the contour track overlaps the infill; a common LOF problem area is the boundary between melt pool tracks and part edges (hatch-to-contour overlap flaw), which can be solved by increasing overlap or using special border scans. LOF can occur if the laser is defocused or misaligned or the powder layer thickness is inconsistent (too thick in spots). Ensuring a uniform powder layer and a well-focused beam will prevent random lack-of-fusion regions. Proper machine calibration and powder recoating are part of quality control here.

Modern LPBF machines may employ photodiodes or cameras to detect signatures of incomplete fusion. For example, a region with unusually low melt pool brightness could indicate a lack of fusion. Some systems can adjust power on the fly or add an extra scan (remelting) if a flaw is detected. At a simpler level, regular density measurements (Archimedes or image based) during parameter development ensure that the chosen parameters yield nearly fully dense (>99.5%) parts, indicating minimal LOF voids.

Standard process development often includes constructing a process map of laser power vs. scan speed for a given material, to identify the "LOF regime." Typically, parts exhibit LOF porosity at too low energy densities (the lower left region of such a map). In contrast, an optimum region yields full density, and extremely high energies lead to other defects (keyhole porosity). LOF defects can be virtually eliminated by operating in the appropriate window and using a robust scan strategy. Indeed, references note that employing a higher energy density and smaller hatch spacing can achieve full consolidation, eliminating lack-of-fusion porosity. Quality control to catch any LOF that does occur involves high-resolution imaging or CT scanning of critical sections of the part, since LOF pores are often irregular and not easily detected by simple density measurements alone.

Balling Phenomenon

Balling is a distinctive phenomenon in LPBF where the molten track breaks into separate beads or droplets rather than forming a continuous weld line. It is classified as a macroscale defect because it disrupts the uniformity of each layer and can be observed directly on the melt track (often with optical microscopy, appearing as a string of beads). Balling leads to a rough, discontinuous surface and often entrains pores between the balled droplets.

Balling is caused by an unstable melt pool driven by poor wetting and capillary effects. When the balance between the molten metal's surface tension and viscosity versus the driving force to spread (wettability) is unfavorable, the liquid tends to minimize its surface energy by contracting into spherical shapes (droplets) rather than wetting the substrate. This typically occurs in the low-energy-density regime. If the laser power is too low or the scan speed is too high (producing a small, rapidly solidifying melt pool), the melt does not have time or volume to stay in a continuous track. Instead, it shrinks into segments, leading to a line of separated beads, hence the term balling. Balling is also exacerbated by the presence of surface oxides or poor substrate wetting, as well as by excessively long melt pools. Visually, balling is quite apparent. The top surface of a balling-affected region will appear irregular and rough, with small metal spheres or globs instead of smooth tracks. Figure 4.11 shows an example of a "balling" defect on a single track, where the track is no longer continuous. In cross section, balling can cause linearly arranged pores (the gaps between balled droplets). Balling introduces porosity and lack of fusion because the droplets do not fuse well with each other or the substrate.

Balling is detrimental in several ways. First, it causes high surface roughness and may compromise the accuracy of each layer, potentially leading to stair-step defects or collision with the recoater. Second, the discontinuities mean some areas are not properly fused, creating LOF defects and associated pores. Mechanically, a surface

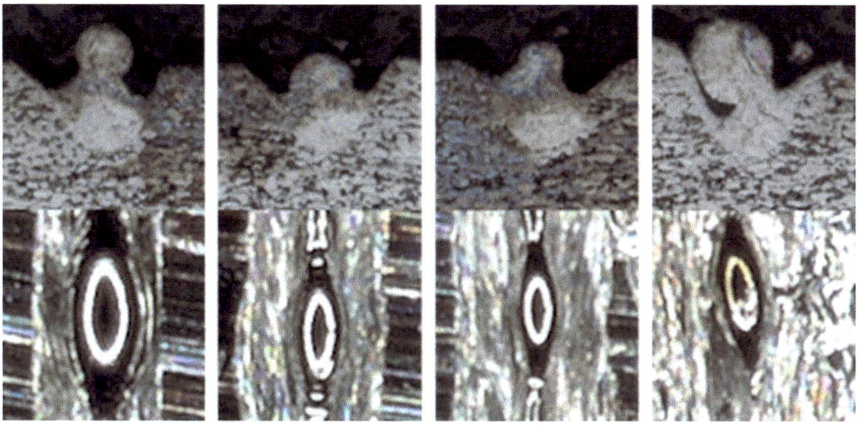

Fig. 4.11 Balling formation correlations with laser power and speed

or internal region afflicted by balling has poor bonding and thus lowers the effective strength. While subsequent passes might remelt a single-balled track in infill, persistent balling will leave strings of pores that act similarly to LOF flaws (i.e., stress concentrators). Balling on surfaces is especially harmful for fatigue performance, as the rough surface and near-surface voids dramatically reduce fatigue strength. Thus, avoiding balling is important for surface quality and ensuring fully dense, strong material.

Balling is influenced by melt pool stability, which depends on parameters and material properties. The primary remedy is to move out of the low-energy regime where balling occurs. Increasing laser power or decreasing scan speed will increase the melt pool volume and promote continuous track formation. There is a consensus that balling occurs when energy density is too low to maintain a stable wetting of the melt pool to the substrate. For example, if one observes balling at 200 W and 1200 mm/s, increasing power or slowing to 800 mm/s often yields a smooth track. Interestingly, some studies (e.g., [43]) observed that high scan speed and power can reduce certain materials' balling. The rationale is that high power compensates for high speed to deliver enough energy, while quickly moving the melt pool can sometimes produce a smoother track if conditions are right. The key is achieving the proper melt pool aspect ratio, which is neither too elongated (which causes balling) nor too deep (which could cause keyholing). Analytical models predict a threshold melt pool length-to-width ratio beyond which balling occurs. A thick powder layer or powders with poor flow can lead to uneven melting and spur balling. Using thinner layers (so the laser energy is more concentrated per unit powder) and ensuring that powder is dry and free of excessive oxides can improve wetting. High oxygen or surface contamination on powder can increase surface tension or prevent good wetting, thus promoting balling; hence, using fresh powder or argon-atomized powder with low oxygen content can help. Some materials are inherently prone to balling because of their fluid properties (e.g., pure metals with low viscosity). Alloys that solidify quickly or have high surface tension may need higher energy input to avoid balling. In some cases, slightly increasing the ambient temperature or using a different inert gas (helium vs. argon can change heat transfer and molten pool behavior) has been explored. However, the most direct control is still through laser parameters.

In practice, balling is often detected early in process development because it is visible. If a certain parameter set produces balling, it is usually marked as a processing limit. For instance, a process map might show that tracks become discontinuous below a certain combination of power/speed (balling), and thus the usable process window is constrained above that line. Implementing a small overlap of melt pools (slight remelt of previous track) also helps ensure continuity and can alleviate mild balling. Suppose balling is observed during a build (via real-time optical monitoring). In that case, a mitigation might be locally increasing energy or adding a rescan (remelting pass) to fuse the balls. Overall, with well-chosen parameters, balling can be avoided.

4.4.1.2 Microscale Defects

Microscale defects are those typically on the order of the microstructure or smaller, often requiring microscopy or CT scanning to observe. The most common microdefects in LPBF are porosity defects, namely gas porosity and keyhole porosity. These are generally distinguished by their shape, size, and formation mechanism. While small, microdefects can have an outsized influence on fatigue and other properties. They are often stochastic (especially gas pores), though their occurrence is tied to process parameters and environment.

Gas Porosity

Gas porosity in LPBF refers to small, usually spherical pores found in the as-built material. These pores are typically a few microns up to perhaps ~100 μm in diameter. They are called gas pores because gas is trapped in the solidifying melt pool. Gas porosity is one of the most ubiquitous defects in metal AM, even in well-optimized processes, because it can originate from inherent sources such as gas within powder particles or gases introduced during melting.

Gas porosity can form via multiple routes, all involving gas entrapment. Metal powders produced by gas atomization can contain internal porosity or gas bubbles (for example, argon entrapped in a particle). The trapped gas is released into the melt pool when such a powder particle melts. A gas pore is left in the solid if it does not escape before solidification. This is a common source of near-spherical pores. Using powders with low internal porosity and proper sizes can mitigate this, but eliminating it is challenging. Many alloys can dissolve a certain amount of gases (like hydrogen and nitrogen) at melt temperatures, but have far lower solubility in the solid state. During rapid solidification, dissolved gas can precipitate out. A classic example is hydrogen porosity in aluminum alloys, in which aluminum melts absorb hydrogen from the atmosphere or moisture, and upon solidification, the excess hydrogen forms bubbles. Maskery et al. described gas pores in AlSi10Mg as spherically shaped (<100 μm) pores caused by hydrogen rejection upon solidification, especially at slower scan speeds that allow more gas segregation. For steels and nickel alloys, oxygen or nitrogen pickup is typically low in the inert environment, but if the chamber purity is poor, some porosity can form this way.

During LPBF, a high-power laser vaporizes some metal, creating a vapor plume above the melt pool. If the inert gas flow is insufficient or the scan geometry causes turbulence, inert gas (argon) or vapor can get trapped in the pool. Rapid solidification can "freeze" that bubble in place. Additionally, any lack of continuous melt flow (as in the balling scenario or spatter entrapment) can trap pockets of gas. Interestingly, the incidence of gas porosity can depend on nonlinear parameters. Some studies have found more gas porosity at extremely high scan speeds (due to turbulent melt pool flow causing bubble entrainment). Maintaining a good inert gas flow and using moderate scanning conditions generally help minimize gas pore formation.

Gas pores are typically nearly spherical (because surface tension shapes the gas bubble in liquid), in contrast to irregular LOF pores. They tend to be isolated and randomly distributed, though they may concentrate in areas like the top of the part (where solidification was very rapid) or where scan tracks ended. Under X-ray CT or microscopy, gas pores appear as round dark spots. They can often be distinguished from keyhole pores by size and shape, since gas pores are smaller and rounder. Figure 4.12 shows micrographs of gas porosity in an LPBF sample (small round voids).

Individually, a tiny gas pore may not severely impact static strength (especially if the part is otherwise dense). However, collections of gas porosity reduce the load-bearing area and can initiate fatigue cracks. It is widely accepted that porosity (including gas pores) can degrade static and dynamic performance, limiting AM part use in critical applications. Fatigue performance is particularly sensitive since a pore acts as a micro-notch where cyclic stress can initiate a crack. Even a <50 μm pore at the surface or just beneath it can nucleate a fatigue crack that grows and

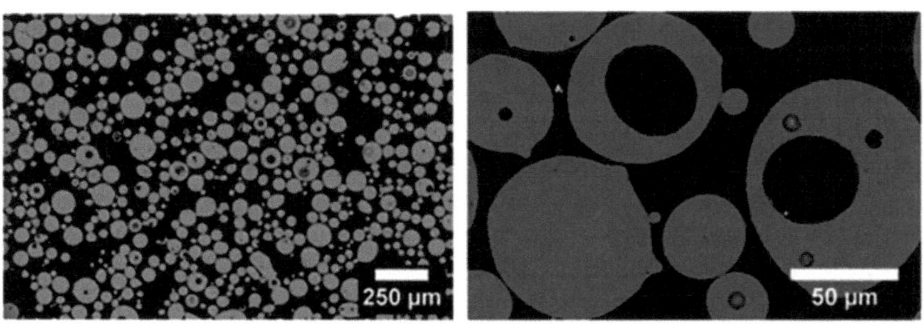

Fig. 4.12 Micrograph of porous powder

causes failure at stresses far below what a pore-free material can withstand. Porosity also reduces tensile ductility since pores concentrate strain and can cause premature localized failure. For example, studies on LPBF 316L show a clear correlation between porosity level and reduction in elongation and fatigue life. That said, gas pores are usually less damaging than LOF defects of similar size because gas pores are smoother (less notch severity) and often smaller. For this reason, high-performance aerospace parts typically require porosity below a very low threshold (often <0.1% by volume).

Starting with low porosity, dry powder mitigates the primary source of gas pores. Vacuum inductively melted powders or those with specific atomization parameters can have fewer entrapped gases. Additionally, storing powder in low-humidity conditions and recycling it carefully (sieving out fines, baking to remove moisture) will reduce gas introduction (especially hydrogen). Furthermore, a balanced set of parameters avoids extremes that cause pore entrapment. If one observes many small pores, one might adjust the scan speed or laser power. Ensuring adequate melt pool lifetime (not too short) and turbulence control (not overly agitating the pool) through parameter tuning is key.

Regarding the build chamber's gas flow, the system should efficiently carry away spatter and plume. Poor gas flow can allow vapor to hover and get trapped. Computational fluid dynamics (CFD) studies and practical tuning (e.g., optimally positioning gas inlet/outlet) can reduce this issue. Some machines use a vacuum or low-pressure environment rather than atmospheric inert gas, which has been shown to virtually eliminate gas pores from entrained air for certain materials (at the cost of more complex machinery). Hot isostatic pressing (HIP) is often employed to close internal porosity and ensure that safety-critical parts are closed. HIP involves high temperature and pressure, which can compress and heal internal gas pores (especially spherical ones) by creep and diffusion. HIP can eliminate gas pores that are fully enclosed, markedly improving fatigue life (note that if they have oxide films or are surface connected, LOF defects may not heal as well). HIP has been shown to improve the high-cycle fatigue performance of LPBF Ti-6Al-4V and steels precisely by removing gas porosity.

Quality control for gas porosity involves X-ray CT scanning or ultrasonic inspection to detect and quantify internal pores. Since gas porosity is often stochastic, process qualification relies on producing samples at given parameters and measuring density (via Archimedes or image analysis) to ensure that it consistently exceeds a threshold (e.g., >99.9% dense). Standards often categorize acceptable pore content and size for a given application. For example, ASTM standards will differentiate between acceptable small spherical porosity (which might be inevitable up to a certain level) and rejectable large or clustered pores. In summary, while eliminating gas porosity is difficult, high-quality materials and well-controlled processing can minimize its occurrence to a level that does not significantly affect performance.

Keyhole Porosity

This is a form of defect in which the deep penetration of the laser source produces spherical pores. Keyhole pores are created by an excess energy point input, either excessively high laser power settings or too much time melting at one spot. Usually with a diameter of less than 100 μm, these pores are visible on micrographs. Figure 4.13 shows a high-resolution SEM image of a keyhole pore formed within a melt pool. Note that the melt pool where the keyhole pore is found is in a distinctly large melt pool region. Keyhole porosity occurs under conditions of excessive energy density, for example, very high laser power or slow scan speed (or a combination, resulting in high energy per unit length). In such cases, the intense laser beam melts the powder and vaporizes some of the metal, creating a localized high-pressure metal vapor. This vapor pushes the melt pool downward, drilling a small keyhole cavity. If the keyhole is stable, it can deepen penetration (which is how deep welds are made in laser welding). However, keyholes tend to be unstable in a dynamic scanning process like LPBF, where they can oscillate and collapse as the laser moves. When a keyhole collapses, it can trap vapor inside the closing liquid, leaving a void. Essentially, a bubble gets encased in the solidifying melt before escaping. The resulting pore often mirrors the shape of the last stage of the keyhole; for instance, it may have a conical or rounded shape with one end pointing toward the build surface.

Keyhole pores typically form deep in the melt pool, near the boundary of fully and partially melted zones. They might be found at the center of a track cross section or slightly below the top surface of a layer. They are often associated with

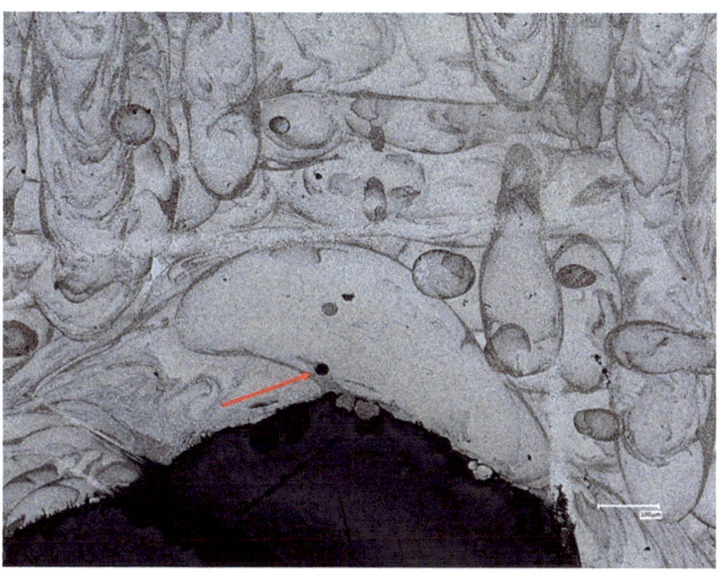

Fig. 4.13 High-resolution SEM image of a keyhole defect

sudden process upsets. For example, a keyhole pore might form there if the laser dwells too long at a point (perhaps at a scan end or corner). In situ X-ray studies have directly observed this and have concluded that, when energy input is high, the melt pool becomes very deep, and then a bubble is seen to pinch off from the keyhole tip and remain as a pore.

A keyhole porosity regime can sometimes be inferred from the process outputs; for example, excessive spatter and a brighter plume are indicators of keyholing (since more vaporization produces more ejected particles and light emission). Keyhole pores are often detected in CT scans or cross sections as larger irregular voids in the parts. They may appear somewhat flattened or tear shaped (with a smoother, curved outline on one side and a sharper point on the other). Because keyhole pores generally occur at high-energy inputs, they might cluster in regions of the part exposed to such conditions, like thick sections (where slow scan or multiple contours were applied) or near supports (if the laser slowed down). Modern standards define keyhole porosity as associated with an instability in the melt pool under high laser energy, distinguishing it from gas porosity. If located near the surface, a keyhole pore can significantly reduce fatigue life, even more so than a gas pore of smaller size. Keyhole pores have been linked to fatigue failures in various studies; for example, one investigation in a nickel alloy found that sporadic keyhole pores (on the order of 200 μm) were initiation sites for fatigue cracks that led to early failure, even when overall density was high. The keyhole porosity also indicates that the process was unstable, which might correlate with other issues (like more spatter or distortion). Thus, avoiding keyhole porosity is essential for achieving consistent, high-quality material. Generally, a single keyhole pore, if large enough, can act like a crack starter under stress. Under static tensile loading, a keyhole pore reduces the load-bearing cross section and can cause localized plasticity or immediate failure if it is large and located at a critical section. Because keyhole pores tend to be isolated (happening only when conditions overshoot), the concern is usually the worst-case pore rather than the overall porosity level.

Preventing keyhole porosity is about not allowing the energy input to exceed the stable melting threshold. Choosing a laser power, spot size, scan speed, etc. is necessary, so the specific point energy (often measured as volumetric energy density, J/mm^3) stays below the keyhole formation threshold. If one increases laser power to improve fusion, one must also appropriately increase scan speed or layer thickness to avoid entering keyhole mode. Empirical and analytical models exist to predict the keyhole threshold for given materials. For example, for stainless steel, a threshold might be around a certain energy density (say ~100 J/mm^3); exceeding that can cause a sudden increase in porosity. Therefore, staying in the conduction-mode melting regime (shallower melt pools) is desired. If keyholing is observed, reducing laser power or increasing speed on subsequent layers in that region is a direct fix. As Taylor and Wicker [6] described, implementing laser ramp-down or shutoff strategies is a practical way to avoid keyhole pores that form at scan end points (where the laser dwells slightly longer). Many machines now ramp down or reduce laser power just before a scan track or turn ends, so the melt pool does not get excessive energy when the laser stops or changes direction. This prevents the brief keyhole that could form when the laser decelerates at a feature edge.

Maintaining process parameters within the recommended range is usually sufficient to avoid keyhole pores. For example, if one were printing Ti-6Al-4V and noticed keyhole pores at 200 W, 600 mm/s, one might increase speed to 800 mm/s or drop to 150 W; the result would be a shallower melt that still fuses but does not keyhole. Research by Cunningham et al. using in situ X-ray showed that once a certain power/speed combination was crossed, the porosity jumped dramatically due to keyhole instability. Thus, engineers determine that threshold in process qualification and ensure that production parameters stay safe.

Quality assurance for keyhole porosity involves scanning for large pores in critical areas. If a CT scan or ultrasonic test finds an anomalously large pore, it could indicate that a keyhole event occurred. In such cases, process logs (if available) can be reviewed to adjust parameters. Design-wise, avoiding sudden changes in geometry that require the laser to slow down (e.g., sharp corners where heat accumulates) can also help reduce keyhole formation. Some standards (like ISO/ASTM 52905 technical report on in-process monitoring) explicitly note that keyhole porosity is typically linked to high laser energy input instability, reinforcing that adherence to proven parameter sets is the primary defense.

4.5 Qualification and Certification of Parts Produced by LPBF

In additive manufacturing, qualification and certification have distinct but complementary meanings. According to ISO/ASTM 52900:2021, qualification is the "process of demonstrating whether an entity is capable of fulfilling specified requirements." LPBF qualification can apply to parts, materials, machines, operators, or processes, proving that these elements consistently meet the needed standards. By contrast, certification is defined (in ISO/IEC 17000:2020) as a "third-party attestation related to an object of conformity assessment," meaning an independent authority formally confirms that a part, process, system, or person conforms to specified requirements. In other words, qualification is often an internal or supplier activity to prove capability. In contrast, certification is an external validation (e.g., by regulatory agencies or standards bodies) that those qualifications meet recognized standards. In LPBF, qualification is significant because it establishes the baseline process reliability and part quality (e.g., verifying a machine or material can produce parts with required properties). Certification builds on this by providing customers, regulators, or end users confidence that the LPBF-produced part or process is acceptable for service, especially in safety-critical applications. Qualification and certification enable trust in LPBF technology by reducing variability and ensuring compliance with industry standards. For instance, an aerospace manufacturer might qualify a new LPBF titanium alloy process through rigorous testing and then pursue certification under an agreed standard (such as AMS or NASA specifications) to allow flight use. In summary, qualification in LPBF is about demonstrating capability and consistency, while certification formally recognizes and attests to that capability under governing standards or regulations.

4.5.1 The IQ/OQ/PQ/MQ Qualification Framework

LPBF process validation often follows the classic IQ/OQ/PQ framework (installation, operational, performance qualification), with an additional focus on material qualification (MQ) for feedstock and resulting material properties. This framework, borrowed from industrial quality systems, provides a structured approach to qualifying AM equipment and processes.

Installation qualification (IQ) ensures that the LPBF machine and its ancillary systems are installed correctly and meet the manufacturer's specifications in the production environment. IQ typically involves documenting that the machine's setup (power, gas supply, software, etc.) adheres to OEM requirements and performing baseline tests after installation. For example, it may include a factory acceptance test (FAT) at the vendor site and a site acceptance test (SAT) upon installation, often by building standard test artifacts to confirm that the machine performs identically before and after shipment. IQ proves that the equipment, facilities, and utilities are adequately prepared for production use.

Operational qualification (OQ) aims to verify that the LPBF process parameters and controls can produce material and parts that meet all specified requirements under normal operating conditions. This stage involves establishing the proper machine settings and process window (laser power, scan speed, layer thickness, etc.) and typically running test builds with coupons (e.g., tensile bars, density samples) to demonstrate repeatability. All critical process variables are characterized and held within control limits. OQ essentially proves that "material specification requirements can be satisfactorily met" with the chosen process parameters on that machine. For instance, an OQ on a new LPBF machine might consist of three back-to-back builds of a coupon package, which are then tested for density, mechanical properties, and geometry to ensure that the machine can consistently produce in-spec parts.

Performance qualification (PQ), also called part qualification, validates that the LPBF process can produce the actual component reliably and with acceptable quality, over time and at production scale. This phase locks in the specific part design, build layout, finalized process parameters, and post-processing steps and then manufactures one or more pilot parts for intensive inspection. A first-article inspection is performed, including dimensional checks and a complete suite of testing (mechanical tests, material analysis) on the part. PQ demonstrates that the part consistently meets all performance and quality criteria with the established process. Successful PQ means that the process is now considered "qualified" for that part family, and future production parts can be made under the same conditions with only routine quality checks. Statistical tools are often employed here to establish process capability (e.g., C_{pk}) and to implement ongoing statistical process control for production.

Material qualification (MQ) is critical given the variability in powder feedstock and the linkage between process and material properties. MQ focuses specifically on qualifying the powder material and resulting printed material properties for use in production. This can include verifying powder chemistry and morphology, establishing acceptable reuse cycles, and generating a comprehensive material property database (density, tensile, fatigue, etc.) for the alloy when processed via LPBF. In many frameworks, MQ is treated as a subset of PQ. For example, one must qualify the material behavior as part of qualifying the overall process. In practice, material qualification requires building and testing many specimens across different builds and powder lots to capture variability. Industry experience shows that this can be a time-consuming and costly process, often taking years to accumulate enough data for high-reliability applications. The payoff is a material allowables dataset or Material Property Suite (MPS) (as termed by NASA) that provides statistically validated design values. A successfully material-qualified LPBF process will have a well-defined allowable strength, ductility, etc. that engineers can rely on in design. In summary, MQ ensures that the feedstock and resultant alloy meet the necessary quality and performance standards (e.g., chemistry within spec, no adverse impurities, consistent microstructure and mechanical properties) before full production. This is especially important when introducing a new alloy or a new powder supplier for LPBF. Once IQ, OQ, PQ, and MQ are all completed, the LPBF process is fully qualified, meaning the machine, material, and process are verified to produce parts that meet requirements, and production can proceed under a controlled state.

4.5.2 *Major Qualification Approaches for LPBF*

Over the past several years, multiple industry standards bodies have published guidelines to standardize how machines, materials, and processes are qualified for LPBF. Notably, NASA, SAE (AMS specifications), and ASTM have developed qualification frameworks with much commonality in intent, but some differences in emphasis. NASA standards (NASA-STD-6030 and 6033): NASA's approach is oriented toward flight-critical space hardware and is accordingly rigorous. NASA-STD-6030 (released 2021) is titled "Additive Manufacturing Requirements for Spaceflight Systems" and provides a comprehensive qualification and certification road map for AM parts used in crewed spacecraft. It introduces a part classification system where parts are categorized as Class A, B, or C based on the consequence of failure. Class A parts have high consequences (failure could be catastrophic or a safety hazard), Class B parts have medium consequences (significant but not catastrophic failures), and Class C parts have low/negligible consequences. This classification drives the stringency of qualification. For example, Class A parts demand the most stringent process controls, testing, and inspection. In contrast, Class C parts have reduced but defined minimum requirements. NASA's standards require an Additive Manufacturing Control Plan (AMCP) and a robust quality management system as the foundation. Machine and process qualification under NASA-STD-6030/6033 entails establishing a qualified material process (QMP) in which a combination of a specific machine, material, process parameters, and post-processing is treated as a "locked-down" process for production. Extensive testing is mandated to populate a Material Property Suite (MPS) for the material-process combination, often requiring tens of builds and many specimens to derive statistically sound design allowables. NASA also emphasizes ongoing controls. For example, equipment and facility qualification is covered in NASA-STD-6033, which addresses maintenance of any associated equipment that could impact the process (laser calibrations, environmental controls, etc.). Personnel training and certification are also required, and NASA has released a standard for qualifying machine operators (ISO/ASTM 52942 for LPBF in aerospace). In summary, NASA's standards prescribe a thorough, data-driven qualification in which the machine must be qualified, the feedstock is qualified, the process is proven stable (typically via witness coupons), and the part itself goes through a stringent acceptance program including pre-production article testing and 100% inspection of critical features. Only after meeting all these can a part be certified for flight. This high level of control is necessary for mission-critical aerospace parts. For instance, NASA requires all Class A (critical) LPBF parts to undergo quantitative nondestructive evaluation (such as volumetric X-ray inspection) with full coverage, per fracture-critical NDE standards, to ensure that no internal defects jeopardize safety.

The aerospace industry (outside NASA) has converged on Aerospace Material Specifications (AMS) standards to qualify additive manufacturing processes for flight parts. Notably, AMS7003 is a process specification for LPBF of metals (focused initially on aerospace Ti-6Al-4V parts), and AMS7032 is a machine qualification standard for fusion-based AM machines. The AMS approach separates the concerns of machine qualification, material control, and process qualification into different documents. AMS7032, for example, outlines how to qualify an LPBF machine for aerospace production where it "mandates several prerequisites" including completing an installation qualification, having trained and qualified operators, implementing controlled facility conditions (e.g., temperature, humidity, dust control), using accredited testing labs for evaluation, and characterizing all key process variables for that machine. In essence, before one even attempts an OQ build, the AMS7032 standard expects the machine to be correctly installed and baseline calibrated (IQ), and the organization to have the necessary quality infrastructure (qualified personnel, procedures for powder handling, etc.). Once those prerequisites are met, operational qualification testing per AMS7032 involves printing a defined set of coupon builds (in one public example, three builds of test coupons were produced consecutively) and subjecting them to an extensive battery of tests. These tests include radiographic inspection (X-ray or CT scans for internal defects), mechanical testing (e.g., tensile tests to measure strength and ductility), metallurgical analysis (microstructure and porosity examination on polished cross sections), and chemical analysis (to verify alloy composition). All results are compared against acceptance criteria (often tied to other AMS material specifications, such as

AMS7031/AMS7011 for powder and properties) to determine if the machine/process is capable. In parallel, AMS7003 provides requirements for ongoing process control during production and establishes process monitoring and maintenance requirements to ensure repeatability. For instance, AMS7003 calls for control of all key process variables (laser power, scan speed, powder layer thickness, etc.) with defined tolerances and monitoring frequency. It also requires specific in situ sensors or calibrations; one source notes "the ASTM F3303 standard requires sensors … to monitor oxygen levels," and AMS7003 would have similar provisions for inert gas purity and other environmental factors. Regular preventative maintenance and recalibration (at least semiannually) are typically required to keep the machine qualified. Unlike NASA's prescriptive class-based approach, the AMS standards implicitly assume that parts will be certified by meeting aerospace quality requirements (often through demonstrating equivalence to wrought properties). Companies using AMS specs will qualify a machine/material combination by AMS7032/7003 and still need the part approved by the customer or airworthiness authority (e.g., obtaining an FAA-approved process for a commercial aircraft part). The AMS standards provide a common language and baseline for such qualifications. They are performance based, focusing on consistent production of material and parts that meet requirements, rather than explicitly categorizing parts by criticality (that categorization is usually done by design authorities using standards like ASTM F3572 or company internal criteria).

ASTM International's Committee F42 has also published standards to guide LPBF qualification. ASTM F3303-18 is a key standard titled "Standard for Additive Manufacturing—Process Characteristics and Performance: Practice for Metal Powder Bed Fusion to Meet Critical Applications." It was developed around the same time as some NASA/MSFC documents and provides a framework to ensure that a PBF process is robust enough for critical applications. In essence, F3303 covers similar ground to AMS7003, outlining recommended practices for machine operation, calibration, feedstock control, process monitoring, and test coupons to qualify a process. For example, F3303 stipulates that the LPBF machine should be installed in a controlled environment (temperature, humidity, vibration) and equipped with sensors to monitor key parameters (it requires sensors in optimal locations to monitor oxygen level in the build chamber, indicating the importance of inert atmosphere control). It also emphasizes documentation and traceability, where the machine's maintenance and calibration records, software version, and configuration should be documented for each qualification. ASTM F3303 provides tables of pre-build checks that an operator must perform before each build (many of these align with NASA MSFC-SPEC-3717, as shown in a side-by-side comparison). These include verifying powder quantity and lot, machine setup (laser optics, recoater condition, build platform ID, and alignment), gas flow, and oxygen content and confirming that process parameters match the approved settings for the material. During qualification, F3303 recommends using test artifacts to evaluate machine performance (for example, density cubes, tensile bars oriented in various directions, surface roughness coupons, and geometry test artifacts to check accuracy). Acceptance criteria might include minimum part density, mechanical properties meeting or exceeding a baseline (such as wrought or a specified percentage of wrought properties), dimensional tolerances, and no critical defects in inspection. Like AMS, ASTM F3303 is meant to be applied in a risk-based situation, given that its title "to Meet Critical Applications" implies that it is geared for high-criticality parts (analogous to Class A/B). For less critical parts, ASTM has separate guidance (and indeed ASTM F3572-22 addresses part classification to allow scaling requirements by criticality). In summary, ASTM's standards approach LPBF qualification by ensuring that the process is well characterized and under control so that any part produced under that process will meet requirements. NASA's approach adds an extra layer of explicitly tailoring the requirements to part criticality and demanding specific risk mitigations for the highest-criticality parts. In contrast, the AMS/ASTM approach is more uniform (but of course, in aerospace, critical parts will inherently need the full rigor). All three frameworks stress the importance of machine qualification, material control, process validation, and quality management. A qualified LPBF process under these standards will have documented evidence of machine capability (calibrations, test builds), material consistency (certified powder with lot traceability and test results), and proven part properties (coupon tests and possibly a demonstration part). This allows organizations to certify parts, for example, by showing compliance to a standard like ASTM F3303 or AMS7003, and obtaining customer/regulatory approval, a critical aerospace LPBF part can enter service with confidence in its reliability.

Despite differences, these standards converge toward a common theme: qualified LPBF production requires controlling every input (machine, material, process) and accumulating evidence (test data, inspections) to demonstrate reliability. Users in aerospace often use a combination of these standards (Table 4.1). For example, a company might follow ASTM F3303 practices to qualify a process internally, then use AMS7032 to align with aerospace customer requirements, and, if it is for NASA, also meet NASA-STD-6030 class-specific criteria. In all cases, the outcome is a robustly qualified process that can be certified for use in critical applications.

4.5.3 Part Families and Criticality Levels by Industry

The part's application and criticality heavily influence the level of qualification rigor and inspection needed for LPBF parts. Different industries categorize parts by how critical their function is to safety or mission success. Structural components (e.g., airframe brackets, engine mounts, load-bearing fittings) are often safety critical in aerospace. Using the classification schemes mentioned earlier, these would typically be Class A or B parts. For example, a flight-critical bracket that, if it fails, could jeopardize aircraft control would have a "high" consequence (Class A in ASTM F3572 terminology or Class A in NASA's system). Such parts demand extensive qualification where material allowables must be established, fatigue life is demonstrated, and 100% inspection (NDE) is usually required. Less critical structural parts (say a secondary bracket whose failure is non-catastrophic but could

Table 4.1 Standards' comparison pertaining to machine qualification

Gas flow/filtration	**ASTM**	**NASA**	**AMS**	**Computer control system**	**ASTM**	**NASA**	**AMS**
Recirculation filter flow rate	✓	✓	✓	Software/firmware lock	✓	✓	✓
Velocity profile	✓			Time		✓	
				Scan strategy	✓		
Chamber conditions	**ASTM**	**NASA**	**AMS**	**Laser**	**ASTM**	**NASA**	**AMS**
Shielding gas composition	✓	✓	✓	Power	✓ ±5%	✓	✓
Gas flow rates for supplement (purge)			✓	Waist location	✓	✓	✓
Moisture/humidity/dew point	✓	✓	✓	Astigmatism			✓
Chamber pressure (decay)	✓		✓	Beam quality factor	✓		✓
Oxygen sensor	✓	✓	✓	Thermal stability	✓		✓
Impurity concentration	✓			Power across build plate	✓	✓	✓
Chamber temperature	✓	✓		Spot size/shape across plate	✓	✓	✓
				Beam profile for each laser	✓	✓	✓
				Pulse characteristic	✓		✓
				On/off delays		✓	
Powder spreading/ management	**ASTM**	**NASA**	**AMS**	**Beam steering**	**ASTM**	**NASA**	**AMS**
Build platform positional accuracy	✓	✓	✓	Travel speed	✓ ±5%	✓	✓
Re-coater speed	✓	✓	✓	Positional accuracy	✓ 0.06 mm	✓	✓
Build platform preheat	✓		✓	Scan field scaling	✓	✓	✓
Layer thickness and uniformity	✓						
Recoater blade material wear	✓						

cause a maintenance issue) might be Class B (medium consequence) and still require aerospace-grade quality but perhaps slightly relaxed inspection frequency or redundant proof testing. The aerospace industry tends to require that any flight hardware produced by AM go through a formal qualification process. However, the depth of testing (e.g., number of coupon tests, extent of NDE) is scaled by criticality. Spaceflight hardware often uses a similar scheme. For example, NASA classifies all LPBF flight parts as A, B, or C. A structural spacecraft part that is mission critical (holds pressure or supports loads) would be Class A and must meet the highest standards. In contrast, a minor nonstructural fitting might be Class C with a leaner qualification approach.

The medical industry classifies devices by risk to the patient. Many additively manufactured implants (spinal cages, hip cups, skull plates, dental implants) are considered high-risk devices because a failure can directly impact a patient's health. For example, a load-bearing orthopedic implant failing could require revision surgery (serious harm), and certain implants (spinal or cardiovascular) could be life-threatening if they fail. Regulators like the FDA categorize implants typically as Class II or Class III medical devices (Class III being the highest-risk, often life-sustaining implants). As such, medical implants made by LPBF must undergo rigorous qualification and validation. The manufacturer must ensure that the LPBF process produces material with appropriate biocompatibility, strength, and fatigue life. Mechanical properties must meet ASTM/ISO standards for the implant type (for instance, ASTM F136 for titanium alloy implants). Fatigue testing is essential for implants like hip stems or spinal cages, which endure millions of loading cycles. LPBF's inherent surface roughness and potential porosity can significantly affect fatigue, so process parameters or surface finishing may need optimization to achieve the required life. Medical implant qualification also involves thorough metallurgical analysis (to ensure no unwanted phases or contamination), and typically, each production part is inspected (at least visually and dimensionally, and often by CT for critical implants) to ensure quality. Once the company is satisfied internally (qualification), a certification/approval step involves regulatory bodies. For example, a technical file can be submitted to the FDA or EU regulators demonstrating that the LPBF implant is as safe and effective as a traditionally made device. The criticality here is patient safety, and, like aerospace, the process gets locked down to where only specified machines, powders, and parameter sets can produce the approved implant. On the other hand, some simpler medical devices (low-risk surgical guides, orthodontic models) made by AM might be Class I (low risk) or exempt devices, which have much lower regulatory burden; these are more akin to "low-criticality" parts in the medical field.

In industries such as consumer products, hobbyist manufacturing, or even specific industrial tooling, there are many parts whose failure would not pose safety risks or significant economic losses. Examples might include a prototype housing, a jig or fixture used on an assembly line that is not load critical, or cosmetic and personalization products. These parts represent the low-criticality end of the spectrum. The qualification requirements are minimal for such parts (comparable to Class D in ASTM F3572, "negligible or no impact" if the part fails). A company may decide that an LPBF machine is "qualified" to make a consumer product if it produces parts that look correct and meet basic dimensional specs, without needing extensive mechanical testing. Certification in the formal sense is usually not required here (since there is no regulatory or safety standard to meet for, say, a decorative item or a one-off prototype). However, manufacturers still implement some quality control to ensure customer satisfaction. For example, they might do sample inspections for dimensions or perform a simple tensile test on coupons from a new powder lot to ensure that the machine functions well. The difference is that cost and speed often prioritize low-risk parts over exhaustive testing. If a print fails or a part is slightly understrength, it might be acceptable or reprinted without significant consequence. That said, even commercial industries benefit from standardized practice. Many users apply a scaled-down version of IQ/OQ/PQ to ensure that their LPBF machines produce consistent results, reducing scrap and improving throughput. In summary, low-risk parts still undergo qualification steps (to dial in machine settings and ensure basic quality). However, the criticality level allows a much leaner approach, and perhaps just a few trials and measurements can be completed compared to the heavy qualification programs for aerospace or medical parts.

4.6 Business Case and Applications for LPBF

LPBF offers significant design and production advantages, but its economic viability depends on several key factors. Major cost drivers include machine and build time costs, material (feedstock), labor/engineering time, and post-processing, including surface finishing, consumables, and machine maintenance. LPBF machines are expensive (often \$500k–\$1 M+ each), and machine depreciation can account for a dominant share of part cost. For example, one case study showed machine costs constituting ~73% of the unit cost for a stainless steel LPBF part [44]. Slow build rates mean long production times; the hourly machine rate and build duration directly impact cost per part. Efficiently filling the build chamber (nesting parts) can improve throughput, but build time remains a limiting factor for high-volume production [45]. Metal powders for LPBF are typically 5–10 times more expensive than equivalent bulk material. Specialized powders (e.g., Ti-6Al-4V or Inconel) can cost \$300–\$600+ per kg, partly due to stringent size and sphericity requirements and low-volume production [45]. For instance, titanium powder commonly used in LPBF costs around \$260–\$450 per kg. These high feedstock costs make material a significant cost contributor, especially for large or heavy parts. On the other hand, LPBF minimizes waste (unused powder can be recycled), so material utilization is efficient compared to machining, which cuts away material.

Engineering effort is required to design for AM (DfAM) and prepare each build. This includes machine setup, parameter optimization, support structure design, and post-build part removal. Engineering and operator labor can be substantial, particularly when introducing a new part. If a build fails (e.g., due to process instability or design error), the lost time and need for redesign add to the cost. Thus, skilled labor and iterative development time are important economic factors. Unlike many traditional processes, LPBF parts usually require extensive post-processing, which adds time and cost. Steps such as cooldown, powder removal, cutting the parts off the build plate, support removal, heat treatments, hot isostatic pressing (HIP) for critical parts, surface finishing (grinding, blasting, polishing), and inspection are often necessary. Post-processing can account for a significant portion of total production cost, in which nearly 27% of the cost of metal AM parts comes from post-processing operations, according to Wohlers Report 2021 [46]. These additional operations may be labor-intensive (e.g., manual support removal) and often require specialized equipment (furnaces, HIP vessels, CNC machines for finish machining), contributing to labor and capital costs.

LPBF machines consume inert gas (usually argon or nitrogen) to shield the powder bed from oxidation. This is an ongoing expense since operating a single machine can use tens of thousands of dollars of argon annually. Other consumables include recoater blades, filters (to capture soot/metal condensate in the gas flow), and wear parts, all needing periodic replacement. Routine maintenance and calibration of the lasers, optics, and mechanical systems are required to keep the machine within specification, which may add to overall operating costs.

These cost factors mean that LPBF per-part costs are relatively insensitive to production volume (there are limited economies of scale, aside from multipart builds). LPBF is most competitive for low-volume, high-complexity, or high-value components, whereas traditional manufacturing would require expensive tooling or multistep fabrication [38]. The technology's value proposition lies in geometric freedom and consolidation of parts rather than sheer throughput. Thus, from a business perspective, the economic sweet spot for LPBF is typically where the design or performance benefits outweigh the higher unit cost. Examples include complex aerospace and medical parts, customized products, and situations where reducing lead time or inventory is critical. In summary, while machine amortization and material costs currently dominate LPBF economics, ongoing improvements (faster multi-laser machines, cheaper powders, post-processing automation) gradually lower cost barriers. Firms must carefully evaluate the "business case" for each candidate part, considering not just production costs, but also the value of weight savings, improved performance, supply chain simplification, and design innovation that LPBF can deliver.

4.6.1 Aerospace Industry Applications

The aerospace sector was an early adopter of metal AM due to its demand for high-performance, lightweight parts in low volumes. LPBF's ability to produce complex, weight-optimized geometries and consolidate assemblies offers enormous benefits for aerospace components. One of the most famous examples is the fuel nozzle in GE's LEAP commercial jet engine. Using LPBF, GE engineers redesigned this fuel nozzle tip as a single piece, consolidating 20 parts into one. The printed nozzle is about 25% lighter than the prior assembly and features intricate internal passages that improve fuel–air mixing (thus enhancing efficiency and durability). By 2018, GE had additionally manufactured tens of thousands of these nozzles for use in service, demonstrating the viability of LPBF for critical, mass-produced engine hardware. This part highlighted a typical aerospace business case. Although the LPBF nozzle was more expensive per unit than a cast one, it enabled weight reduction and performance gains that yield life cycle cost savings (e.g., fuel burn reduction) and eliminated numerous braze/weld joints, boosting reliability.

Beyond fuel injectors, LPBF is being applied to structural and engine components throughout aerospace. For instance, General Electric's new turboprop and turboshaft engines incorporate LPBF parts to reduce weight and part count. In the GE T901 helicopter engine, an oil sump assembly that was initially over 50 pieces was redesigned and printed as a single component, resulting in a 20% weight reduction [47]. Such part consolidation cuts weight and avoids assembly labor and potential failure points at joints. Topology optimization is often used with LPBF to redesign aircraft brackets, mounts, and ribs with organic shapes that remove unnecessary material. Airbus achieved a notable example with a topology-optimized cabin bracket produced by LPBF, which was significantly lighter while meeting load requirements. Even satellite and space launch companies print components like antenna mounts and rocket engine parts to maximize weight savings and design freedom. Every kilogram saved in aerospace has immense value in fuel or payload capacity.

A key driver in aerospace is the performance enhancement that complex LPBF geometries enable. For example, heat exchangers and heat sinks within jet engines or avionics can be printed with intricate cooling channels to improve thermal management (GE has reported printing complex heat exchanger units for its engines, integrating them into tight spaces impossible to machine). Additionally, LPBF allows using high-temperature superalloys that are hard to machine. This is particularly useful for turbine components, as NASA and others have experimented with printing nickel alloy turbine blades and vanes with internal cooling passages. While challenging, such parts could improve engine efficiency by permitting novel cooling schemes or lighter designs.

From a business case perspective, aerospace companies justify the higher cost of LPBF parts through system-level benefits. Weight reduction directly translates to fuel savings over an aircraft's life. Part count reduction simplifies assembly and maintenance (fewer parts to track and replace). Additionally, LPBF can shorten development cycles since designers can iterate and test aerospace components much faster than traditional casting or forging, accelerating innovation. There are also supply chain advantages, especially with on-demand printing of spare parts (or low-volume legacy parts), which can reduce inventory costs and lead times for aircraft maintenance. For example, Boeing has used LPBF to produce replacement environmental control system ducts and other spares for older military aircraft, avoiding reopening obsolete production lines. It should be noted that aerospace is highly regulated, so extensive qualification is required for any flight-critical AM part. Components often undergo post-print HIP and heat treatments to ensure defect-free performance and must meet stringent inspection criteria [47]. Despite these hurdles, the aerospace sector has steadily progressed from small noncritical items to flight-critical hardware made by LPBF. As of the last few years, multiple aircraft engines in commercial service contain LPBF parts, and aircraft structural parts (like brackets) have flown after certification.

4.6.2 Medical and Biomedical Applications

The medical industry, primarily orthopedic and dental implant manufacturing, has embraced LPBF due to its unique ability to produce patient-specific shapes and controlled porous structures. Orthopedic implants (such as hip replacements, spinal cages, knee components, and bone plates) benefit significantly from LPBF's design freedom. Traditionally, implants are made in standard shapes and often fully dense, leading to issues like stress shielding (where a stiff implant carries too much load, causing bone loss). With LPBF, engineers can design implants with lattice or trabecular structures that mimic the porosity of bone. These porous surfaces and internal lattices encourage osseointegration, meaning bone can grow into the implant, creating a stronger biological fixation. Studies have shown significantly improved bone ingrowth for LPBF-printed implants compared to conventional solid implants. For example, [48] Dall'Ava et al. [48] found that additively manufactured porous acetabular cups achieved ~63% bone attachment area versus 37% for a traditional implant, indicating superior osseointegration. Clinically, this can translate to longer-lasting implants and better patient outcomes.

Another advantage is customization. Using CT or MRI scan data, surgeons can design patient-specific implants that match an individual's anatomy. LPBF then fabricates the implant to that exact geometry, something impractical with standard manufacturing. This is especially valuable for complex reconstructive surgeries (e.g., cranial plates, jawbone replacements, tumor resections) where off-the-shelf implants might not fit. Dozens of custom titanium implants, from skull patches to rib replacements, have been successfully implemented using LPBF, often in cases where no other solution existed. Even for standard implants like hip or knee replacements, LPBF allows mass-customized solutions (adjusting geometry for each patient's physiology), which could improve comfort and biomechanics.

Regarding the business case, LPBF enables higher-performance implants and can streamline the supply chain for medical devices. Companies like Stryker have developed proprietary LPBF titanium porous structures (e.g., "Tritanium" cages for spinal fusion) and obtained their FDA clearances. These implants feature a high porosity, rough surface texture, and optimized stiffness to promote bone ingrowth and reduce stress shielding. They have seen widespread adoption in spine surgery, demonstrating that surgeons and patients are willing to embrace 3D printed implants when the clinical benefits are clear. By 2019, more than 100,000 LPBF-printed acetabular hip cups and spinal cages had been implanted globally [49], reflecting growing confidence in the technology. Medical device firms justify the higher manufacturing cost because these designs improve patient outcomes and often command premium pricing. Moreover, LPBF can reduce the need for inventory of many sizes. Instead of stocking a range of implant sizes, a hospital could print on demand or have a near-just-in-time supply of the exact needed variant.

Dental applications also thrive with LPBF. Customized dental crowns, bridges, and orthodontic appliances are printed in cobalt-chrome or titanium, achieving excellent fit and detail. Dental labs commonly use metal 3D printing to produce frameworks for removable partial dentures or implant abutments. The fine resolution of LPBF and the ability to create complex shapes (like the mesh of a partial denture) make it suitable, and it can be more efficient than casting for small-batch production. The medical field also benefits from reduced lead times, and a patient-specific implant can go from design to surgery in days or weeks, which is crucial for treating complex cases promptly.

Regulatory bodies have kept pace with this innovation. The U.S. FDA has issued guidance on 3D printed medical devices and has cleared numerous LPBF-made implants via the standard approval pathways [50]. As long as manufacturers validate their processes and ensure quality (through standards like ASTM F3303 for metal AM part qualification), printed implants are treated similarly to forged or cast ones. In summary, LPBF has opened new frontiers in the medical industry by enabling personalized, anatomically optimized implants with improved biological integration. The high value of medical devices justifies the costs, improved success rates, and patient satisfaction with the designs.

4.6.3 Heat Exchangers and Thermal Management

Heat exchangers are critical in many industries, and LPBF has unlocked radical new designs for these components. Traditional heat exchangers (HXs) are often limited to simple geometries (straight tubes, plates, fins) due to manufacturing constraints like machining and brazing. With LPBF, engineers can create complex internal channels, lattice structures, and organic geometries that maximize heat transfer surface area while minimizing pressure drop and volume. This has led to a new generation of compact, high-performance heat exchangers that were previously unattainable.

One striking example is the work of Gerstler and Erno [51], who designed a multi-furcating fluid passage geometry for a fuel-cooled oil cooler and fabricated it with LPBF. Their novel HX featured a network of branching channels impossible to make conventionally. The result was a 66% reduction in weight and 50% reduction in volume compared to a standard design, yet it achieved equivalent heat transfer performance and pressure drop [51]. Such weight and space savings are extremely attractive in aerospace and automotive cooling systems. Similarly, researchers have used LPBF to build air-to-air heat exchangers with gyroid or lattice cores, demonstrating high effectiveness in ultracompact form factors [52]. These complex internal geometries promote turbulent flow and enhance thermal contact, often leading to performance far beyond what finned tubes can do.

Another application is conformal cooling channels for high heat flux components. For instance, in electronic cooling or mold tooling inserts, LPBF can create curved cooling channels that closely follow the shape of the part being cooled. This yields more uniform temperature control. In power electronics and battery systems, companies have printed cold plates with intricate internal fin arrays that improve heat dissipation within tight spaces. Conflux Technology in Australia demonstrated a 3D printed water–oil heat exchanger for automotive use that used an internal gyroid lattice core. It achieved the required heat rejection with a pressure drop 82% lower than a conventional design, and a 39% reduction in weight, while being produced in a single piece [53]. Such improvements can directly translate to better vehicle performance and energy efficiency.

The business case for LPBF heat exchangers lies in their unparalleled performance and assembly consolidation. A conventional heat exchanger might require many parts (fins, tubes, manifolds) joined by welding or brazing. LPBF can integrate the entire exchanger (headers, core, etc.) into one monolithic structure, eliminating joints that can leak or fail. For example, Siemens reported printing a leakproof gas turbine heat exchanger that combined what used to be 10+ parts into one, drastically simplifying production and maintenance. Additionally, the ability to tailor channel geometries means that designers can optimize the balance between heat transfer and fluid flow losses with a previously impossible freedom. In aerospace, where heat exchangers for oil cooling or intercooling must fit into irregular, small volumes, LPBF provides a solution that meets thermal requirements within spatial constraints. The improved cooling can enable engines to run at higher power or efficiency, again a trade-off of higher component cost for a more significant system benefit.

That said, designing for LPBF in this domain requires addressing challenges. Notably, very thin walls or fine features can push process limits, and removing unfused powder from complex internal passages can be difficult. Engineers must incorporate powder removal channels or use sacrificial support strategies for internal features. Despite this, multiple companies have moved toward commercializing AM heat exchangers, primarily for aerospace, high-performance automotive, and electronics cooling. The compact size and weight reduction can justify the weight-sensitive and high-value applications. As LPBF machines scale up in build volume, even larger heat exchangers (for industrial processes, HVAC, etc.) could be printed. In summary, LPBF is revolutionizing heat exchanger design by allowing forms that optimize thermal performance in ways conventional manufacturing cannot. This offers industries the potential for more efficient thermal management systems, albeit at a premium cost, often offset by the performance gains.

4.6.4 Luxury Goods and Jewelry Applications

LPBF has also found a niche in the luxury goods sector, particularly in jewelry and high-end consumer products, where customization and intricate design are prized. Traditional jewelry manufacturing (lost-wax casting, fabrication) imposes constraints on geometry and is labor-intensive for complex one-off pieces. LPBF, by contrast, allows jewelers and designers to create highly complex, latticelike or filigree structures in precious metals (gold, platinum, silver) directly from a digital model. This capability enables levels of creativity and design complexity that were previously impractical or extremely costly by hand.

One clear advantage is mass customization. Brands and designers can offer bespoke jewelry tailored to each customer's preferences (personalized text, unique organic shapes, exact sizing) without needing new tooling for each piece. Thangamani et al. [54] note that, unlike traditional methods, where complexity and cost are directly correlated, LPBF significantly decouples this relationship. This means that a highly ornate design with interlocking parts or internal detail can be produced in one step. In contrast, conventionally, it might require masterful craftsmanship and assembly of many tiny components. For example, luxury watchmakers have used LPBF to produce watch cases or metal bracelet links with intricate internal lattice frameworks that reduce weight and create novel aesthetics, which is impossible to achieve by CNC machining alone.

The economic considerations for jewelry LPBF hinge on production speed and material savings. AM can reduce material waste since precious metal powders that are not fused in the build can be reused, unlike the significant scrap in subtractive methods. It can also shorten the time from design to finished product. One study compared manufacturing a batch of jewelry items via traditional casting versus LPBF and found that the 3D printed jewelry could be delivered in a few weeks versus potentially months for the conventional route [54]. This agility allows designers to rapidly prototype new pieces, verify the design, and go into small-scale production without the delay of making molds or dies. Small-batch production is where LPBF shines. For example, producing 50 units of a complex new ring design can be more cost-effective with printing than investing in casting equipment for the same small run. As noted by industry reports, LPBF becomes competitive for high-complexity jewelry in low-to-medium volumes. In contrast, traditional methods still hold an economic edge for the mass production of simple designs.

Several jewelry manufacturers have adopted LPBF for both prototyping and final production. Cooksongold, Inc., for instance, offers direct printing in 18k gold and has collaborated on one-off fine jewelry pieces with elaborate geometries. Fashion designers have also used LPBF in haute couture, creating ornate metal lacework for dresses or accessories that push the boundaries of form. Beyond jewelry, other luxury or lifestyle products leveraging LPBF include bespoke eyewear frames, intricate pen bodies, and custom decorative hardware for interior design. The ability to iterate designs quickly is valuable since artists can tweak and reprint a digital model, rather than re-carving a wax model or retooling a mold, fostering innovation and bespoke services.

However, challenges remain, such as achieving the high-quality surface finish expected in jewelry. LPBF parts typically emerge with a matte, slightly rough finish and require polishing. Fortunately, precious metals are relatively soft and can be hand-polished or electro-polished to a mirror shine. A typical production workflow is to print the detailed geometry and have a jeweler do final finishing touches, marrying technology with traditional craftsmanship. As machine and material costs come down, we can expect broader use of LPBF in luxury goods, particularly as consumers increasingly demand unique, personalized products. The value added by personalization and complex beauty often justifies the higher cost of AM production in this sector, aligning well with the high margins of luxury products.

4.6.5 Automotive and Motorsports Applications

LPBF is primarily utilized in high-performance and niche segments in the automotive industry, emphasizing lightweight design, rapid prototyping, and part consolidation rather than high-volume production. For mainstream automotive manufacturing, the cycle times and costs of LPBF are generally not yet competitive with casting or machining for thousands of parts. However, for motorsports, supercars, and advanced prototyping, LPBF has become an indispensable tool.

Motorsports (e.g., Formula 1, endurance racing, motorsport motorcycles) leverage LPBF to produce complex components with extreme weight optimization under tight development timelines. Formula 1 teams, for instance, constantly iterate aerodynamic components and engine parts between races. Metal LPBF allows them to create one-off or few-off parts (like titanium hydraulic manifolds, suspension rockers, or gearbox housings) with internal channels and organic geometries that would be impossible to machine. These parts can be printed and tested within days, accelerating the design cycle, which is crucial in competitive racing. The performance gains justify the cost since shaving a few hundred grams or packaging a part more tightly can provide a competitive edge. It is common for F1 cars to contain dozens of LPBF metal parts, from brackets to cooling elements, and many polymer AM parts for aero testing. This rapid development and implementation of new designs in motorsports showcase the business case of LPBF as an enabler of innovation when time is a critical factor.

Manufacturers have used LPBF to achieve designs that enhance performance and exclusivity in the supercar and luxury automotive realm. A prominent example is Bugatti's development of a 3D printed titanium brake caliper for its Chiron hypercar. The caliper, produced in 2018 as a prototype, was the largest functional titanium LPBF part and was about 40% lighter than the aluminum counterpart while maintaining the required strength. Though not yet employed in full production due to print time and cost (each took ~45 h to print), it demonstrated the potential for significant unsprung weight reduction in a critical safety component. Bugatti viewed this as an investment in future production methods for ultralow-volume, high-performance parts. Similarly, Koenigsegg has mentioned using LPBF to create components of its high-end cars' engines and exhaust systems that integrate multiple functions into one piece, reducing assembly complexity and weight.

Perhaps the most celebrated success is in engine components, which occurred in 2020 when Porsche revealed that it had LPBF-printed metal pistons for the 911 GT2 RS's engine in a development project. These printed pistons were 10% lighter than forged pistons and featured an integrated cooling oil channel in the crown, which was a design not achievable by conventional means. Testing showed that the engine could run at higher RPMs and lower temperatures with these lighter, better-cooled pistons, unlocking up to 30 additional horsepower from the twin-turbo engine. This is a remarkable improvement in an already highly optimized engine, achieved solely through design changes made possible by LPBF. While these pistons were experimental, Porsche's project with partners proved that LPBF could meet the quality and performance needs of the engine internals. The business case here is clear: extra horsepower and reliability are valuable enough for sports cars to consider advanced manufacturing despite higher unit costs. It also opens the door for producing spare parts for classic cars or limited editions without expensive tooling.

Beyond these examples, LPBF is widely used for automotive prototyping and tooling. Before committing to expensive casting dies, automakers will print functional prototypes of parts like intake manifolds, pump housings, or shift mechanisms to evaluate fit and performance. This speeds up R&D and can reveal necessary design tweaks early. Some companies also print assembly jigs, robotic end effectors, or custom tooling in metal to help with manufacturing processes, benefiting from quick turnaround and design freedom (e.g., conformal cooling channels in molds for plastic parts as mentioned earlier). For aftermarket and custom car applications, LPBF has enabled the creation of custom brackets, tailored exhaust tips, and even stylistic components in small quantities that would be infeasible to mass produce.

In the high-end automotive segment, where production volumes are low (dozens to hundreds of cars) and the emphasis on innovation and differentiation is high, LPBF's cost is more easily justified. The weight savings can improve acceleration, handling, or electric vehicle range; the part consolidation can improve reliability and simplify assembly; and the ability to create complex shapes allows cutting-edge design aesthetics (for example, LPBF lattice wheels or grille structures). As a result, we see an emerging trend of boutique car makers and concept cars featuring 3D printed metal parts (for instance, a few production models have included printed metal mirror brackets or decorative trim).

Overall, in the automotive domain, LPBF is a strategic tool for innovation. Its use aligns with high-value outcomes such as winning races, achieving record performance, or offering bespoke features to premium customers. While not yet a solution for mass-producing millions of car parts, it is already invaluable in shortening development cycles and enabling small-scale production of ultra-optimized components that elevate automotive engineering beyond the limits of traditional manufacturing. As machine speeds increase and costs drop, it is conceivable that more widespread automotive adoption could follow, especially for electric vehicle components and complex cooling or structural elements where LPBF's strengths directly translate to vehicle efficiency.

4.6.6 Oil and Gas Industry Applications

The oil and gas sector, including petrochemicals and power generation (turbines), is increasingly exploring LPBF to produce robust, complex components that can withstand harsh operating conditions. Historically, this industry is conservative and focused on proven methods. However, the potential benefits of additive manufacturing, such as part consolidation, reduced lead times for spares, and design-driven performance improvements, are driving adoption in critical areas.

One successful example is in downhole tools used in oil well drilling and logging. Baker Hughes (a major oilfield service company) reengineered a formation-testing tool's buffer tube using LPBF. The buffer tube is a pressure-bearing subassembly that isolates hydraulic fluid from the healthy fluids during sampling. By redesigning it for AM, Baker Hughes could consolidate what was originally a seven-piece assembly with multiple seals and joints into a single monolithic component. The LPBF-fabricated buffer tube incorporated internal fluid channels and eliminated all welds and threaded connections. This led to improved reliability. After thousands of hours of field operation, the 3D printed buffer tubes showed zero failures, whereas the conventional assembly had known failure modes at the joints. The development cycle was also fast, going from concept to field-ready part in about 1 month (an unusually short timeline in oil tool development). This case highlights the part consolidation and durability gains that justify LPBF in O&G. That is, even though the part's manufacturing cost was higher, it removed weak points and improved tool uptime, which is extremely valuable in costly drilling operations.

LPBF also enables design improvements for valves, pumps, and other fluid flow components in the oil/gas and petrochemical industry. Complex internal geometries, such as curved channels or lattice structures, can be printed inside valve bodies or pump impellers to optimize flow or reduce turbulence. For instance, engineers have designed novel choke valves with tortuous internal pathways that attenuate fluid pressure in stages, minimizing the erosion damage typically caused by high-pressure drops. Major oil companies like Total have publicly stated that the design freedom of AM is a "major advantage" for creating more efficient, consolidated parts for refineries and rigs. Additionally, LPBF can produce parts in corrosion-resistant superalloys (e.g., nickel-based Alloy 625 or 316L stainless steel) with complex hollow structures, which is beneficial for heat exchangers or separators in offshore platforms where weight is at a premium and the marine environment is aggressive.

Another compelling use case is on-demand spare parts and repair. Oil and gas operations often face long lead times when replacing broken components, especially if they are older or customized equipment. LPBF offers the possibility of digitizing the inventory. In other words, keep a digital library of parts that can be printed when needed, rather than stocking physical spares for decades. This is particularly attractive for remote or offshore installations with limited storage space and highly costly downtime. For example, a failed pump impeller on an offshore platform could be 3D printed in a matter of days at a nearby service center, instead of waiting weeks for a new casting to arrive from a factory. Some oil companies have begun qualifying LPBF processes to produce such spare parts on demand, under controlled conditions, to ensure that they meet the rigorous certification standards. Early trials include printing parts like pressure sensor housings, pipe connectors, and turbomachinery components and subjecting them to the same testing as conventionally made ones.

A noteworthy example is a tool manufactured by Baker Hughes, which showed that eliminating joints can remove failure modes, thus increasing the reliability of downhole operations, which avoids costly delays. In refining and chemical plants, a single innovative AM part that cuts maintenance frequency or energy usage can justify its cost quickly. Additionally, AM can facilitate designs for condition monitoring with embedded channels for sensors or inspection, which traditionally would be complex to manufacture. A printed part might include an internal cavity for fiber-optic monitoring of strain in a pipeline component, offering operators new data to prevent failures.

Despite the promise, adoption in oil and gas is gradual due to qualification and safety concerns. Components for this sector must often adhere to standards from organizations like American Petroleum Institute (API) and ASME for pressure vessels. Thus, much effort is going into establishing that LPBF parts have equivalent or superior properties (fatigue strength, fracture toughness, corrosion resistance) compared to forgings. Moreover, hot isostatic pressing and heat treatments can alleviate internal porosity and residual stresses, yielding parts that meet code requirements. The long-term vision is that certified AM processes could produce on-site replacement parts or novel designs that improve efficiency (for example, lighter drill bits or optimized burners for gas turbines). Siemens Energy already uses AM to refurbish and even newly manufacture certain gas turbine burners and fuel nozzles, taking advantage of part consolidation to improve cooling and fuel flow.

In summary, the oil and gas industry's use of LPBF is driven by a combination of design innovation (better performance, fewer parts) and supply chain optimization (on-demand production, reduced inventory). While still emerging, there are clear instances where the high cost of a 3D printed part is offset by the value it delivers, whether that is preventing a platform from shutting down or extending the life of a critical component. As standards catch up and more field data is gathered, we can expect LPBF to become a more common tool in the oil and gas manufacturing toolkit, especially for complex, low-volume components where conventional manufacturing struggles to deliver the functionality needed.

4.7 Trends and Future Outlook in LPBF

4.7.1 Larger Build Volumes and Multi-Laser Systems

One clear trend is the scaling up of LPBF machines in terms of build volume and laser count. Manufacturers are designing printers with significantly larger build chambers, enabling the production of much bigger components or dense batches of parts in a single run. Additionally, modern systems are increasingly equipped with multiple lasers (ranging from two on midsized machines to as many as a dozen or more on cutting-edge platforms) working simultaneously. By having several laser beams sintering powder in parallel, the productivity of LPBF is dramatically increased. Large parts that once took days to print can now be completed far faster, and multiple smaller parts can be printed at once without lengthening the build time.

This multi-laser approach pushes LPBF closer to actual mass production capabilities, making it more competitive with conventional manufacturing in throughput.

However, system complexity inevitably rises as build volumes grow and more lasers are added. Coordinating numerous laser optics and scan paths across a wide powder bed requires advanced calibration and control to ensure uniform results. Overlapping laser scan regions must be carefully managed so that seams between areas melted by different lasers have consistent microstructures and strength. Larger machines also demand more sophisticated inert gas flow and filtration to maintain a stable environment over a large build area, since removing soot and spatter becomes more challenging in a meter-scale chamber. Despite these complexities, the industry consensus is that such trade-offs are worthwhile. Engineers are meeting the challenge with innovative solutions (like intelligent scan algorithms and improved real-time monitoring) to keep quality high. The net result is that future LPBF systems will be far more extensive and faster, unlocking the ability to manufacture large industrial parts (for example, sizable aerospace components or tooling) in one piece and accelerating production runs of smaller parts. This scaling up of LPBF hardware is a foundational step toward broader adoption in heavy industries.

4.7.2 Advanced Laser Sources and Beam Control

Another pivotal area of innovation is laser sources. Traditionally, LPBF machines have relied on single-mode continuous-wave fiber lasers with fixed Gaussian beams at a standard infrared wavelength (~1060 nm). Now, we are seeing multimode beam shaping and other beam control techniques emerging to push performance to new levels. In practical terms, this means that future machines can dynamically adjust the laser's spot size and intensity profile on the fly. For example, a laser might use a small, focused spot to resolve fine details and switch to a broader or ring-shaped beam to fill bulk areas with minimal defects rapidly. This ability, coupled with increasing laser power (with industrial systems now moving toward 1 kW and even 2 kW lasers per beam), is expected to translate directly into higher deposition rates. Beyond beam shaping, there is growing interest in using alternative laser wavelengths to expand LPBF's material capabilities. High-reflectivity metals like pure copper and certain aluminum alloys have always been challenging to process with standard infrared lasers (which those materials tend to reflect rather than absorb). Some next-generation LPBF machines employ green or blue laser sources with shorter wavelengths to address this.

4.7.3 *Expanding Material Portfolio*

The palette of materials available for LPBF continues to grow as researchers and companies develop powders and processes for new alloys. Early LPBF applications were limited to a handful of well-understood materials (such as tool steels, stainless steels, Ti-6Al-4V titanium, nickel superalloys like Inconel 718, and a few aluminum-silicon alloys). The future, however, points to a much broader library of LPBF-compatible materials, including many high-performance alloys that were previously unweldable or difficult to process. For example, advanced aluminum grades (traditionally valued for aerospace) are being modified or paired with tailored process parameters to overcome issues like hot cracking, so that their high strength can be realized in printed form. Specialty nickel-based and titanium alloys designed for higher temperature or fatigue performance are likewise being adapted for LPBF, opening the door to printing parts for more demanding engine and structural applications. Refractory metals and exotic alloys such as tungsten, tantalum, or high-entropy alloys are seeing experimental success. As the understanding of microstructure control in LPBF improves, we can also expect new alloys explicitly engineered for additive manufacturing, yielding superior properties because they are optimized for the rapid solidification environment of LPBF rather than for casting or forging.

In addition to single-material improvements, there is intense interest in multi-material printing or producing parts that combine dissimilar alloys in a single build. Creating a graded transition from one metal to another within a component could enable, for instance, a part that transitions from a tough steel to a highly conductive copper in different zones, marrying the strengths of both. Achieving this in LPBF is challenging (due to different melting points and thermal expansion rates between metals). However, progress is being made through techniques like sequential powder deposition or multiple feed hoppers that swap materials during a build. We will likely see early applications of dissimilar alloy printing in areas like tooling (e.g., mold inserts with copper cooling channels inside a steel block) and aerospace (functionally graded structures that handle heat and stress optimally). As material flexibility increases, the potential applications for LPBF multiply, driving further interest and investment in the technology.

Equally important is the push for more reproducible and standardized powder feedstock. As LPBF moves into mainstream production, consistency in powder quality from batch to batch and supplier to supplier becomes critical. The future will bring more refined standards for powder characteristics so that certified powders produce reliable results across different machines and sites. Industry groups and standards organizations are working toward reference specifications and improved testing methods for powders. Better powder standardization will aid in developing trusted material property data for LPBF parts, since mechanical properties can be more reliably correlated when the feedstock is consistent.

4.7.4 Evolving Landscape of Qualification and Certification

No component can be delivered in safety-critical industries like aerospace, medical, and energy without rigorous assurance that it meets quality and performance requirements. The early adopters of LPBF in such fields often had to develop their bespoke qualification programs, which involved extensive testing of printed specimens, in-depth audits of each machine and process, and one-off certifications for each new part application. This approach, while necessary at the time, is time-consuming and expensive, and it does not scale well as additive manufacturing becomes more widespread. The future of LPBF thus hinges on more streamlined and standardized qualification frameworks that instill confidence in the technology globally.

One significant development is the rise of standardization efforts led by collaborations between industry, academia, and government agencies. Standards organizations are publishing comprehensive guidelines for additive manufacturing processes and materials (for example, unified standards for machine calibration, material testing, and design allowables specific to LPBF). These provide a standard reference so companies do not have to start from scratch. At the same time, certification bodies and regulatory agencies are increasingly working with the AM industry to create clear pathways for certifying LPBF-produced parts. We see initiatives where multiple companies and government labs share data and best practices to establish qualification protocols that everyone can trust. For instance, there are joint projects to develop standard test artifacts that can be printed to evaluate a machine's performance or to define baseline material property datasets for commonly used alloys that all stakeholders recognize.

Moreover, international cooperation is strengthening global trust in LPBF. This means that a part qualified in one region (say, meeting European aerospace additive standards) can be more readily accepted in another (like U.S. FAA requirements), thanks to harmonized standards and mutual recognition agreements. Organizations worldwide are aligning their requirements to smooth out the certification process for manufacturers operating across borders. These measures contribute to a future where LPBF is not an outlier technology but an entirely credible manufacturing method. In such a future, customers, regulators, and manufacturers will have a high degree of confidence that LPBF-produced parts are reliable and meet all necessary criteria, accelerating the adoption of LPBF in critical production environments.

References

1. ISO/ASTM (2015) Additive manufacturing—general principles—terminology (ISO/ASTM 52900:2015)
2. Hilton P, Gray Z (2022) Practical considerations for industrial LPBF machine calibration. In: Proceedings of the 33rd annual international solid freeform fabrication symposium. University of Texas at Austin, Austin, TX, pp 1–9

3. Khorasani AM, Gibson I, Ghaderi R, Awang M (2020) A review of technological improvements in laser-based powder bed fusion of metal printers. Int J Adv Manuf Technol 108(1–2):191–209. https://doi.org/10.1007/s00170-020-05495-1
4. Brown B, Lough C, Wilson D, Newkirk J, Liou F (2024) Atmosphere effects in laser powder bed fusion: a review. Materials 17(22):5549. https://doi.org/10.3390/ma17225549
5. Chouwdhury, et al., (2022) https://doi.org/10.1016/j.jmrt.2022.07.121
6. Taylor P, Wicker R (2022) Control methods for synchronizing laser on/off timing with galvo motion in metal powder bed fusion. Procedia Manuf 53:112–119
7. Tenbrock et al. (2021). https://doi.org/10.1016/j.addma.2020.101810
8. ASTM International (2018) Standard practice for additive manufacturing—process characteristics and performance: laser beam powder bed fusion for critical applications (ASTM F3303-18)
9. Zhang G, Liu X, Guan R (2019) Vacuum vs. argon environment in LPBF of aluminum alloys. Mater Lett 254:119297. https://doi.org/10.1016/j.matlet.2019.119297
10. Dietrich et al. (2020) https://doi.org/10.1016/j.addma.2019.100980
11. Luzon P, Andersson J, Koike M, Okazaki Y (2020) Effect of oxygen content on Ti-6Al-4V implants manufactured by powder bed fusion. J Biomed Mater Res B Appl Biomater 108(6):2576–2586. https://doi.org/10.1002/jbm.b.34583
12. Mirz M, Liu S, Li C, Zhao M, Zhang Y, Hong R (2022) Effects of oxygen content in powder bed fusion of Ti6Al4V: from microstructure to mechanical performance. Mater Sci Eng A 831:142246. https://doi.org/10.1016/j.msea.2021.142246
13. Pauzon C, Leung CLA, Marussi S, Atwood RC, Lee PD (2021) In situ X-ray imaging of the formation and ejection of pore defects in LPBF. Acta Mater 207:116662. https://doi.org/10.1016/j.actamat.2021.116662
14. Ladewig A, Schlick G, Fisser M, Schulze V, Schneider G (2016) The shielding gas flow influences the removal of process by-products in the selective laser melting process. Addit Manuf 10:1–9. https://doi.org/10.1016/j.addma.2016.01.004
15. Bidare P, Bitharas I, Ward RM, Attallah MM, Moore AJ (2018) Fluid and particle dynamics in LPBF. Acta Mater 142:107–120. https://doi.org/10.1016/j.actamat.2017.09.051
16. Anwar and Pham (2017) https://doi.org/10.1016/j.jmatprotec.2016.10.10.015
17. Elkins CJ, Mireles J, Estrada HH, Morgan DW, Taylor HC, Wicker RB (2023) Resolving the three-dimensional flow field within commercial metal additive manufacturing machines: application of experimental magnetic resonance velocimetry. Addit Manuf 73:103651. https://doi.org/10.1016/j.addma.2023.103651
18. Reijonen A, Pikka R, Salminen A (2020) Examining the effects of a lack of shielding gas flow in LPBF. J Manuf Process 56:111–117. https://doi.org/10.1016/j.jmapro.2020.04.065
19. Shen K, Panwisawa M, Bhagavatula G (2020) Optimizing argon flow velocity to minimize spatter in LPBF. CIRP J Manuf Sci Technol 31:229–239. https://doi.org/10.1016/j.cirpj.2020.02.006
20. Berez B, Smith R, Jones T (2022) Location-dependent defect accumulation in multi-LPBF builds. Addit Manuf 49:102903. https://doi.org/10.1016/j.addma.2022.102903
21. Lyons K, Lane B, Whitenton E (2020) Gas flow characterization in laser-based additive manufacturing systems using hot-wire anemometry. J Res Natl Inst Stand Technol 125:125001. https://doi.org/10.6028/jres.125.001
22. Tang Y, Pan W, Zavala C, Yang R (2022) Effects of gas flow speed on melt pool stability in multi-LPBF. Addit Manuf 54:102759. https://doi.org/10.1016/j.addma.2022.102759
23. Ladewig A, Noll B, Wehmöller M, Maier HJ (2018) Influence of recoater strategies on surface conditions during LPBF: a comparative study. Addit Manuf 24:522–530. https://doi.org/10.1016/j.addma.2018.09.016
24. Meier F, Haberland C, Schmid H-J, Drossel W (2019) Powder spreading and compaction in LPBF: the effect of particle size distribution and shape. Addit Manuf 30:100861. https://doi.org/10.1016/j.addma.2019.100861

25. Xiang S, Zhou X, Tang H (2021) Particle-scale modeling of powder spreading and its effect on microstructural uniformity in LPBF. Comput Mater Sci 194:110410. https://doi.org/10.1016/j.commatsci.2021.110410
26. Zhang H, Wu Y, Qi H, Gu D (2022) Discrete element modeling of roller-based powder spreading in LPBF. Powder Technol 399:117–130. https://doi.org/10.1016/j.powtec.2022.04.045
27. Mitterlehner T, Buchmayr B, Scherkl P (2022) Recoating defects in LPBF: a parametric study on layer thickness, powder type, and blade velocity. Addit Manuf 49:102499. https://doi.org/10.1016/j.addma.2022.102499
28. Snow Z, Nassar AR, Reutzel EW (2019) Effects of recoater velocity on powder bed distribution in LPBF. Addit Manuf 29:100804. https://doi.org/10.1016/j.addma.2019.100804
29. Herzog D, Seyda V, Wycisk E, Emmelmann C (2016) Additive manufacturing of metals. Acta Mater 117:371–392. https://doi.org/10.1016/j.actamat.2016.07.019
30. Sehhat MH, Mahdian A (2021) Effects of recoater type on mechanical properties in selective laser melting of metal powders. Addit Manuf Lett 1:100007. https://doi.org/10.1016/j.addlet.2021.100007
31. Grasso M, Colosimo BM (2017) Process defects and in situ monitoring methods in metal powder bed fusion: a review. Meas Sci Technol 28(4):044005. https://doi.org/10.1088/1361-6501/aa5c4f
32. Scime L, Beuth J (2019) Using machine learning to identify in-situ melt pool signatures indicative of flaw formation in a LPBF additive manufacturing process. Addit Manuf 25:151–165. https://doi.org/10.1016/j.addma.2018.11.010
33. Honigmann P, Sharma N, Okolo B, Popp U, Msallem B, Thieringer FM (2018) Patient-specific surgical implants made of 3D printed PEEK: material, technology, and scope of surgical application. Biomed Res Int 2018:4520636. https://doi.org/10.1155/2018/4520636
34. Kalpakjian S, Schmid S (2014) Manufacturing engineering and technology, 7th edn. Pearson
35. Lass EA, Stoudt MR, Williams ME, Katz MB, Levine LE, Phan TQ, Gnaeupel-Herold TH, Ng DS (2017) Formation of the Ni_3Nb δ-phase in stress-relieved Inconel 625 was produced via laser powder-bed fusion. Metall Mater Trans A 48(11):5547–5558. https://doi.org/10.1007/s11661-017-4304-6
36. Tang M, Pistorius PC, Beuth JL (2016) Rapid solidification: selective laser melting of AlSi10Mg. JOM 68(3):960–966. https://doi.org/10.1007/s11837-015-1763-3
37. DebRoy T, Wei HL, Zuback JS, Mukherjee T, Elmer JW, Milewski JO, Beese AM, Wilson-Heid A, De A, Zhang W (2018) Additive manufacturing of metallic components—process, structure and properties. Prog Mater Sci 92:112–224. https://doi.org/10.1016/j.pmatsci.2017.10.001
38. Frazier WE (2014) Metal additive manufacturing: a review. J Mater Eng Perform 23(6):1917–1928. https://doi.org/10.1007/s11665-014-0958-z
39. Seifi M, Salem A, Beuth J, Harrysson O, Lewandowski JJ (2016) Overview of materials qualification needs for metal additive manufacturing. JOM 68(3):747–764. https://doi.org/10.1007/s11837-015-1810-0
40. Gibson I, Rosen DW, Stucker B (2015) Additive manufacturing technologies, 2nd edn. Springer
41. ASM International (2019) ASM handbook, Volume 24: Additive manufacturing processes. ASM International
42. Agrawal AK, Thoma DJ (2022) High-throughput surface characterization to identify porosity defects in additively manufactured 316L stainless steel. Addit Manuf Lett 3:100093. https://doi.org/10.1016/j.addlet.2022.100093
43. Kruth JP, Froyen L, Van Vaerenbergh J, Mercelis P, Rombouts M, Lauwers B (2004) Selective laser melting of iron-based powder. J Mater Process Technol 149(1–3):616–622. https://doi.org/10.1016/j.jmatprotec.2003.11.051
44. Lindemann C, Jahnke U, Moi M, Koch R (2015) Towards a sustainable and economic selection of part candidates for additive manufacturing. Rapid Prototyp J 21(2):144–152. https://doi.org/10.1108/RPJ-12-2013-0122

45. Khanna N, Salvi H, Karaş B, Fairoz I, Shokrani A (2024) Cost modelling for powder bed fusion and directed energy deposition additive manufacturing. J Manuf Mater Process 8(4):142. https://doi.org/10.3390/jmmp8040142
46. Belforte D (2021) Wohlers Associates publishes report on post-processing in additive manufacturing [Press release], 18 Oct 2021. Laser Focus World. https://www.laserfocusworld.com/industrial-laser-solutions/article/14222033/wohlers-associates-publishes-report-on-post-processing-in-additive-manufacturing
47. James MS (2021) Additive manufacturing and the U.S. Army's improved turbine engine, 16 Jul 2021. Army.mil. https://www.army.mil/article/248519
48. Dall'Ava L, Hothi H, Henckel J, Di Laura A, Tirabosco R, Eskelinen A, Skinner J, Hart A (2021) Osseointegration of retrieved 3D-printed, off-the-shelf acetabular implants. Bone Joint Res 10(7):388–400. https://doi.org/10.1302/2046-3758.10.7.BJR-2020-0462.R1
49. Wang W, Ning J, Liang SY (2021) Analytical prediction of balling, lack-of-fusion, and keyholing thresholds in PBF-LB. Appl Sci 11(24):12053. https://doi.org/10.3390/app112412053
50. FDA (2017] FDA.gov here: https://www.fda.gov/files/medical%20devices/published/Technical-Considerations-for-Additive-Manufactured-Medical-Devices---Guidance-for-Industry-and-Food-and-Drug-Administration-Staff.pdf
51. Gerstler WD, Erno D (2017) Introduction of an additively manufactured multi-furcating heat exchanger. In: Proceedings of the 16th IEEE intersociety conference on thermal and thermomechanical phenomena in electronic systems (ITherm, 2017). IEEE, pp 624–633
52. Johnson LK, Richburg C, Lew M, Ledoux WR, Aubin PM, Rombokas E (2019) 3D printed lattice microstructures to mimic soft biological materials. Bioinspir Biomim 14(1):016001. https://doi.org/10.1088/1748-3190/aae10a
53. Conflux Technology (2022) Qualification and certification of additive manufactured heat exchangers [Industry article], 24 Jun 2022. https://www.confluxtechnology.com/qualification-and-certification-of-additive-manufactured-heat-exchangers/
54. Thangamani G, Felicioni S, Padovano E, Biamino S, Lombardi M, Ugues D, Fino P, Bondioli F (2024) A comprehensive review of laser powder bed fusion in jewelry: technologies, materials, and post-processing with future perspective. Metals 14(8):897. https://doi.org/10.3390/met14080897

Chapter 5
Microstructure, Mechanical Properties, and Design Considerations for Additive Manufacturing

Abstract Microstructure has a direct influence on the mechanical properties of metallic alloys.

5.1 Introduction

Microstructure has a direct influence on the mechanical properties of metallic materials. An alloy's microstructure is a result of its inherent chemistry, manufacturing process, and heat treatment. Historically, alloys have been developed and optimized to meet their intended requirements in conjunction with a specific manufacturing process, such as casting, forging, or sheet forming. Additive Manufacturing (AM) is a relatively new process that offers significant benefits for rapid design and implementation and is currently being considered by many industries including aerospace. Many of the already existing aerospace alloys are capable of being processed into viable components by a variety of AM methods. Current expectation is that research in AM processing will lead ultimately to development of unique microstructures that will have a significant impact on the alloy's mechanical properties.

This paper discusses the microstructure and properties of traditional aerospace alloys processed by AM with respect to potential use in gas turbine engine applications. In the past few years, Honeywell has evaluated the AM capability of a wide variety of aerospace alloys. Mechanical property data were generated to determine process capability in comparison to the corresponding cast or wrought versions and to identify potential failure modes in their design and applications. The materials were manufactured in a laboratory environment in the absence of any AM material specifications and controls. Currently, AM lacks an established industrial supply base and the required standards, specifications, and quality system to meet aerospace requirements.

K. Hsu et al., *Additive Manufacturing of Metals*, Springer Series in Advanced Manufacturing, https://doi.org/10.1007/978-3-032-04961-2_5

Table 5.1 A partial list of PBLFT alloys investigated at Honeywell Aerospace

Name	Final heat treatment parameters
IN-718	1352 K/1 h + 991 K/8 h + 894 K/8 h
IN-625	1255 K/1 h
Hast-X	1422 K/1 h
Co-Cr	1464 K/1 h
Ti-6-4	1033 K/2 h
15-5PH	1339 K/3 min + 825 K/4 h
316L	1339 K/1 h
AlSi10Mg	802 K/5 h + 433 K/12 h

5.2 Specimen Manufacturing

Specimens of various alloys were manufactured as cylindrical rods or rectangular plates using powder bed laser fusion technology (PBLFT). Three build directions were investigated, which were vertical, horizontal, and 45° with respect to the horizontal build platform. After building, the specimens remained on the build plates and were subjected to a stress relief treatment to mitigate possible cracking from as-built residual stresses. After stress relief, the blanks were excised from the plates and hot isostatically pressed (HIP). HIP was employed to reduce the occurrence of defects such as pores or possible unbonded regions which could affect fatigue strength. Following HIP, the specimens were heat treated in a manner consistent with their corresponding wrought versions.

Nearly a dozen such alloys have been studied at Honeywell in the past 3 years. Table 5.1 provides a list of materials and the applied post-hip heat treatments. Test samples were machined into standard specimens and tested per ASTM recommended procedures. Some specimens were tested with the additive-manufactured as-built surface intact in the gage section in order to ascertain its influence on fatigue strength.

5.3 Design Considerations

Component stresses at the temperature of application dictate the proper use of a material. The expected failure mode against which the design must have an acceptable margin of safety may be the initiation of a crack in low-cycle fatigue (LCF), adequate high-cycle fatigue (HCF) strength in resonance, creep rupture (CR) at elevated temperatures, or simply metal loss or recession due to oxidation. Typical applications envisaged for the PBLFT materials, at least initially, are those where the alloys are used in cast or wrought forms, but not in applications that are safety-critical. For applications where strength is the main design consideration, IN-718, 15-5PH, and Ti-6-4 are viable candidates. For higher-temperature applications, Co-Cr, IN-625, Hast-X, and IN-718 are the likely choices.

LCF is a phenomenon which occurs mostly at stress concentrations and at stress ranges >75% of the yield strength. The LCF design cycle is typically representative of the loads and temperatures associated with the typical start-to-stop cycle in a turbine engine. The component may need to be certified with an extremely low probability of failure (<10^{-9}/cycle) in the cyclic range of 10^3–10^5 repeated loads.

HCF, on the other hand, occurs in response to vibratory loads at a fairly high cyclic frequency (10^2–10^4 Hz) and at low stress ranges. The design requirement would be absence of failure due to random or resonant vibratory loads during the component's certified life.

The phenomenon of creep occurs at relatively higher temperatures. Avoidance of change in shape or dimension or rupture over the life of the component so as not to compromise its intended function is the goal.

Damage tolerance signifies the ability of the material to sustain failure due to propagation of a crack initiated by any of the above scenarios through an inspection period. Here crack growth and fracture toughness properties dictate proper usage.

Oxidation is an important consideration at temperatures well above 922 K especially in components such as turbine cases, combustor liners, and accessories such as fuel injectors.

In addition to the information on microstructure, mechanical property data and fractographic information will be presented below. Where available, the data will be compared to published averages of the mechanical properties expected from wrought or cast versions.

5.4 Grain Size

All alloys in Table 5.1 except IN-625, Hast-X, Co-Cr, and 316 L are precipitation strengthened. The rest derive their strength from solid solution hardening. In both cases, grain size can have a significant influence on mechanical strength. Superalloys such as IN-718 are the most influenced by grain size as they have a planar mode of deformation in which slip is distributed heterogeneously within a grain and concentrated on fewer slip planes.

In processing of wrought products, recrystallization may take place during hot working or subsequent solution heat treatment. The resultant grain size is determined by the level of prior hot or cold work, the level of thermal energy available for grain growth, and the resistance to grain growth offered by grain boundary pinning particles. A dispersion of fine incoherent particles can limit the grain growth by the phenomenon called Zener pinning [1]. Since almost all of the materials shown in Table 5.1 have carbon as a constituent, carbide particles can appear as a fine dispersion in AM products. In addition, any remnant oxygen in the build atmosphere may create a dispersion of fine oxides in the matrix under certain conditions. If nitrogen is used as a build atmosphere, it can contribute to further inhibition of grain growth by formation of fine nitride particles. Although in materials like IN-718 the HIP and solution temperatures are well above the solvus temperature of the strengthening precipitates, these fine carbide, nitride, or oxide particles present even at the super solvus temperature of the precipitates can limit grain growth to ASTM 3–5. In IN-625 and Co-Cr, vendor-to-vendor differences have been noticed in grain sizes brought about by differences in distribution of these grain boundary pinning inclusions.

5.5 Tensile Properties

Tensile properties of all alloys in the table were determined at selected temperatures. In general, they all compared well with the corresponding wrought material and were better than their cast versions. IN-718 was the most characterized of all as it is one of the most used alloys in gas turbine engine applications. Tensile, LCF, HCF, CR, fracture toughness, and cyclic crack growth data were generated for IN-718 at a variety of temperatures. A typical micrograph of PBLFT IN-718 is shown in Fig. 5.1 which had a grain size of ASTM 4. The particles which were identified as carbides of niobium or titanium are seen as a fine dispersion in the background, and the distribution is fairly uniform across. Grain size and shape were found to be uniform in both transverse and longitudinal metallographic sections of the test bars. There were no discernible differences in microstructure between the bars built in vertical, horizontal, or 45° orientations, which attests to the capability of PBLFT process to impart homogeneous microstructure. Measurement of elastic

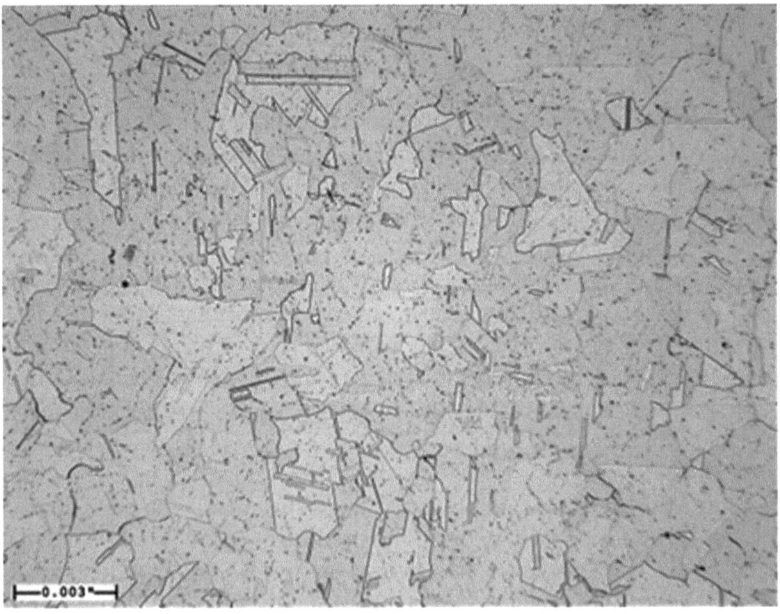

Fig. 5.1 Microstructure of PBLFT IN-718

moduli, which is a good indication of texture, showed no significant differences between horizontally built, vertically built, or those built in 45° orientations. At 589 K, they were 185, 173, and 194 GPa, whereas for a fine-grained wrought IN-718 [2], the reported value is 184 GPa. These results indicate insignificant crystallographic texture in PBLFT IN-718 even when the grain size was coarse compared to typical wrought IN-718 (ASTM 8–10) [2].

Figure 5.2 shows tensile strengths of PBLFT IN-718 for three orientations with respect to the build plate. Horizontal bars were tested at 297, 589, 700, 811, and 922 K, whereas vertical and 45° orientations were tested only at 700 and 922 K. The data are comparable to expected strengths for a grain size of ASTM 4. There were a total of 72 tensile bars tested. YS and UTS refer to 0.2% yield and ultimate tensile strengths, respectively. The variations between different orientations are not discernible on the average, which points to uniformity of texture across different build directions. The tensile elongation values were found to be slightly higher for the PBLFT product than for fine-grained wrought product [2]. When compared to cast IN-718, tensile strengths and ductility of PBLFT IN-718 were far superior. Statistical analysis of the YS and UTS at 700 and 922 K (24 values for each temperature) that contained all three orientations showed a standard error of estimate of 33 MPa for YS and 27 MPa for UTS at *either* temperature. The fact that these values are somewhat lower than those observed in typical wrought IN-718 and much lower than those in cast versions speaks well of the uniformity that can be expected in a PBLFT process.

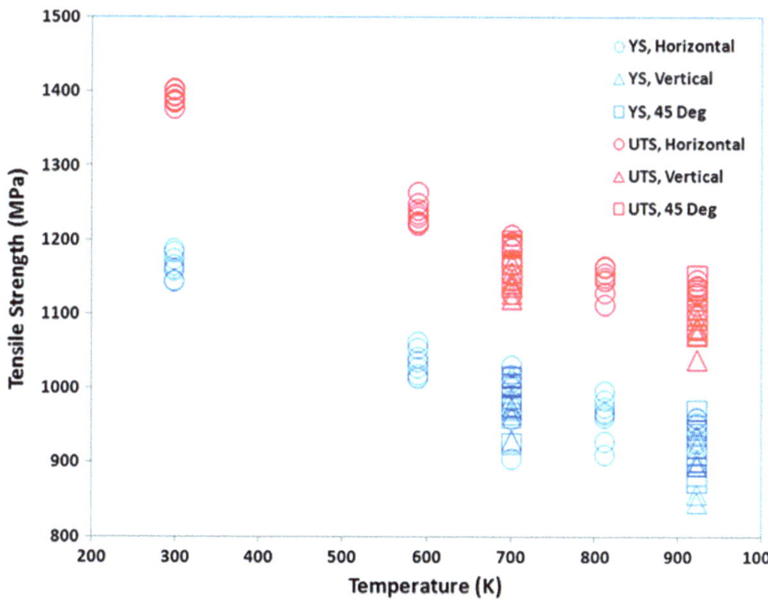

Fig. 5.2 Comparison of tensile strengths of PBLFT IN-718 for various build orientations

For Ti-6-4 alloy, the tensile strengths of PBLFT product compared well with the wrought version as seen in Fig. 5.3. The average lines for YS and UTS are taken from CINDAS [3]. This closeness to wrought behavior may be due first to the small grain size associated with PBLFT processing and subsequently to the fine and uniform lamellar structure (alpha-beta) observed in Fig. 5.4. The ductilities were higher than those for the wrought. Measurements of elastic moduli in horizontal, vertical, and 45° orientations showed no significant texture. For Ti-6-4 also, the tensile strengths and ductility were superior to the cast versions. The scatter in YS and UTS values, 31 and 15 MPa, at 297 K for 24 specimens equally distributed among the three orientations was similar to or lower than those observed for typical wrought products.

The PBLFT process is amenable to duplicating wrought properties in PH (precipitation hardening) steels like 15-5PH. Figure 5.5 shows a comparison. There are a total of 48 specimens with equal distribution with respect to the four temperatures and each temperature set containing equal distributions of the three orientations. The different orientations are not distinguished in the plot. The average properties in original brochures for the wrought version are shown in the figure by dashed lines [4].

For solid solution alloys such as Hast-X, IN-625, and Co-Cr, the tensile properties approached those of the standard wrought product and, in some cases, even exceeded reported values. This is a direct result of inhibition of grain growth by dispersoids present in the microstructure. Figure 5.6 shows a plot of tensile

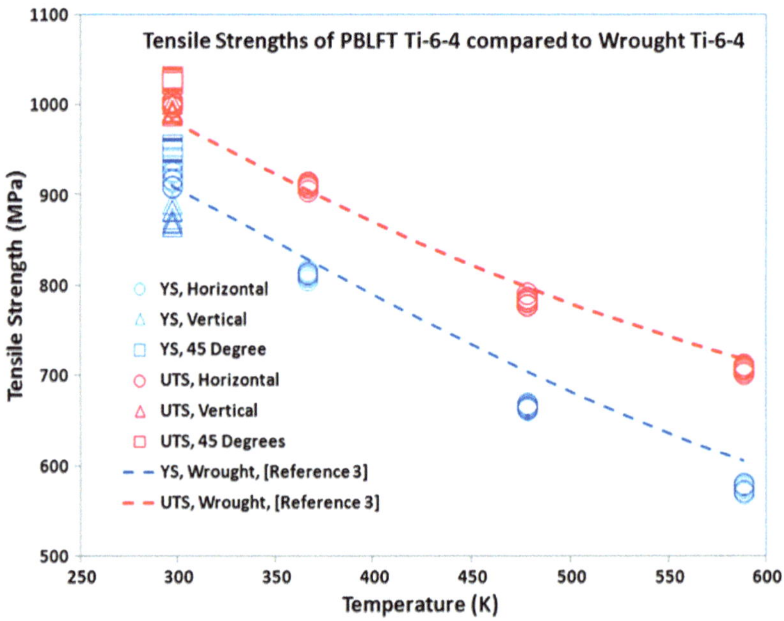

Fig. 5.3 Comparison of tensile strengths of PBLFT Ti-6-4 to wrought Ti-6-4

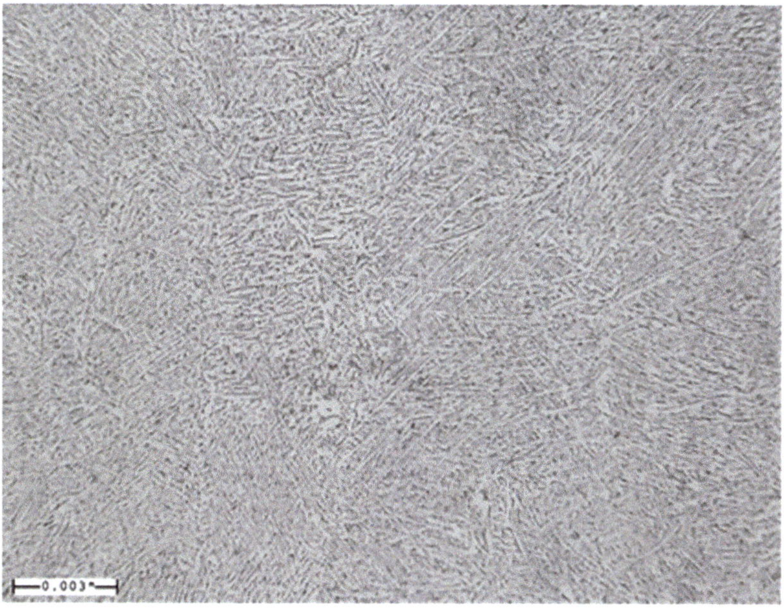

Fig. 5.4 Microstructure of PBLFT Ti-6-4

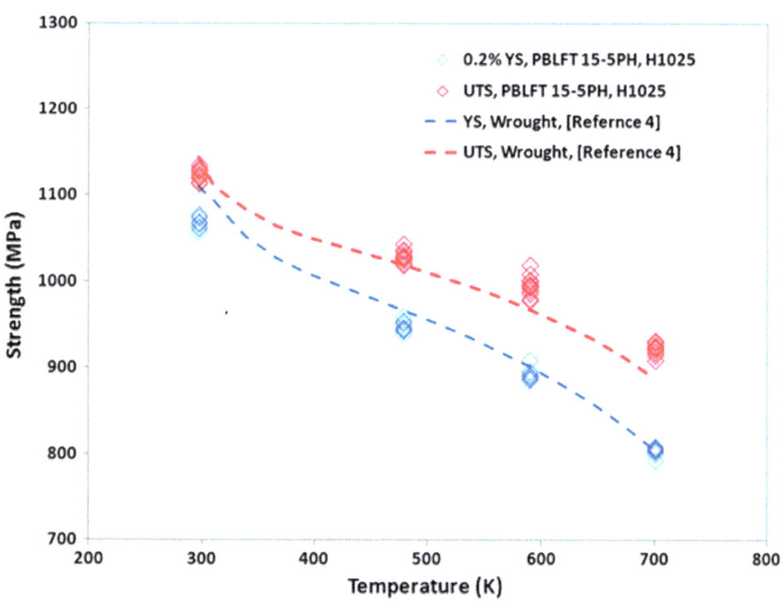

Fig. 5.5 Comparison of tensile strengths of PBLFT 15-5PH to wrought 15-5PH

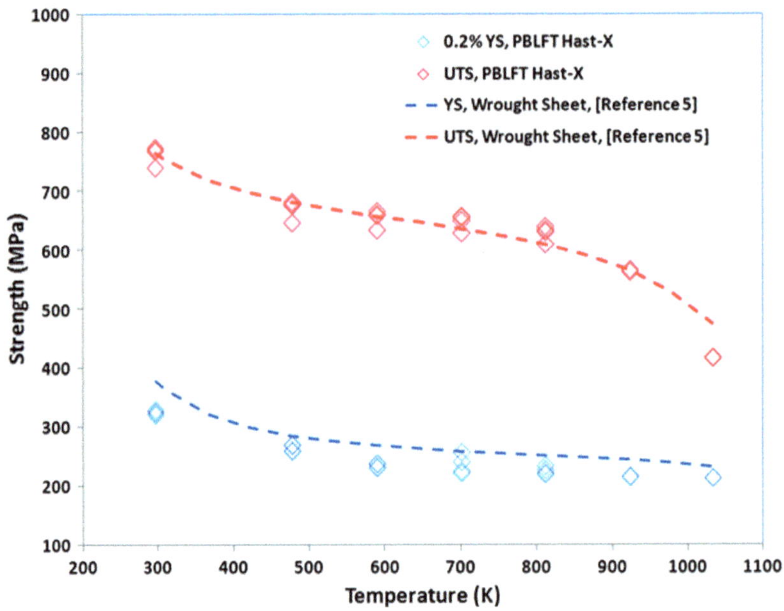

Fig. 5.6 Comparison of tensile strengths of PBLFT Hast-X to wrought Hast-X sheet

properties of Hast-X compared to the reported wrought values in Haynes' brochure [5]. The data contained the three orientations at each temperature. Similar results were observed for IN-625 and Co-Cr. For Co-Cr, the comparison was made against properties of wrought Haynes 188 which is a cobalt-based alloy widely used in high-temperature applications in gas turbine engines. For AlSi10Mg alloy, tensile strengths of the PBLFT version were found to be superior to those for cast aluminum alloy 356 and equivalent to wrought aluminum 6061-T6 at temperatures lower than 366 K.

5.6 Fatigue Properties

For fatigue testing of all alloys except for IN-718, the specimen gage was machined to remove the as-built surface. In the case of IN-718, the major part of the characterization was conducted on machined gage surfaces but a selected number of specimens were tested with the as-built surface in the gage section to ascertain the debit owing to defects present in the as-built layer.

Figure 5.7 shows a LCF plot of PBLFT IN-718 where the stress range on the ordinate is normalized to a stress ratio, R (min/max), of −1.0 by means of a stress ratio normalization factor similar to the Smith–Watson–Topper parameter [6]. The data covers strain ratios of 0.0 and −1.0 and three orientations of the bars with respect to the build plate. The LCF strengths seem to be independent of build

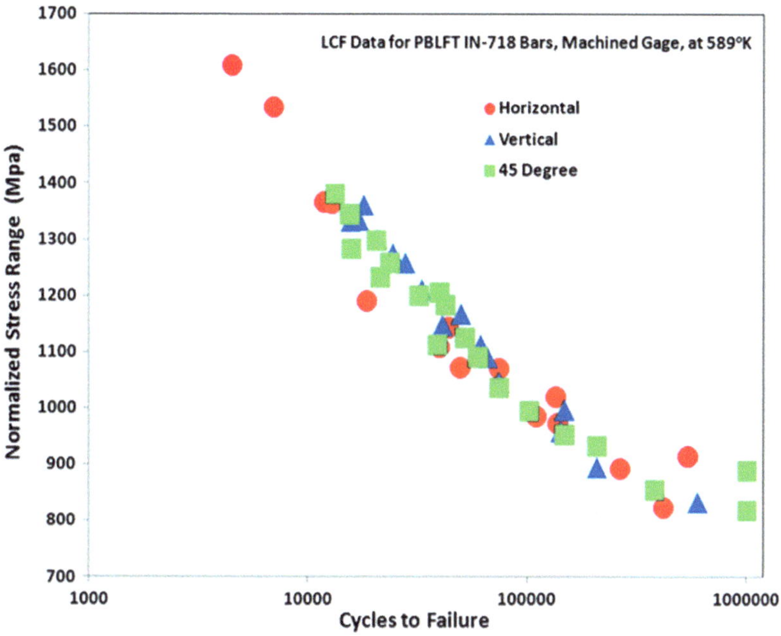

Fig. 5.7 LCF data at 589 K for PBLFT IN-718 with machined gage section

orientations. Also, a single stress ratio correction was found to be applicable for all three build orientations. This again speaks well of the uniformity of microstructure and lack of texture in all three build orientations. When the data are compared to those reported in the literature [7] for finer-grained, higher-strength wrought IN-718 (ASTM 8–10), the difference is explainable simply by the grain size effect. Figure 5.8 shows the debit in LCF strength when the as-built surface is not removed from the gage. It is interesting to note that the effect is not evident in completely reversed cycles ($R = -1.0$). At more positive strain ratios, the debit is well recognizable. Note that all data shown in Fig. 5.8 are from vertically built bars.

Figures 5.9 and 5.10 show scanning electron micrographs of typical crack initiation in LCF for as-machined and as-built surfaces, respectively. Note the subsurface crystallographic Stage-I initiation in the machined surface and a more frontal, surface-initiated, noncrystallographic Stage-II type for the as-built surface. The morphology in Fig. 5.9 is typical when the crack initiation is not associated with or influenced by defects. In Fig. 5.10, *the solid brush line* drawn near the bottom of the fracture (*not to be mistaken for a crack*) demarcates the shape and location of the defect front in the plane of fracture from which fatigue initiated. This debit due to defects in as-built surface was also seen in HCF data generated at 589 K. HCF strength for specimens with machined gage surfaces was found to be equivalent to that for the wrought version. This may be related to smaller slip distances involved in HCF crack initiation which may be influenced by dispersoids. An important point to note in both LCF and HCF behavior of PBLFT IN-718 is that even the strength

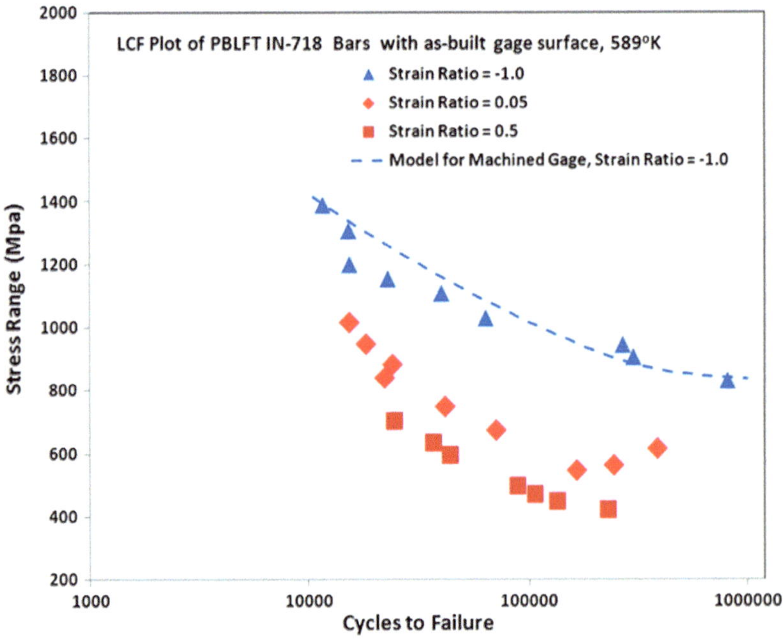

Fig. 5.8 LCF data at 589 K for PBLFT IN-718 with as-built gage surface, all vertically built bars

Fig. 5.9 SEM micrograph of crack initiation with machined gage section, vertically built, Nf = 206,544

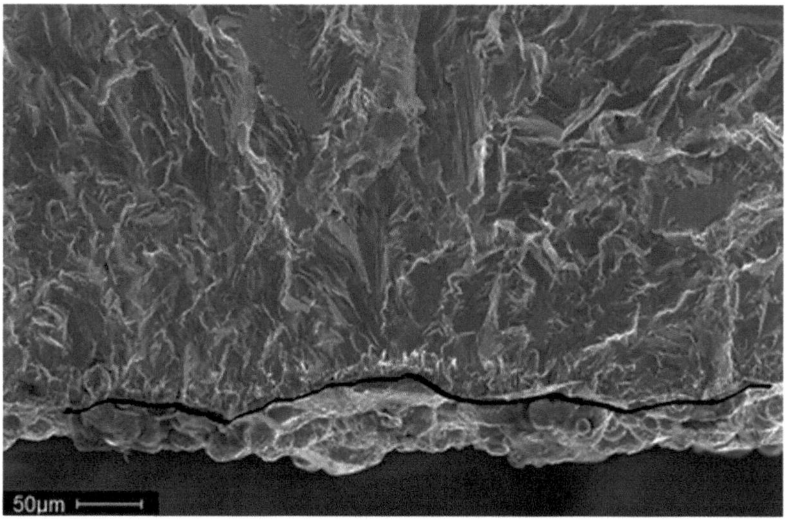

Fig. 5.10 SEM micrograph of crack initiation with as-built gage section, vertically built, Nf = 21,855. Note the rough as-built surface at the bottom edge and the solid brush line indicating the LCF initiation front

with the as-built surface present in the gage was far superior to those of the cast versions. With new methods ever evolving in precision surface machining techniques, the as-built surface debit in LCF in PBLFT products may become a thing of the past.

The fatigue strength of Ti-6-4 was found to be equivalent to that for wrought and much superior to the cast version. The former is attributed to the fine nature of the microstructure and the latter to the very large grain sizes of the cast. HCF strength of PBLFT Ti-6-4 was higher than that for wrought especially at higher mean stresses. For 15-5PH and 316L, the LCF strength at 297 K fell between average and minimum (defined by −3Sigma) of wrought models. For PBLFT IN-625 and Hast-X, the fatigue strengths were found to be equivalent to those for wrought. For Co-Cr, the LCF strength was far superior to that for wrought Hast-X, HA-230, or the cobalt-based wrought alloy HA-188.

5.7 Creep

Creep properties were measured for PBLFT IN-718. For both rupture and times to various percentages of creep strains, the PBLFT version was superior to wrought or cast versions. The superiority to the cast version was unexpected. Figure 5.11 shows a comparison of rupture times. The PBLFT version significantly exceeded the cast version in capability at 977 and 1033 K. This superiority in creep and rupture strengths of PBLFT IN-718 may also be attributed to the effective pinning of grain boundaries by dispersions. There were no orientation-related differences in properties between different build directions.

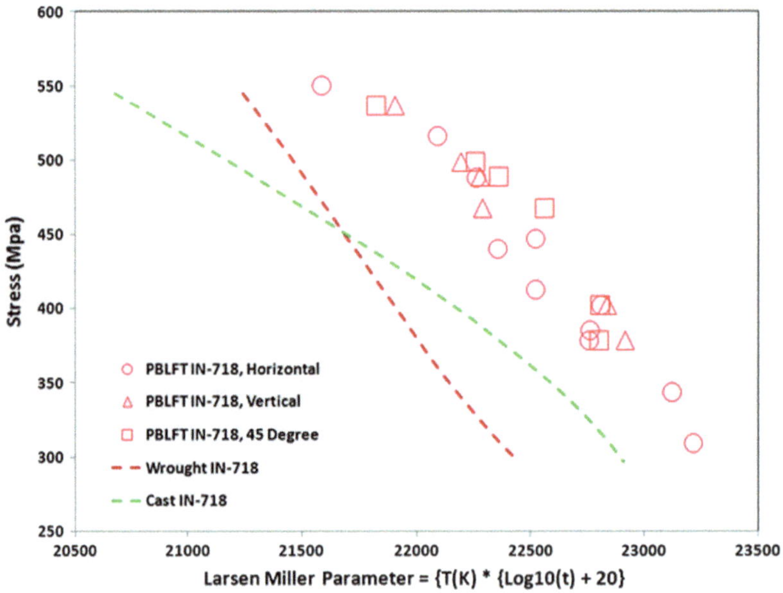

Fig. 5.11 Larson–Miller plots of CR for PBLFT IN-718

5.8 Fracture Tolerance

Fracture toughness measured at room temperature for various orientations of the crack for different build orientations for PBLFT IN-718 showed the same values as those reported for the wrought fine-grained version in the literature. The crack growth rates measured for PBLFT IN-718 at 589 K were lower than those of the fine-grained wrought version. This difference may be attributed to the larger grain size of the PBLFT product.

5.9 Influence of Dispersoids

A uniform distribution of fine incoherent particles in an alloy matrix will have an influence on keeping grain size from becoming too large in spite of heat treatments conducted at a high enough temperature to put into solution all other strengthening phases. This is somewhat akin to the behavior observed in super solvus processed PM (powder metallurgy) nickel-base superalloys like LC PM Astroloy, Alloy-10, PM Rene95, and others. The grain-growth-inhibiting particles in the PM alloys are a fine dispersion of carbides or oxides created during powder manufacture. In the case of AM-processed IN-718, these particles are primarily carbides of niobium or titanium.

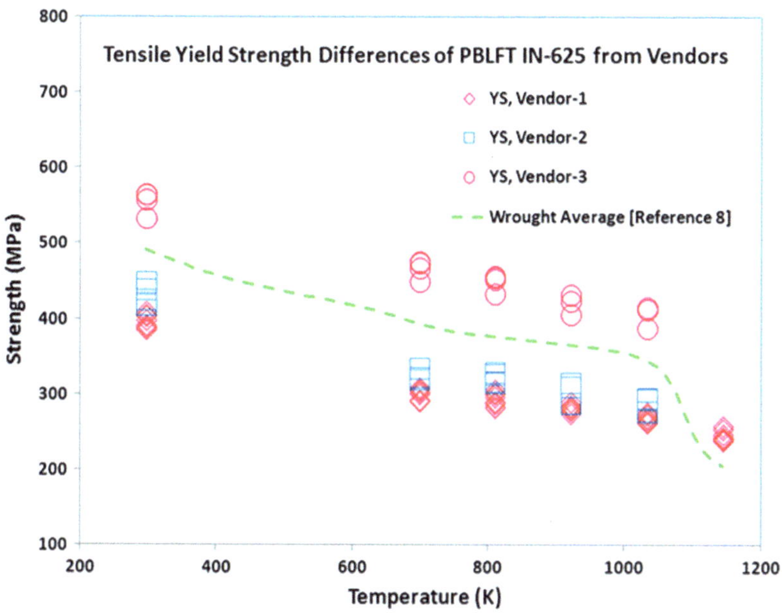

Fig. 5.12 Vendor differences in tensile yield strength of PBLFT IN-625

Vendor-to-vendor differences have been noted in tensile yield strength for IN-625, Co-Cr, and stainless steel 316L. Figure 5.12 shows the observation on PBLFT IN-625 in which the data are compared to standard wrought IN-625 [8]. The effect is explainable based on grain size differences and influence of dispersoids. In the case of Co-Cr, our investigation has shown that introduction of nitrogen (versus argon) into the build atmosphere can produce the same effect, with nitrogen build atmosphere contributing higher yield strength and higher creep strength. In addition, variations in processing parameters or the initial chemistry of the powder itself could affect grain growth.

5.10 Electron Beam Technology

AM using electron beam technology (EBT) has begun to get attention for production of aerospace components. IN-718 has been used as a vehicle at Honeywell for understanding the unique microstructure and properties produced by EBT. Because the thermal intensity of the beam is higher and the base plate has to be held at a high enough temperature (1200–1350 K) during manufacture, the vertical growth of the successively deposited layers seems to occur highly epitaxially. As a result, there is a strong and significant texture developed in the vertical (build) direction. LCF tests

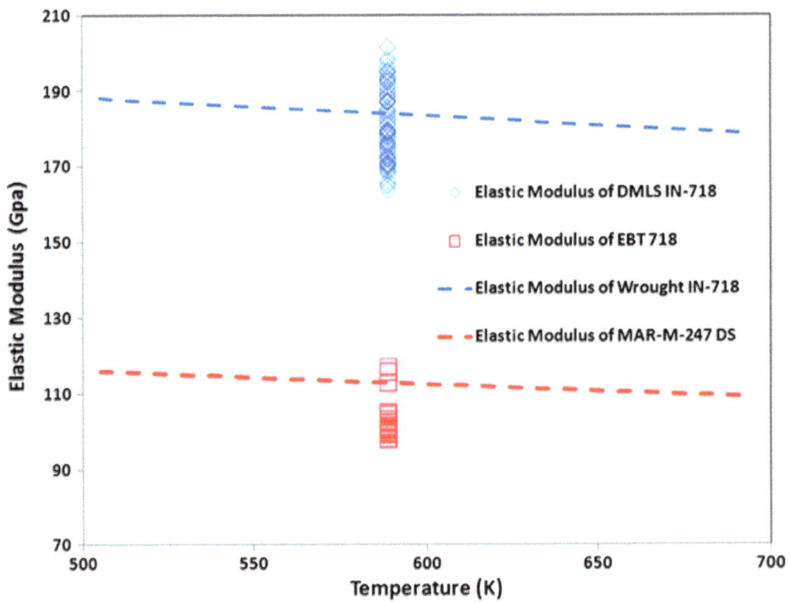

Fig. 5.13 Comparison of elastic moduli of EBT IN-718 and PBLFT IN-718

have shown that the modulus is close to that for directionally solidified nickel-base superalloys like DS MAR-M-247 in the growth direction, which is also close to the modulus value for nickel-base superalloy single crystals in the ⟨001⟩ direction. Figure 5.13 shows the data for heat-treated EBT IN-718. Modulus of PBLFT IN-718, however, is close to that for wrought IN-718.

Microstructure of EBT IN-718 consists of grains with a high aspect ratio in the build direction as shown in Fig. 5.14. This texture was seen to exist even in the as-built condition. Electron back-scattered diffraction (EBSD) orientation map of these grains is shown in Fig. 5.15. The strong red coloring [9] is indicative of the preferred [100] crystallographic direction in the build orientation (vertical in the figure).

In spite of the strong texture, the tensile and LCF strengths of EBT IN-718 were close to those for the PBLFT material. Figure 5.16 shows a comparison of the LCF data. The EBT data fall on top of the PBLFT version. Because of the low elastic modulus in the build direction, EBT process may be suitable for situations where thermally induced stresses are predominant in the build direction of a component. The fractography of EBT bars resembled those observed for DS MAR-M-247 with significant Stage-I crystallographic faceted crack initiation.

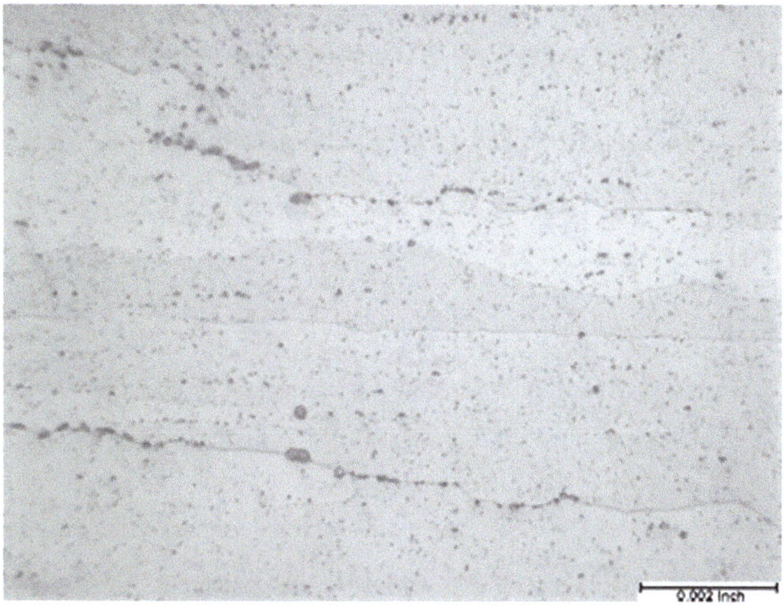

Fig. 5.14 Longitudinal section (*Horizontal direction in the micrograph, also the Build Direction*) of heat-treated EBM IN-718 showing grains with a high aspect ratio

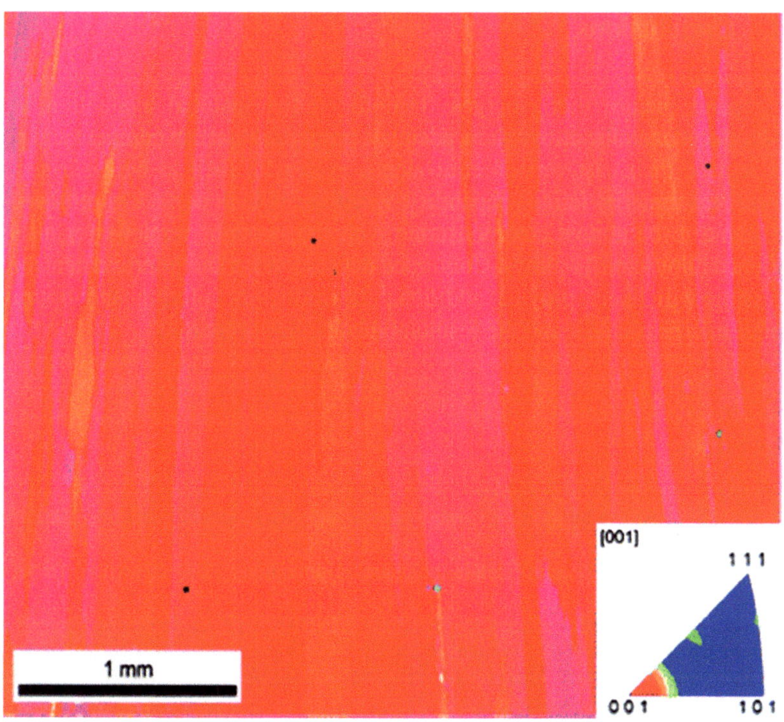

Fig. 5.15 EBSD crystallographic orientation map of EBT IN-718 (*The vertical orientation here is the build direction, and corresponds with the horizontal direction in Fig. 5.14*)

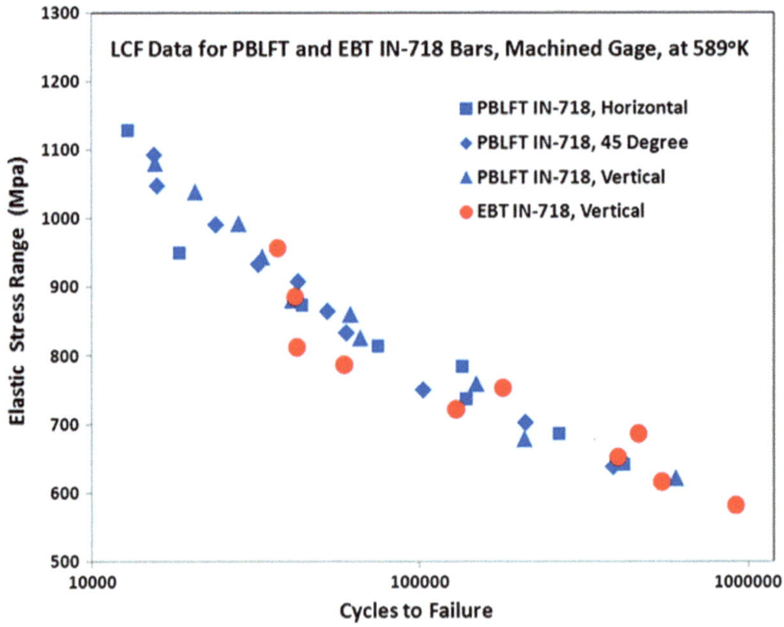

Fig. 5.16 LCF data of EBT IN-718 compared to PBLFT IN-718 at 589 K, strain ratio of 0.0

5.11 Conclusions

The general picture that emerges from the initial material evaluation is that AM-produced metallic alloys show adequate properties that can compete and potentially replace their cast-version equivalents, even with their as-built surface. The inherent AM material fatigue capability, when the as-built surface is removed, has been found to be almost commensurate with the wrought versions for all solid-solutioned alloys and Ti-alloys. Continued developments and improvements in AM processing will likely provide additional flexibility in tailoring and optimizing material microstructures and the resultant mechanical properties.

References

1. Hillert M (1988) Inhibition of grain growth by second-phase particles. Acta Metall 36(12):3177–3181
2. Inconel Alloy 718, Special metals brochure. Huntington, 3200 Riverside Drive, WV 25705-1771
3. Aerospace and high performance alloys database, CINDAS. Data digitized from Figure 3.3.1.8, Ti-6-4 Chapter, Section 3707, November 2000
4. Product bulletin no. S-6, Armco 15-5 PH VAC CE precipitation hardening steel. Armco Stainless Steel Products, Baltimore, MD 21203

5. Haynes brochure, Hastelloy X Alloy. Haynes International, 1020 West Park Avenue, Kokomo, IN 46904
6. Smith KN, Watson P, Topper TH (1970) A stress-strain function for the fatigue of metals. J Mater ASTM 5:767–778
7. Korth GE (1983) Mechanical properties test data of Alloy 718 for liquid metal fast breeder reactor applications. Report EGG-2229, Idaho National Engineering Laboratory, January 1983
8. Haynes brochure, IN-625 Alloy. Haynes International, 1020 West Park Avenue, Kokomo, IN 46904
9. EBSD work was conducted at Arizona State University under the guidance of Professor Amaneh Tasooji and the students of Capstone Project, May 2016

Chapter 6
Electron Beam Technology

Abstract Additive manufacturing (AM) or three-dimensional (3D) printing is the process of creating 3D objects or products, layer by layer, from a 3D digital model. The 3D digital model can be created from CAD, CT, MRI, or laser scanning. AM is the reverse process of traditional manufacturing technologies, such as machining, in which one starts with a block of material, and subsequently, the material is removed in a subtractive process forming the final desired part.

6.1 Additive Manufacturing

Additive manufacturing (AM) or three-dimensional (3D) printing is the process of creating 3D objects or products, layer by layer, from a 3D digital model. The 3D digital model can be created from CAD, CT, MRI, or laser scanning. AM is the reverse process of traditional manufacturing technologies, such as machining, in which one starts with a block of material, and subsequently, the material is removed in a subtractive process forming the final desired part. Alternatively, AM starts with a blank platform, and material is added in a controlled method where required—in each layer—until the final part is formed [1].

In order to help standardize AM in the United States, the ASTM F42 Committee on AM Technologies was formed in 2009 and categorized AM technologies into seven categories including vat photopolymerization, material extrusion, powder bed fusion, material jetting, binder jetting, sheet lamination, and directed energy deposition (F42 Committee 2012). In order to provide sufficient background for the research completed in this dissertation, this chapter provides more information on the metal AM processes that are commonly used today in AM.

K. Hsu et al., *Additive Manufacturing of Metals*, Springer Series in Advanced Manufacturing, https://doi.org/10.1007/978-3-032-04961-2_6

6.1.1 Powder Bed Fusion

Powder bed fusion systems use thermal energy to melt the powder into the desired pattern. The majority of powder bed fusion systems use a laser to melt polymer or metal powder to fabricate the 3D structures layer by layer (F42 Committee 2012). Powder bed fusion technology is the most common metal-based technology used to manufacture end-use engineered products, many of which are being used in aerospace, defense, and medical applications.

6.1.2 Electron Beam Melting

The electron beam melting (EBM) AM process builds complex parts out of metal powders within a high temperature and vacuum environment. EBM has been used in final production for many years, and more than 50,000 EBM-built devices have been implanted for the medical industry. In recent years, EBM has started to gain momentum in the aerospace industry to manufacture production components. In order to expand the EBM market, new materials such as Alloy 718 have been developed and characterized in the EBM technology. Alloy 718 is a nickel-chromium-based super alloy ideal for high temperature and corrosive environments, with excellent mechanical properties at elevated temperatures.

Electron beam melting (EBM) is an AM technology that selectively consolidates metal powders such as titanium, Inco 718, and cobalt alloys to fabricate 3D structures. Compared to conventional manufacturing processes, the EBM process is capable of fabricating low-volume, high-value articles at reduced lead times. The technology was invented in 1993 in Sweden at the University of Technology in Gothenburg. Arcam was founded in 1997 and sold its first commercial system in 2002. The company currently has over 200 systems installed around the world [2, 3].

The process includes the focusing of an electron beam at discrete areas within a powder bed composed of metal particles (average size of 45–105 μm) to produce melting, followed by re-solidification that enables fabrication of complex geometries. The EBM system (Fig. 6.1) consists of an electron beam gun, vacuum chamber (~10–4 torr), build tank, and powder distribution mechanisms (powder hoppers and rake). Within the electron beam gun, a tungsten filament is heated to emit electrons accelerated at high voltage (60 kV), resulting in an electron beam (carrying high kinetic energy) that is focused by electromagnetic lenses. This system is capable of electron beam scan speeds of up to 8000 m/s, electron beam positioning accuracy of ±0.025 mm, and layer thicknesses in the range 0.05–0.2 mm [4–6].

Part fabrication is initiated by the uniform distribution of powder over a start plate as seen in Fig. 6.2a. The powder layer is preheated by the electron beam followed by a sequence of line-by-line scanning that melts the loosely joined powder. The powder is spread and packed, only by the raking system; therefore, the powder is disorganized with gaps. As a result, before melting, the layer thickness is about

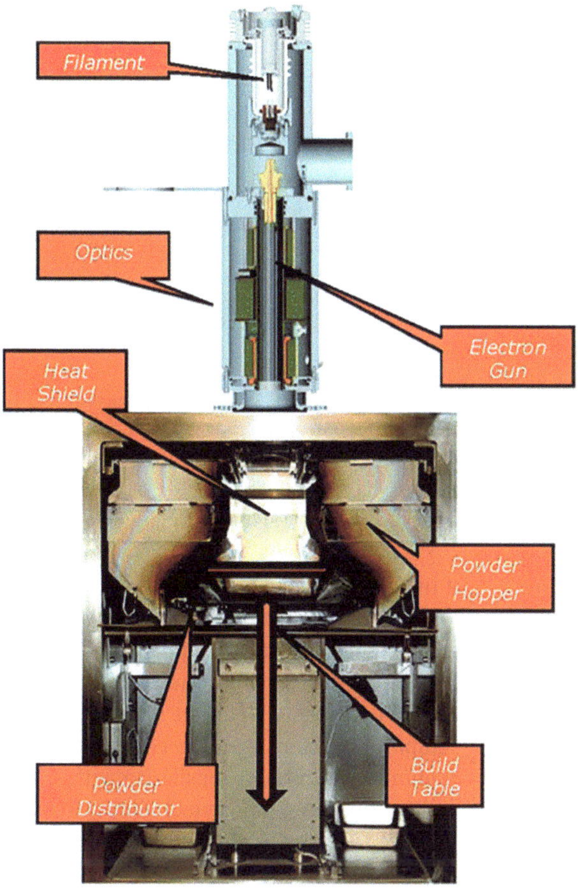

Fig. 6.1 EBM system

2–3 times greater than it should be. When the powder is melted, it reduces to the correct layer thickness (Fig. 6.2b). After completion of a layer, the process platform is lowered a distance equivalent to one layer thickness (0.05–0.2 mm). A new layer of powder is applied, and the process is repeated until the full build is complete. Completion of the process is followed by a helium-assisted cool-down sequence lasting about 6 h, depending on the build size and material used.

6.1.3 Materials

Titanium and associated alloys have been used in industry for over 50 years—becoming more popular in recent decades [7]. Titanium has been most successful in areas where the high strength-to-weight ratio provides an advantage over aluminum and

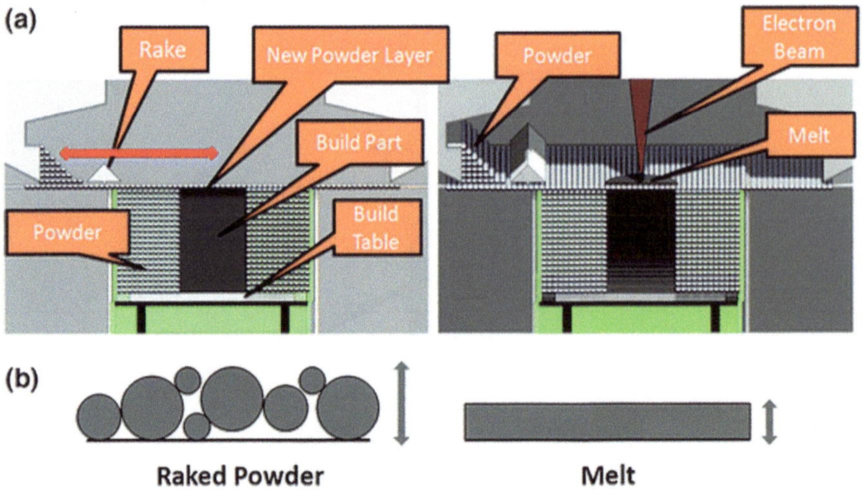

Fig. 6.2 (**a**) EBM raking and melting process. (**b**) Powder layer thickness before and after

steels. Other advantages of titanium include the ability to be biocompatible and corrosion resistant [8]. The new lighter and more efficient airplane designs require stronger frames and thus the obvious replacement for aluminum structural components is titanium; since aluminum and composites cannot provide the required structural support that titanium can provide. As composites become more popular in the aerospace industry, more designers employ titanium airframe components [9]. Commercial (99.2% pure) grades of titanium have ultimate tensile strength of about 63,000 psi (434 MPa), equal to that of common, low-grade steel alloys, but with the dramatic advantage of being 45% lighter. Titanium is 60% more dense than aluminum, but more than twice as strong as the most commonly used 6061-T6 aluminum alloy [10].

Titanium is the ninth most abundant element in the Earth's crust (0.63% by mass) and the seventh most abundant metal. Current titanium production methods used to produce both titanium ingot and wrought products have naturally high manufacturing costs, due in part to the strong chemical bonding between titanium and the associated interstitial elements such as carbon, hydrogen, nitrogen, and oxygen. The cost increases significantly due to the energy required to separate these interstitial elements [11]. Ti-6AL-4V (Ti64) is the most popular titanium alloy for medical and aerospace applications [12], and the costs for titanium have increased due to recent high demand from these industries.

Alloy 718 is a solution annealed and precipitation hardened material designed for high-temperature applications. Alloy 718 is hardened by the precipitation of secondary phases (e.g., gamma prime and gamma double-prime) into the metal matrix. This alloy is the workhorse of nickel superalloys, and it is widely used for gas turbine components and cryogenic storage tanks, jet engines, pump bodies and parts, rocket motors and thrust reversers, nuclear fuel element spacers, hot extrusion tooling, and other applications requiring oxidation and corrosion resistance as well as strength at elevated temperatures.

6.1.4 Powder Metallurgy Requirements for EBM

Before attempting to develop parameters using an EBM system, powder must be characterized to determine if it will be a good candidate for the technology. The powder having flowability (it must be able to flow 25 s/50 g like Arcam's supplied powder), high apparent density (>50% of the density of solid material), no internal porosity (from the production process), and no small particles (<0.010 mm) is preferred.

In order to determine the flowability of a powder, B212-09 ASTM International Standards were followed. The powder was allowed to flow freely and unaided through a Hall flowmeter funnel from ACuPowder International (Union, NJ) (shown in Fig. 6.3a), consisting of a calibrated orifice and a cylindrical brass cup with a nominal capacity of 25 cm^3 and a distance of ~25 mm from the bottom of the funnel to the top of the density cup [13]. Apparent density, flow rate, and percent density change were obtained for different powders and compared to Ti64 powder provided by Arcam. High flowability is recommended to improve the hopper's feeding and to obtain a more stable raking—both providing smoother powder layers. If the percent density change of the material to be analyzed is comparable to or higher than the value obtained for Ti64, the microstructure can be analyzed and the particle size distribution can be measured. Powder examination can be performed through the use of a stereomicroscope to identify the shape of the particles, the presence of satellites, any foreign particles, or contamination. Figure 6.3b shows a scanning electron microscope (SEM) micrograph of tantalum powder with a 35% density change

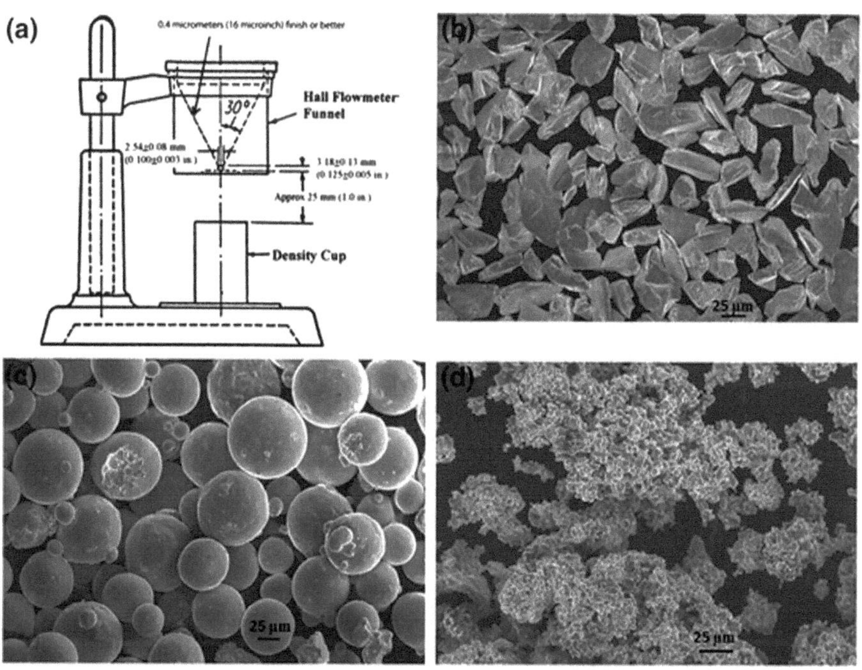

Fig. 6.3 (**a**) Hall flowmeter funnel. (**b**) Tantalum powder. (**c**) Ti64 powder. (**d**) CPTi powder

where angular morphology can be observed. Figure 6.3c depicts Arcam's Ti64 powder with a 55% density change, and Fig. 6.3d illustrates a sponge-like commercially pure titanium (CPTi) powder with a 16% density change. Higher apparent densities will provide better heat conduction and therefore reduce the risk of sample swelling or overheating. A higher value also improves the quality of supports. It can be observed that powders with undesirable morphologies (angular or irregular) have similar flowing capabilities [14].

Another aspect to consider when introducing a new powder to the EBM system is its internal porosity. Internal porosity is represented by spherical voids located along the inside of a single powder particle and can be analyzed by looking at a sample's cross section.

6.2 Powder Manufacturing

6.2.1 Gas Atomization

The process traditionally utilized to obtain spherical powder is gas atomization, which has existed since 1872 and was first patented by Marriot of Huddersfield. Since then, several designs have been used including different versions known as "free fall," "confined" or closed nozzles, and "internal mixing" [15]. This process, as depicted in Fig. 6.4, consists of dispersing liquid metal by a high velocity jet of air, nitrogen, argon, or helium. Many materials can be spheroidized by this process, such as copper and copper alloys, aluminum and its alloys, magnesium, zinc, titanium and its alloys, nickel-based alloys, cobalt-based alloys, lead/tin solder, precious metals, etc. Inert gases must be used when atomizing reactive metals such as titanium and superalloys, to avoid oxidation [16]. Figure 6.4 shows an SEM image obtained from a Ti64 alloy that was gas atomized and purchased for EBM fabrication. In the case of titanium, the metal travels as a falling stream with the aid of gravity for 100–200 mm. Maximizing the gas's velocity and density when it meets the metal stream allows one to achieve particle sizes in the range of 40–60 μm [10]. The schematic depicts a confined gas atomization system.

6.2.2 Induction Plasma Atomization

Induction plasma processes consist of in-flight heating and melting of feed material particles by plasma, followed by solidification under controlled conditions. Depending on the size and apparent density of the treated powder, the time of flight of the particle is controlled such that the molten droplets have sufficient time to complete solidification before reaching the bottom of the reactor chamber (www.tekna.com). The particle morphology and size distribution can be controlled by modifying the nozzle design, particle exiting velocity from the nozzle, and drop solidification distance [17].

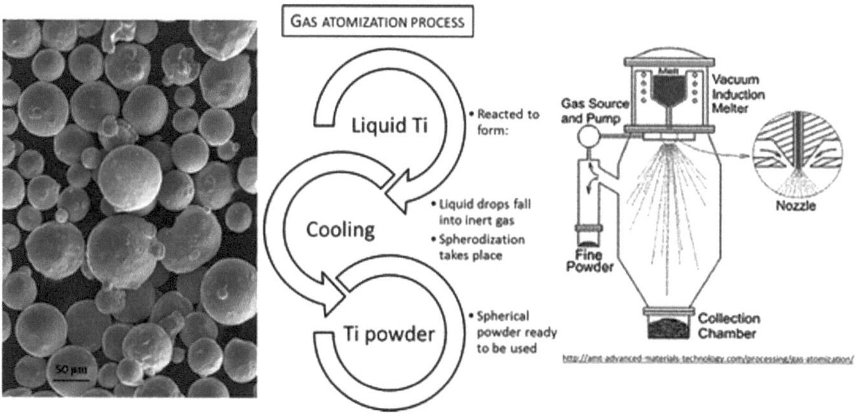

Fig. 6.4 SEM image of Ti64powder obtained from gas atomized powder. Process procedure and diagram inserted next to image

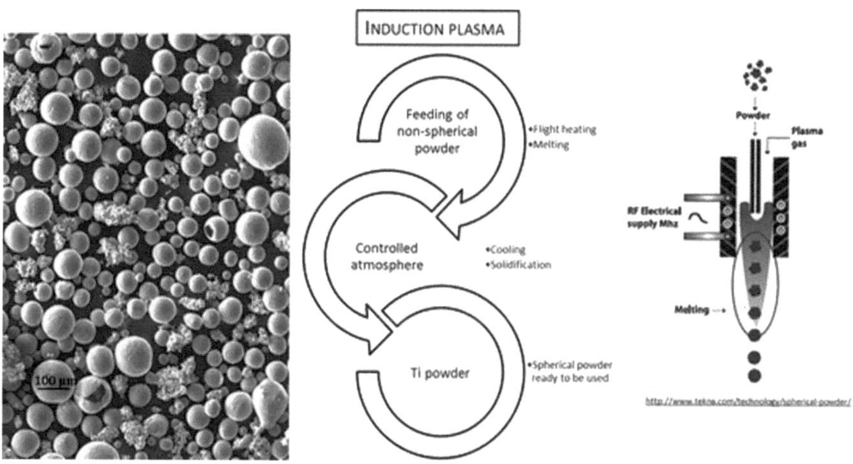

Fig. 6.5 SEM image of Armstrong Ti powder after induction plasma process. Diagram and schematic of the process illustrated next to image

The induction plasma process is used to improve powder characteristics such as increased flowability, decreased porosity, increased powder density, and enhanced powder purity. Flowability is increased when the particles are more spherical, making them easier to flow. Porosity is reduced when the material is remelted and resolidified. Powder tapped density increases since the re-spheroidized particles are more uniform, improving packing. The melting process can also be used to improve powder purity through the selective/reactive vaporization of impurities by increasing the plasma melting temperature [18]. Figure 6.5 shows re-spheroidized International Titanium Powder-Cristal Global (ITP) CPTi powder and the induction plasma process.

6.2.3 Armstrong Process

The Armstrong process provides high purity metal and alloy powders at a lower cost and is particularly well suited for titanium. The main aspect of this process is the elimination of the traditional Kroll process which is responsible for producing the unintentional "sponge" morphology. Alternatively, the Armstrong process starts producing powder to be consolidated subsequently, and therefore, powder manufacturing cost is reduced through the minimization of processing steps [19]. The Armstrong process is owned by ITP and can be used for a wide range of metals, alloys, and ceramics (www.itponline.com). Among the many benefits of this process are high purity, controllable oxygen content, uniform grain structure, and cost reduction [20, 21]. More importantly, ITP powder has been shown to meet ASTM grade 2 and ASTM grade 5 properties for CPTi and Ti64 [22]. Ti and Ti64 powders obtained from the Armstrong process have been utilized in the production of roll compacted sheets and vacuum hot pressed plates with appropriate properties [23]. Figure 6.6 shows an SEM image of the sponge-like titanium obtained via the Armstrong process as well as the sequence of steps necessary to obtain such powder. Titanium tetrachloride (TiCl4) is reacted with sodium to form pure titanium and sodium chloride (NaCl).

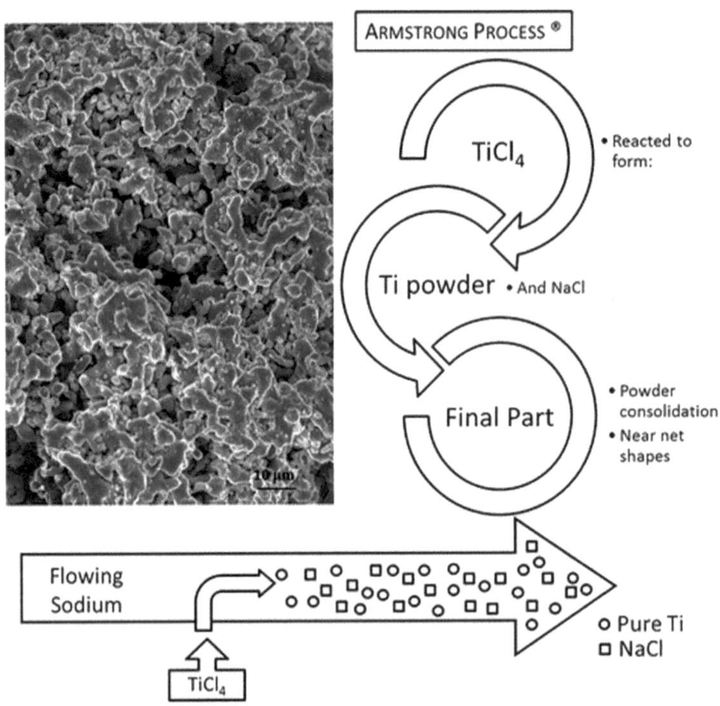

Fig. 6.6 SEM image of pure Ti sponge powder obtained from the Armstrong process. Diagrams illustrating the process are shown adjacent to image

Titanium powder from the Armstrong process is comparable to CPTi ASTM Grade 2 with regard to oxygen and carbon content since both are within the ASTM range. In Armstrong research, Armstrong Ti powder was compacted by a specialized vacuum hot press (VHP) unit, and tensile testing was performed. UTS and YS are higher than typical CPTi Grade 2 due to the all-alpha, fine, and equiaxed grain structure obtained, ~2–4 Å [24].

6.2.4 Hydride–Dehydride

The hydride–dehydride (HDH) process has the capability to produce titanium, zirconium, vanadium, and tantalum powders. The process, as seen in Fig. 6.7, consists of performing a reversible reaction on the material by hydrating titanium at elevated temperatures (~650 °C) for long periods of time, allowing the brittle phase of the titanium to be crushed and milled for finer powder size. The crushed powder is then reheated under high vacuum at ~350 °C to remove the hydrogen [25]. The result is an angular shaped (nonspherical) powder that can be used successfully for fabrication in traditional powder production. The traditional methods for utilizing HDH powder are press sintering, metal injection molding, cold isostatic pressing, and hot isostatic pressing [26].

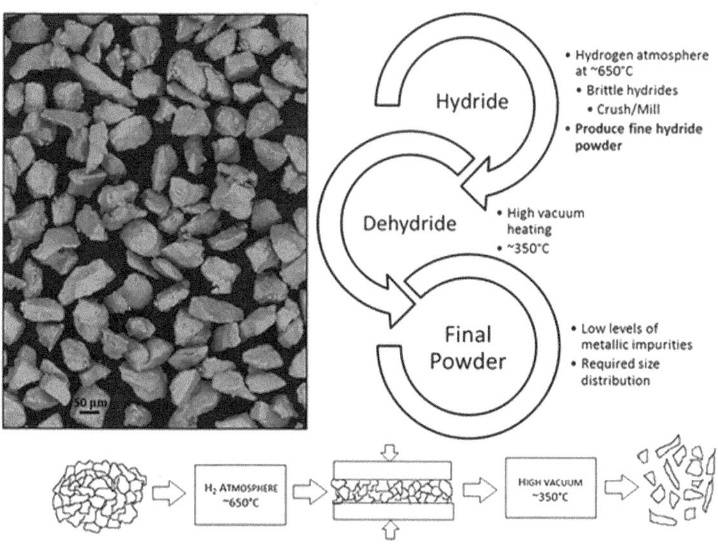

Fig. 6.7 SEM image of pure Ti64 powder obtained from the HDH process. Diagrams illustrating the HDH process are shown adjacent to image

6.3 Powder Characterization

Prior to parameter development using the EBM system, the powder must be characterized to determine if it is a good candidate for the technology. The steps in Fig. 6.8 were followed for initial powder analysis. In order to determine if the powder was safe to handle and use in the EBM system, a minimum ignition energy test was performed following standard BS EN 13821:2002 (Determination of minimum ignition energy of dust/air mixtures) [25]. The test measured the ease of ignition of a dust cloud by electrical and electrostatic discharge. Electrodes were connected to a circuit that produced a spark with a known energy. A cloud of powder was blown past the spark. If the powder ignited, the energy was reduced until there was no reaction. Arcam recommends powder with 45–105 μm particle distribution that have minimum ignition energy of 0.5 J.

The powder must have high flowability (it must be able to flow 25 s/50 g, like Arcam's supplied powder) to be able to rake correctly and flow in the machine hoppers. Following ASTM standard B213, a Hall flow meter was used to test flowability [13]. The powder must also have high apparent density (apparent density must be >50% of the density of solid material) to be able to melt the powder and obtain fully dense parts with current scanning methods. Following standard ASTM B212, a Hall flow meter and a density cup were used to test apparent density [13]. The particle size distribution was verified as Arcam recommends that particle distribution be between 45 and 105 μm with a normal distribution curve. The powder must not contain small particles (<0.010 mm) since these can become a fi and health hazard. Next, the powder was mounted and polished to check for internal porosity (from the gas atomization production process). Internal porosity can become problematic with various materials, like Ti64, that do not allow entrapped gas to flow out of the part as the powder melts and solidifies.

Microstructure analysis of the powder was performed to obtain a basis of comparison before and after the powder was processed with EBM technology. Chemical analysis was also performed to compare the powder before and after melting by the

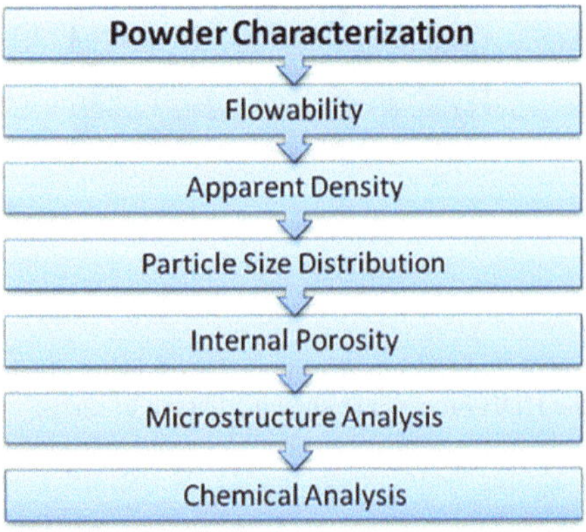

Fig. 6.8 Process flow diagrams showing steps for initial powder analysis

EBM process, since the electron gun's energy can evaporate a percentage of the lower melting element in the material. This can put the material out of specification, making the end part non-compliant. These powder analyses can save a lot of time in determining if the powder will be usable in EBM systems, stopping one from using a powder that has the incorrect packing or flowability density.

An Alloy 718 powder selection study has been done with ARCAM and ORNL. The study involved powders from nine different powder manufacturers using four different manufacturing processes. The morphology of the gas atomized powders to the rotary atomized, plasma atomized, and the plasma rotated electrode.

6.4 Parameter Development

In order to start the material parameter process development (Fig. 6.9), a smoke test was performed in the build chamber with cold powder. Smoke takes place when the charge distribution density exceeds the critical limit of the powder. As a result, an electrical discharge occurs since the powder particles repel each other, causing a powder explosion inside the chamber, as seen in Fig. 6.10. The smoke test consisted of using various machine parameter settings of line order, line offset, beam speed, and focus offset to determine which settings were within the "smoke-free" parameter window. Once the parameter window was identified, the parameters could be modified safely. If the parameters are within the window, the build should be smoke free.

Next, a sintering test was conducted to identify the temperature required to preheat the build platform to sinter powder underneath the platform. The sintering test consisted of preheating a clean build platform inside the machine to a constant temperature and holding it for 20 min. The build was stopped, and the chamber was opened to check if the powder underneath the build platform was sintered. If the powder was not sintered, the test was repeated at a higher temperature. The powder must be sintered for platform stability when raking as well as for thermal conductivity. The advantage of the sintering test is that it provides an indication of the

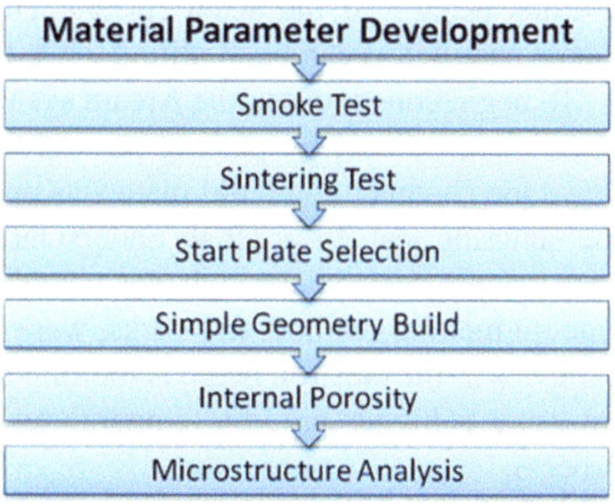

Fig. 6.9 Process flow diagrams showing steps for initial process development

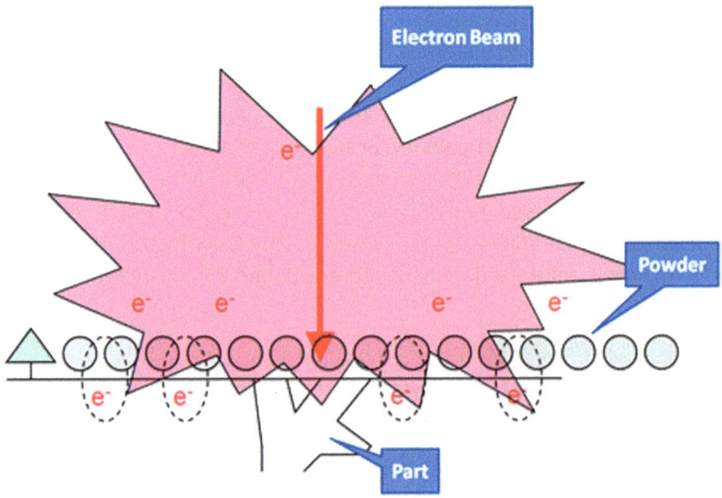

Fig. 6.10 The charge distribution density exceeds the critical limit of the powder, due to the electron beam, and the powder repels each other creating an explosion inside the build chamber

current required for preheating to maintain a stable temperature during the build. The start plate material must be compatible with the new powder material since the start plate holds the powder in place during a build. Stainless steel has been the most successful start plate material as it is compatible with a wide selection of build materials.

Once the correct start temperature and start plate materials were identified, the first process build was prepared. The first build must be a simple geometry, like blocks with no negative surfaces or supports. Initially, the Ti64 themes provided by Arcam can be used as a good starting point. There are three main build themes for every build: the preheat theme, the melt theme, and the support theme. For every build step, there is a theme that provides the parameters to the machine. The initial themes were modified to reflect the findings in the smoke test, so that they were within the parameter window, to eliminate smoke during the build. Next, the preheating theme was modified by taking the average current obtained in the sintering test and using it as the beam current. The advantage of the Arcam system is that theme parameters can be modified while the system is building.

To begin construction, the Arcam system must be prepared. Next, the build start plate was insulated by placing powder underneath and leveled. After vacuuming down the chamber, the build plate was preheated to the temperature determined by the sintering test. In order to understand and control how the system builds, the process was viewed through the chamber window. Parameters that produced a smooth melting, without fireworks, were found. Fireworks occur when the powder explodes as the electron beam is melting it. Fireworks can be eliminated or reduced by using a higher process temperature or increasing the current in the preheat process.

Powder sintering was controlled by the preheating theme parameters: min/max current, focus offset, and number of preheating repetitions. The preheating current controls how much energy is delivered during one cycle run of the heating process. The preheat focus offset controls the electron beam diameter during preheating. The focus offset parameter provides the area where energy is being distributed and can be set to deliver maximum energy without hard-sintering the material. The number of repetitions conveys how many times the build is preheated and is used to raise and stabilize the bed temperature. These parameters control the temperature and stabilize the powder before the melting process begins.

The melt process is when the powder is liquefied layer by layer forming the final part. During this process, the powder surface must look even and smooth to obtain fully dense parts. The melt process was controlled by the melt speed function and focus offset. The speed function value controlled the melting speed in automatic mode. Higher values provide higher speed and deliver less energy when melting. Once again the focus offset was modified to control the melt providing an even and smooth top surface. Focus offset calibration was done by modifying the focus offset in each cube to obtain the best top surface finish, as seen in previously built samples in Fig. 6.11, where the best surface is the one circled in red. The value selected was used as the focus value for the rest of the builds.

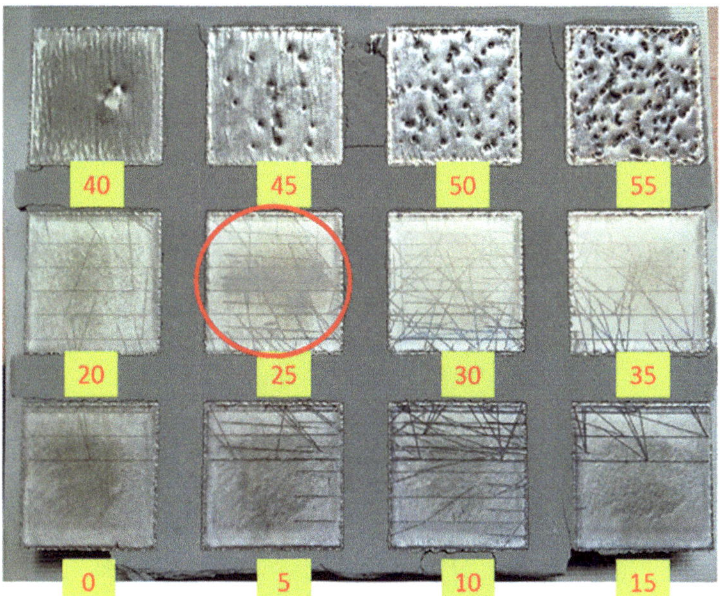

Fig. 6.11 Focus offset test for selecting best focus offset (*circled in red*)

6.5 Build Setup and Process

Once the parameters were determined to be sufficient, a verification build was manufactured using a 150 × 150 mm 316 stainless steel start plate. The verification build (Figs. 6.12 and 6.13) consisted of four rectangular blocks of 100 × 25 mm and 32 cylinders of 10 mm in diameter and 105 mm in height. Additionally, four cubes are also included for chemistry and microstructure analysis. The stainless steel plate was given an initial plate preheat temperature of 1000 °C and a sinter temperature of ≈20 min. Overall, the build time was ≈30 h with a ≈6 h period required for cooldown of the build. For the present build, a layer thickness of 75 μm was chosen as a compromise between build speed and surface quality.

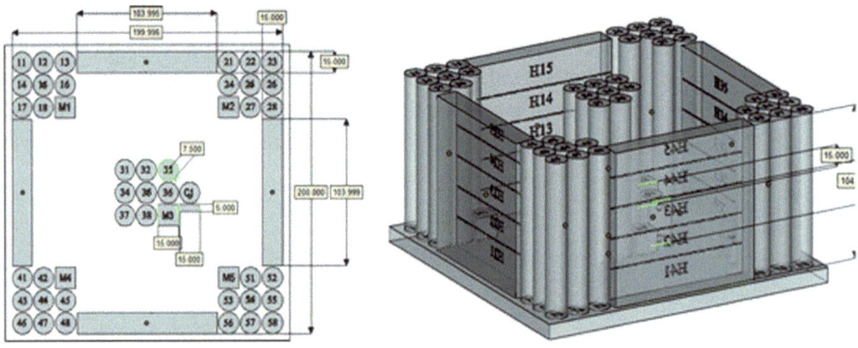

Fig. 6.12 Arcam verification build 1

Fig. 6.13 Verification build 1 made in a 150 × 150 mm start plate

References

1. Gibson I, Rosen DW, Stucker B (2010) Additive manufacturing technologies: rapid prototyping to direct digital manufacturing. Springer, Berlin
2. www.arcam.com
3. Wohlers T (2012) Wohlers report 2012. May. https://lirias.kuleuven.be/handle/123456789/372635
4. Gaytan SM, Murr LE, Medina F, Martinez E, Lopez MI, Wicker RB (2009) Advanced metal powder based manufacturing of complex components by electron beam melting. Mater Technol (Adv Perform Mater) 24(3):180–190. https://doi.org/10.1179/106678509X12475882446133
5. Murr LE, Gaytan SM, Medina F, Lopez H, Martinez E, Machado BI, Hernandez DH et al (2010) Next-generation biomedical implants using additive manufacturing of complex, cellular and functional mesh arrays. Philos Trans R Soc A Math Phys Eng Sci 368(1917):1999–2032. https://doi.org/10.1098/rsta.2010.0010
6. Murr LE, Martinez E, Gaytan SM, Ramirez DA, Machado BI, Shindo PW, Martinez JL et al (2011) Microstructural architecture, microstructures, and mechanical properties for a nickel-base superalloy fabricated by electron beam melting. Metall Mater Trans A 42(11):3491–3508. https://doi.org/10.1007/s11661-011-0748-2
7. Boyer RR, Briggs RD (2005) The use of β titanium alloys in the aerospace industry. J Mater Eng Perform 14(6):681–685. https://doi.org/10.1361/105994905X75448
8. Azevedo CRF, Rodrigues D, Beneduce Neto F (2003) Ti–Al–V powder metallurgy (PM) via the hydrogenation–dehydrogenation (HDH) process. J Alloys Compd 353(1–2):217–227. https://doi.org/10.1016/S0925-8388(02)01297-5
9. Peters M, Kumpfert J, Ward C, Leyens C (2003) Titanium alloys for aerospace applications. Adv Eng Mater 5(6):419–427. https://doi.org/10.1002/adem.200310095
10. Cui C, Hu BM, Zhao L, Liu S (2011) Titanium alloy production technology, market prospects and industry development. Mater Des 32(3):1684–1691. https://doi.org/10.1016/j.matdes.2010.09.011
11. Ivasishin OM, Anokhin VM, Demidik AN, Savvakin DG (2000) Cost-effective blended elemental powder metallurgy of titanium alloys for transportation application. Key Eng Mater 188:55–62. https://doi.org/10.4028/www.scientific.net/KEM.188.55
12. Baufeld B, Brandl E, van der Biest O (2011) Wire based additive layer manufacturing: comparison of microstructure and mechanical properties of Ti–6Al–4V components fabricated by laser-beam deposition and shaped metal deposition. J Mater Process Technol 211(6):1146–1158. https://doi.org/10.1016/j.jmatprotec.2011.01.018
13. Standard ASTM (2007) B212-99 'Standard test method for apparent density of free-flowing metal powders using the hall flowmeter funnel'. Annu Book ASTM Stand 2:23–25
14. Santomaso A, Lazzaro P, Canu P (2003) Powder flowability and density ratios: the impact of granules packing. Chem Eng Sci 58(13):2857–2874
15. Lee PW, Trudel Y, German RM, Ferguson BL, Eisen WB, Mover K, Madan D, Sanderow H, Lampman SR, Davidson GM (1998) Powder metal technologies and applications. Prepared under the direction of the ASM International Handbook Committee. ASM Handb 7:1146
16. Neikov OD, Naboychenko S, Mourachova IB, Gopienko VG, Frishberg IV, Lotsko DV (2009) Handbook of non-ferrous metal powders: technologies and applications. Elsevier, Amsterdam
17. Boulos M (2004) Plasma power can make better powders. Met Powder Rep 59(5):16–21. https://doi.org/10.1016/S0026-0657(04)00153-5
18. www.tekna.com
19. Stone N, Cantin D, Gibson M, Kearney T, Sri L, David R, Wilson R, Yousuff M, Rajakumar R, Rogers K (2009) Implementing the direct powder route for titanium mill product: continuous production of CP sheet. Mater Sci Forum 618–619:139–142. https://doi.org/10.4028/www.scientific.net/MSF.618-619.139

20. Chen W, Yamamoto Y, Peter WH, Gorti SB, Sabau AS, Clark MB, Nunn SD et al (2011) Cold compaction study of armstrong process® Ti–6Al–4V powders. Powder Technol 214(2):194–199. https://doi.org/10.1016/j.powtec.2011.08.007
21. Froes FH (1998) The production of low-cost titanium powders. JOM 50(9):41–43. https://doi.org/10.1007/s11837-998-0413-4
22. Peter WH, Blue CA, Scorey CR, Ernst W, McKernan JM, Kiggans JO, Rivard JDK, Yu C (2007) Non-melt processing of 'low-cost', armstrong titanium and titanium alloy powders. In: Proceedings of the 3rd international conference on light metals technology. Saint-Sauveur, Quebec, pp 24–26
23. Rivard JDK, Blue CA, Harper DC, Kiggans JO, Menchhofer PA, Mayotte JR, Jacobsen L, Kogut D (2005) The thermomechanical processing of titanium and Ti64 thin gage sheet and plate. JOM 57(11):58–61. https://doi.org/10.1007/s11837-005-0029-x
24. Eylon D, Ernst WA, Kramer DP (2013) Ultra-fine titanium microstructure development by rapid hot-compaction of Armstrong-process powder for improved mechanical properties and superplasticity. http://www.itponline.com/docs/Eylon_Plansee_2009_TiPM_Paper.pdf. Accessed 30 April
25. McCracken CG, Motchenbacher C, Barbis DP (2013) Review of titanium. Powder-production methods. Int J Powder Metall 46(5):19–26
26. McCracken CG, Barbis DP, Deeter RC (2011) Key characteristics of hydride-dehydride titanium powder. Powder Metall 54(3):180–183. https://doi.org/10.1179/174329011X13045076771849

Chapter 7
Additive Manufacturing Supply Chain

7.1 Overview of the Current State of Making Metal Parts Using Additive Manufacturing

Everything being said about how additive manufacturing will change supply chains is coming true:

- A new distributed manufacturing capability is being built, allowing us to make parts where they are needed in the world.
- A large ecosystem of partners is being put in place to make additive-manufactured parts:
  - OEM machine manufacturers
  - AM part manufacturers
  - Post-AM processing: heat-treating, machining, etc.
  - Material manufacturers
  - Supply chain services partners—for manufacturing, post-processing, and material services
- We will not need to make and store parts in inventory in anticipation of demand.
- We will be able to send a digital file to a "print location" and have the part "on demand."
- DfAM—Design for Additive Manufacturing allows for the design of whole systems, reducing the number of parts, manufacturing steps, and material usage and allowing for a more robust total cost perspective (vs. the traditional manufacturing cost-per-part mantra).
- The new AM parts supply chains will provide reduced environmental impacts.

K. Hsu et al., *Additive Manufacturing of Metals*, Springer Series in Advanced Manufacturing, https://doi.org/10.1007/978-3-032-04961-2_7

The overall additive manufacturing industry (products and services combined) has grown from ~$5B in 2015 to ~$20B at the end of 2023.[1]

The powder bed fusion and other powdered metal technologies have also grown significantly in the last 5 years from ~1800 in 2017 to ~3800 systems sold in 2023. The metals material market is ~19% of the $3.65 billion materials market, or approximately $717 mm at the end of 2023.[2]

Large part manufacturers have already gone out to their manufacturing supply chain to get parts made. Supply-chain partners have stepped up and made the investment in equipment and expertise to be able to extend the manufacturing supply chain beyond the four walls of the original part manufacturer. Volumes are increasing beyond R&D and prototyping and turning into industrialized parts manufacturing. Industry is driving toward the "industrialization" of AM.

Increased volume drives increased performance needs of the supply chain. However, the industry is at a point today where there are lots of idle machines waiting on orders, material, and other services to produce the parts that are being demanded. In some cases, manufacturers are waiting months for material—to print a part that can be made in hours. Until this is solved, industry growth will be constrained.

7.2 Overview of the Future State of Making Metal Parts Using Additive Manufacturing

In the future, larger volumes of additively manufactured parts and systems will be produced in many more locations. We already see the industrialization of AM in aircraft parts. There are those in the automotive industry talking about moving "millions of pounds of additive materials and parts through a manufacturing plant" to support auto part and system builds https://www.hp.com/us-en/printers/3d-printers/products/metal-jet.html. "New" industries (to additive) like oil and gas (old industries applying new additive capabilities) are looking to build parts and systems using additive manufacturing so they can avoid castings and forgings (and the long lead times). Other industries such as aircraft, industrial manufacturing, and the military are also considering additive to bypass castings and forgings. America Makes has done a number of projects showing the impact of using additive to replace or supplement the casting and forging supply chains. The military is demonstrating printing entire tanks and talking about printing large sections of ships. Customized medical devices and implants are made by the hundreds of thousands (today). The focus of the future will be bringing the technology closer to the physician, called point-of-care (POC) manufacturing.[3] This will drive a more "distributed manufacturing" supply chain. Building construction will be done at the site and

[1] Wohler's Report—2024 3D Printing and Additive Manufacturing State of the Industry, p. 103.

[2] Ibid., p. 114.

[3] Ibid., p. 36.

allow designs that would have been too costly and inefficient with old construction methods. There are companies planning to print on the Moon. https://www.cbsnews.com/news/3d-printing-on-earth-and-moon-60-minutes/.

In the future, large additive manufacturing volumes will be made across many distributed manufacturing sites, and manufacturers will expect their materials to be available "just-in-time." The future AM supply chains will have the same expectations of service and performance that large industrial manufacturing supply chains drive today. Additive manufacturing sites will have a schedule of parts to be made tomorrow and will tell their supply chain partner today what materials need to be delivered to the floor tomorrow. That material, which will be within a few hours (or closer) of the manufacturer, will be delivered according to the production schedule so there is no manufacturing downtime. This is called the "Dell Model" in the "old school" supply chain. It is how high-customization, high-volume manufacturing works—this is where we are headed for additive manufacturing: (https://www.everything-supplychain.com/dell-push-pull-supply-chain-strategy-9-min-video-20k-views/).

However, there is one significant new capability that needs to be added to the additive manufacturing ecosystem. Because we will not be moving parts around the world, as production will take place at the "point of need," we will need to move and position ***materials*** around the world. This new distributed manufacturing supply chain will drive new materials supply chains. These materials supply chains are "greenfield"—they need to be built, as they will be derived based on the distributed manufacturing location demands.

The additive manufacturing industry will be critical to our competitive advantage but is currently constrained by the fact that there are no underlying material supply chains built yet to support its growth properly. In addition, material comes from a very limited number of locations that are challenging from a location and commerce perspective, increasing the risk of these supply chains and material supplies.[4]

This is not a technical problem. This can all be addressed through supply chain strategies.

7.3 Supply Chain Changes Coming as a Result of Additive Manufacturing

Design for additive will drive "part" cost down by reducing the ***number of parts*** that need to be made and assembled. In traditional manufacturing, there are numerous examples of "parts" that require dozens of separate components to be produced and assembled. Additive manufacturing enables the design of "parts" as a single

---

[4] A Comprehensive Risk Assessment and Evaluation Model: Proposing a Risk Priority Continuum; Stanley E. Griffis and Judith M. Whipple; Transportation Journal, Vol. 51, No. 4, 2012 Copyright © 2012 The Pennsylvania State University, University Park, PA. Permission to reference given by: Judith M. Whipple, Interim Dean, Bowersox-Thull Endowed Professor in Logistics and Supply Chain Management, Broad College of Business, Michigan State University, whipple@broad.msu.edu.

"system," and producing a single "part" reduces the ***number of parts*** in the supply chain from hundreds to single digits, thereby reducing the total cost of that system.

A great example is the partnership between General Atomics and Divergent Technologies producing an unmanned vehicle system that has gone from needing to manufacture and assemble many dozens of parts, to additively manufacturing four pieces that are automatically assembled into the entire system. The total cost reduction is represented in Chap. 9.

As was predicted in the first edition of this book, the fundamental strategy of additive manufacturing reducing total cost holds true. In 2017, it was proposed:

> The additive strategy of holding a print file, sending it to a manufacturer, and printing on demand will be a major economic disruptor to those industries that rely on aftermarket revenues. For companies that design, produce, and sell components to military and other industrial organizations, the day is coming when they will only sell a license to print a part. Multi-month delivery schedules will be replaced with printing a part in just a few hours.
>
> When companies are able to email or electronically move CAD files across oceans and time zones and print the components closer to the end user, delivery schedules can be shortened and the costs associated with the delivery of a part will be reduced.
>
> Likewise, companies that normally would inventory higher value (and presumably higher margin) components will see those inventory levels (and those profits) reduced. In the coming future, components will be produced when the customer orders the parts, much closer to the point of need/use. The United States Defense Logistics Agency is beginning to implement this type of strategy across the trillion-dollar defense industry. It is believed other nations, industries, and industrial manufacturers will follow suit with this type of activity.
>
> The movement of goods (components) costs money and time. It should be noted that the real goal of any supply chain is to reduce inventory levels and costs and increase inventory turns. These three things are key elements of any supply chain management effort.
>
> The concept of print-on-demand means that inventories will not be sitting in large buildings waiting for an order. This will reduce costs and help companies manage cash flows and reduce the amount of money tied up in capital (inventory).
>
> The old traditional supply chain model is focused on low costs and high volumes (such as automotive or the appliance industry). Additive manufacturing will create a new industry model with a focus on low volumes, no inventory of parts, and quick profits. This industry can be located anywhere in the world due to most manufacturing elements being global commodities, able to be deployed globally, and influenced very little by labor expenses.[5]

Shifting the inventory burden to the materials (a lower value item in the supply chain) is a classic total cost of supply chain cost reduction strategy. In the future, we will inventory materials (lower value), not parts (higher value), thus reducing the amount of capital tied up in supply chain inventory.

---

[5] Exact text from Edition 1.

7.4 Materials Development

Materials development, especially for powdered metals, is still at a relatively nascent stage in terms of industry maturity. Many part manufacturers are doing a tremendous amount of R&D and experimenting with materials. New materials are and will continue to be developed specifically for additive manufacturing.

The key to supporting the industrialization of additive manufacturing (increasing the volume of parts being made) will be to rapidly move materials up the curve from R&D to industrial manufacturing volumes.

Today, part manufacturers are working with material developers and manufacturers to develop individual, specialized specifications, looking for some type of competitive advantage. While advantages might be uncovered through various material development efforts, this is actually a constraint on the additive manufacturing industry today.

Until part manufacturers commit to certain materials that support industrial volumes, the material manufacturers are stuck in a cycle of make-to-order manufacturing.

7.5 Materials Manufacturing

Today (2025), the industry is stuck in an early/immature market stage that drives most material production to be done on a MTO (Make-to-Order) basis. Material manufacturers are waiting until they get an actual order from a part manufacturer for the material they want.

There are several impacts on the additive market as a result:

First, the material manufacturer is not making a material unless it is "sold" (as evidenced by the PO for the material). Some of the reasoning behind this is the part manufacturer is ordering "special" materials that cannot be sold to anyone else. So, the materials companies are not going to be stuck holding material that is "special" for one particular company if the company doesn't use it.

Second, because it is MTO, every order generally goes to the "back of the line" in terms of the production schedule. When the volumes aren't particularly high compared to the capacity to manufacture, it doesn't matter too much. But now that the volumes are starting to get to industrialization levels (beyond R&D and testing), these small, frequent orders continue to simply be "put in line" by the material manufacturers—resulting in longer and longer lead times for getting material.

What do long lead times mean? Machine downtime and delays for those wanting to make parts. We now have industrial-sized machines, capable of printing industrial volumes (for additive), sitting idle waiting for material. Why? Because manufacturers are not planning their material demands, not using purchasing vehicles that assure ordering and payment, and not coalescing around "standard" materials;

and material manufacturers are reluctant to invest in capital to produce more material.

The future atomization capacity investment that will be needed to grow the entire industry could be staved off for a little while, if part manufacturers helped the material manufacturers be more efficient. How? By ordering in volumes (even of their "special" material) that make it enticing for the material manufacturer to move up that order → more volume = more revenue = increased manufacturing efficiency. This should also bring the cost per pound/kg of this material down for the part manufacturers, because it is lowering manufacturing costs and bringing revenue in sooner for the material manufacturers.

However, ordering, producing, and shipping in higher volume likely means the end of shipping directly to the customers in small quantities from the point of manufacturer. A more mature supply chain structure needs to be put in place. Putting more mature supply chains in place that support a more mature manufacturing stance will help part manufacturers operate more efficiently and get value from their very expensive pieces of capital, and material manufacturers optimize atomization capacities. The goal is to keep part-making machines running, by delivering material to the plant floor every day based on what is needed to support that day's manufacturing schedule (called JIT, Just-in-Time, in the supply chain world). JIT doesn't happen with small quantities shipping across the globe direct-to-customers months after the material was ordered.

Many of the large manufacturing companies (materials and part manufacturers alike) understand this. They know how the supply chains work for their large traditional manufacturing operations. They have sophisticated, dialed-in, efficient manufacturing operations and supply chains to support them.

But the supply chains for additive manufacturing are going to look different because of AM's distributed manufacturing. AM is going to drive having lots of manufacturing locations at the point of need for a part. Thus, material is going to need to be delivered to each of these points of manufacture. That is different than today's large, high-volume manufacturing locations that produce millions of parts in a single location, and then ship those parts, with material being delivered to that single manufacturing location.

Picture what supply chain professionals call the "food cold chain" that supports restaurants and institutional food preparation locations. These supply chains are designed around a highly distributed manufacturing situation with materials that are specific, specialized, and have special storage and handling requirements. This is a supply chain with lots of points of manufacture (every restaurant and industrial food location such as offices, schools, etc.). They "manufacture" at the point of need. They produce a lot of different items (parts) with lots of different ingredients, every day.

If you think about all the individual restaurants or institutions as individual, highly distributed "points of manufacture," they individually might not consume and produce that much volume on a daily basis. Also, their "materials" (food/

ingredients) are typically somewhat unique, might change on a daily basis, and require specialized storage and handling requirements (temperature, cleanliness, separate storage, etc.). The individual "manufacturing sites" also don't want to store a lot of materials on-site (it is expensive and takes up space). So, they receive multiple deliveries a week, from smaller, distributed, local storage locations, containing just enough materials needed to "manufacture" their unique items at the point of manufacture over the next couple of days. Now think of all of those individual locations grossed up to geographic volumes (a major city, or Southern California, for instance). All of a sudden, the collective volume of what needs to be stored and delivered (with special storage and handling requirements), is at a point where each of the individual restaurants (point of manufacture) doesn't want to manage or own that supply chain. And, their suppliers certainly don't want to be shipping direct from their point of manufacture (which could be hundreds or thousands of miles away).

Additive manufacturing for much of its future volume is going to look and behave this way. The problem is the supply chain to support it hasn't been built yet.

The additive industry is still at the "MTO" (make-to-order) stage of materials ordering and delivery. Every single part manufacturer is putting in an individual order for their "unique" materials/ingredients, directly with the material manufacturer. This is all well and good if the volumes involved are small and can be worked into the overall material manufacturer manufacturing schedule on an MTO basis. You might have to wait a few weeks for your material order.

However, additive manufacturing volumes are increasing—which means at some point, the economics of the MTO model for materials doesn't work anymore—because customers are now waiting months for their materials, because material manufacturers don't have the capacity for all the small quantity orders. The number of small shipments is becoming expensive. The increased lead times and the associated AM machine downtime are putting economic pressure on the current supply chain and constraining the growth of additive manufacturing as an industry. The problem of a 6+-month lead time on materials is a supply chain problem and can be solved with supply chain business models and practices.

But because the AM supply chain is distributed, the economics don't support the materials simply "piggy-backing" on the current materials and manufacturing supply chains. Being a distributed manufacturing supply chain, it is literally physically different, which drives different economics and supply chain solutions. Thus, the need to design and build the needed green-field supply chains that will support the continued expansion of additive manufacturing as an industry.

So, let's get more detailed about what is happening today. How are materials flowing today? What is the physical flow? How are orders and shipments taking place? What are companies driving in terms of material development and specifications?

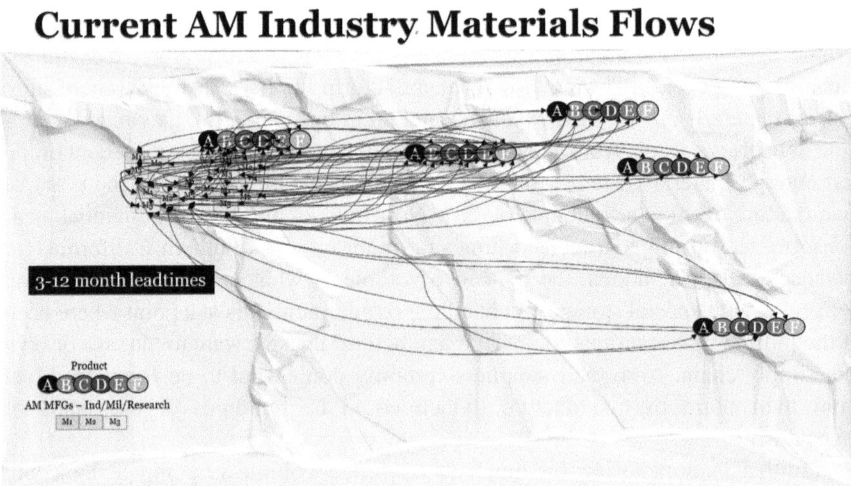

In 2025, most materials flow along single "point-to-point" flows. A single company orders a small amount of material (of their "special" material), and sends that order to the material manufacturer; the material is produced, and then shipped directly to the company that ordered the material. This happens multiple times a month or year.

A diverse set of unique material specifications, plus relatively low-volume production requirements, has driven material suppliers in the supply chain to organize their operations for first-in-first-out (FIFO) make-to-order (MTO) service. This results in unintended small-lot production, escalating material prices and lengthening lead times, reducing overall supply assurance, and jeopardizing research and production schedules.

This causes the material manufacturers to wait until they have an order in hand to produce material. When they have unused atomization capacity, this does not cause much of a problem for them or the part manufacturer waiting for the material. The order gets scheduled, within a presumably agreeable amount of time, the material gets manufactured, and is shipped directly to the point of use.

However, with the increasing volume of material demand, atomization capacity has shrunk, and the material manufacturer can no longer simply "fit in" an order. That Make-to-Order (MTO) for that special material now goes to the "end of the line" to get made—now several months away as their manufacturing capacity is used up with lots of other orders ahead of it.

7.6 Future State: Building the Extended Ecosystem Supply Chain

In the future, as part-manufacturing volumes increase in the distributed manufacturing supply chains that are being built, an extended ecosystem of supply chain partners will also need to further develop:

- Machine OEMs
- Material manufacturers
- Post-processing partners for finishing, machining, etc.
- Machine/printing part and consumables partners (gases, build plates, machine repair parts and services, etc.)
- Supply chain and logistics services to handle aggregation of materials, parts, supplies, etc.

Speed will need to increase. The speed of being able to send a part file, the speed of building the finished part, and the speed of having material available, so that expensive OEM equipment is not sitting idle and customers are not waiting for parts, will be critical to the growth of additive manufacturing. Time is money.

This means manufacturing and material planning processes need to be set up to enable the procurement activities required to ensure the machines can run and materials are available.

Many additive parts will still require post-build processing. It doesn't help if the distributed manufacturing of an additive part means the part now has to be moved to multiple locations to be finished, especially if the movements require going long distances. That puts the industry back to moving parts again.

The new distributed manufacturing supply chain for additive manufacturing in North America means the post-processing ecosystem also needs to follow the new supply chain. Post-processing services will need to be set up next to or in concentrated locations close to or within geographic centers of additive manufacturing. These centers are already becoming clear:

- Southwest US (Phoenix–Southern California)
- Central Texas and mid-South
- Mid-Atlantic states
- Northeast/New England
- Midwest US
- Northwest US

At a minimum, in order to speed part completion, ecosystem partners are going to need to be within a day's transit of the point of additive manufacture (preferably less). Once you get beyond 2 days, you add tremendous logistics cost and part lead time. It would be preferable for all operations to be "side-by-side," or just a few minutes down the road. You want to minimize the number of times and distance a part is moving before completion.

This does not mean all operations and equipment need to be owned and/or operated by the part manufacturer. Ecosystem needs lend themselves well to partnering with small companies looking to provide these services. They can even be co-located with the part manufacturer.

New materials will certainly continue to be developed specifically for additive. The question will be how quickly the industry can coalesce around material specifications. The purpose of this is to drive material manufacturing volume and movement efficiencies. Every time a different material has to be made by the manufacturer, it requires extra capacity—either through dedicated machines and processes, or through high manufacturing turns. This increases manufacturing cost and cost of the materials.

To lower cost and decrease material lead times, the sooner the industry can start ordering, manufacturing, and shipping in volume, the sooner material will be available in quantities needed to industrialize additive metal parts.

To do this, part manufacturers need to be ordering material in higher volumes—"We need to be ordering 10,000kg of material, not 1,000kg of material!" as one manufacturing engineer stated at an industry conference. They are right! Material volumes need to be aggregated and planned across multiple users in geographies so that material movement costs can be reduced, and manufacturing lead times are reduced.

By aggregating demand from AM manufacturers, higher volume production requirements and larger lot production for material manufacturers lower material manufacturing costs (and presumably prices). It will also shorten material lead times, improving overall material availability and production schedules.

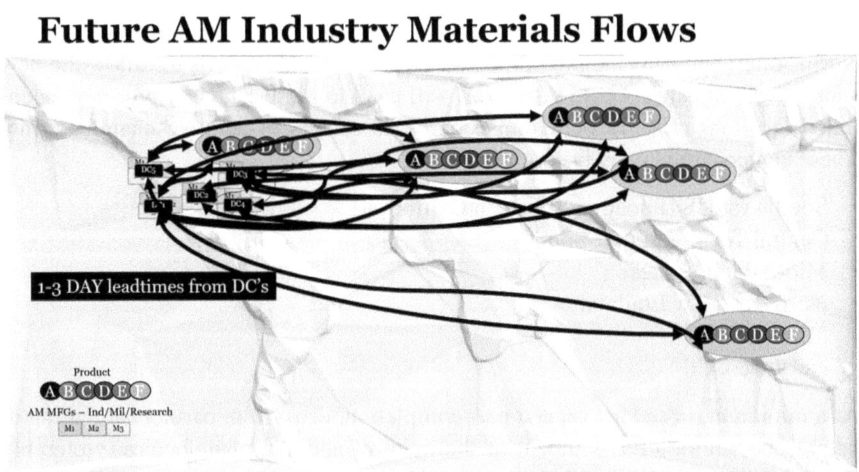

7.7 Reducing Material Variability

In addition to improving manufacturing efficiency and reducing cost (material cost, machine downtime, etc.), supply chain solutions can reduce material variability. Changing the way we procure and transport materials will help ensure material quality in the following ways:

- Increasing the volume of batches made and shipping material in single batches and single shipments reduces material variability
- Reducing the number of batches of powder produced reduces variability

This also helps reduce material manufacturing costs for the material providers at the same time.

Shifting from the make-to-order small volumes shipped directly to make-to-stock operations—ordering in quantities, shipping and storing in volume, and having materials available when needed—reduces variability and cost. The lower manufacturing cost, lower transportation/shipment cost, fewer material quality issues, and better part manufacturing efficiency (lower cost) will mean an actual lower total manufacturing and supply chain of the additive part/system.

7.8 Green-Field Supply Chain Development

As manufacturing and material volumes continue to increase, the ecosystem supply chain will continue to evolve over the next decade or two. More and more distributed manufacturing locations will drive materials and services supply chains to set up support for the many locations.

Also, the manufacture of materials will continue to develop. It is anticipated that the US will eventually bring the atomization of metals, and more production of metals, to the US.

Did you know that if you combined all the Ti718 volume consumed in 2024 in California alone, the entire North American atomization capacity cannot currently keep up? The expansion of metal additive manufacturing is actually constrained by the material manufacturing bottleneck of atomization. That investment will come when part manufacturers mature from the MTO to MTS order behaviors. This is why we will need to continue to build the green-field supply chains needed to move and position those increased volumes of materials coming in from foreign locations (Europe and Asia mostly) and eventually manufactured in the United States.

In order to produce additive "parts" effectively and efficiently, material needs to be available to the production floor in a matter of hours, or at most a few days. This is why material supply chains are coming.

There is also the question of vertical integration. Who is going to decide to invest in the atomization capacity that will be needed? Today's material companies?

Today's additive part manufacturers? There is an interesting long-term play here for who will own the capacity, and therefore the materials markets.

In the future, additive manufacturers will expect to plan parts manufacturing and keep their additive machines running because materials have also been planned and are positioned to be delivered "as needed" with very little lead time. Part manufacturers will also not want to use up valuable manufacturing capacity space (capital) storing large quantities of materials. They will let their supply chain partners know what materials are needed tomorrow and expect the materials to be delivered to the manufacturing floor—tomorrow.

7.9 Sustainability

In addition, the fact green-field supply chains need to be built provides the advantage of building in sustainability from the very beginning, especially in the materials and consumables reuse/recycle/upcycle space. One of the first things that needs to be built in the ecosystem is the partnerships to drive the recovery, recycle, and reuse of materials. The supply chain partners that are delivering "virgin" materials in the first place also need to partner with the companies that are looking to reclaim/reuse used materials. Some of these partners will be outside the additive manufacturing space.

Additive manufacturers are already demonstrating how AM can be a more sustainable manufacturing practice. EOS, a machine OEM, has a framework for determining if making a part via AM is a more sustainable and lower-impact solution than traditional manufacturing. Additive manufacturers are also already asking for recycling and reuse solutions, especially related to used materials. The volume of parts, and therefore the volume of materials needing to be recycled, resold, or reused for other applications, is increasing. The question is, what partners in the supply chains, still to be developed, will provide these services?

- https://www.weforum.org/agenda/2023/06/how-3d-printing-helps-accelerate-sustainable-manufacturing/
- https://www.eos.info/06_corporate-and-hr/corporate/corporate-social-responsibility/eos_sustainability_report_2022.pdf

7.10 In Summary

The development of new, green-field supply chains supporting the additive manufacturing industry is going to support industrialized manufacture of parts. Additive will be both a supplement and a replacement of traditional manufacturing depending on the application and the part design. Additive is going to drive the development and manufacture of new materials. Partners will work on positioning those materials around the world to be consumed "Just-in-Time." The development of the services ecosystem will be needed around the world, likely consolidated in additive manufacturing regions, to support making a finished part.

Industry growth will continue to be constrained until these additional supply/value chain models are put in place. Additive manufacturing has a true opportunity to build with the future in mind. Additive brings an opportunity for a total cost perspective in manufacturing—a lower total cost. New additive supply chains provide the possibility of sustainability being built-in to the business models. Additive will drive a lower impact on the environment, thus adding to its total cost argument over the long term.

Chapter 8
Supplier and Machine Qualification

Abstract Laser powder-bed fusion can allow for a faster and more efficient supply of aviation parts. L-PFMs must adhere to the high standards of airworthiness held by the FAA and EASA. Regulatory agencies have adopted a performance-based regulation of additive components. Components created through additive manufacturing must *meet or exceed* the requirements of conventionally made parts.

It can be an expensive and time-intensive process, costing upwards of hundreds of thousands of dollars and several years to complete.

Additive manufacturing can reduce this cost and shorten this schedule by providing game-changing solutions for standardization and qualification. When a company qualifies a component for airworthiness, they must qualify "the process" as well as "the machine" along with "the part."

The medical industry introduced a process of quality assurance referred to as installation qualification (IQ), operational qualification (OQ), and performance qualification (PQ). To explain these three efforts, please refer to the following:

1. Installation qualification is used to ensure that all equipment are installed correctly.
2. Operational qualification is the process of ensuring that the equipment and processes are operating within specified limits.
3. Performance qualification is validating the production process as designed.

Some aerospace companies have begun using IQ, OQ, and PQ nomenclature, but mostly, the aerospace markets use a fixed process/first article inspection approach.

The intent of this chapter is to guide the reader through the steps needed to meet the correct standards for a 3D printing process and machine.

K. Hsu et al., *Additive Manufacturing of Metals*, Springer Series in Advanced Manufacturing, https://doi.org/10.1007/978-3-032-04961-2_8

8.1 An Introduction to Airworthiness

The aviation market (defense, aerospace, and commercial) is a risk-averse industry. Extreme care and diligence must always be applied to the manufacturing of any component, regardless of how it is produced, when the safety of people (and particularly the public) is involved.

Every aerospace company employs people who work in the area of flight safety (FS). The additive manufacturing team's job is to use empirical data or what might be referred to as "performance-based data" to show the FS team that the printing process produces components that meet the aforementioned standards. They must also show that the process produces repeatable and predictable builds over long periods. Performance-based data over time is what regulatory agencies are looking for.

8.1.1 Qualifying a Supplier

In the process of supplier qualification, measurables are more than just the technology associated with 3D printing of the component. Supplier development organizations want to ensure that the company selected is managed properly in the areas of accounting and finance as well as engineering standards.

8.1.1.1 General

Business areas that should be reviewed when qualifying a vender will include but not limited to (1) finance and cost management, (2) continuous improvement, and (3) health and safety.

The manufacturing areas to be audited are going to be but not limited to (1) additive manufacturing, (2) experience with alloy to be printed, (3) suppler base (particularly with powder source and heat treat), and (4) statistical process control.

If a company is going to 3D print components, it makes sense to first ask what experience that company has in printing, and not just printing but also what alloy (powder) they are printing and with what machines. When the need involves an aluminum alloy, a company with 3–5 years' experience in printing with nickel alloys is not as competent as a company with 3–5 years with aluminum. When printing with aluminum, how a machine is cleaned and how the build room is cared for with regard to humidity can be totally different from a room where nickel alloys are stored and printed.

It is not just the years of experience a small company might have but also the type of experience that needs to be considered. It is important to stratify a company with its experience with complex versus simple geometries. Can the company print thin walls as well as thicker walls?

In addition, one needs to examine the process controls associated with additive manufacturing. Topics to address should be statistical process control, how does the company perform daily production meetings, and does the company have a process for noncompliance of parts, to name a few.

Just because a company can print with one alloy does not mean that the same company can print with any alloy. The quality controls associated with aluminum are different than when printing with nickel or titanium. For this reason, a vendor should be reviewed based on alloy.

8.1.1.2 Finance

Financial security and cost management are critical. No company can afford to have a key supplier of critical technology go out of business because of poor financial management. The supplier must make financial planning and cost management of critical importance.

8.1.1.3 3D Printing Experience

The number of years of experience a company has will dictate the type of work that the company will be allowed to bid on. Considerations will also focus on whether the parts are complex or simple and if the components have thin walls or thick walls. Maybe most importantly, the supplier will be evaluated on how well and to what detail the company implements additive manufacturing process controls.

8.1.1.4 Materials

What material technology is located in-house or in the supplier's supply chain is going to be critical. A 3D supplier needs to know how to order, store, and inspect the powder used in an L-PBF machine.

If a supplier can assist in helping their customer produce and maintain or grow the material property database, then that supplier will become a long-term partner.

8.2 Qualifying a Supplier's Machine and Process

Before a printed component is placed on an engine, the printing process must be fixed, and all known process variations must be removed. In essence, the printing process, which includes the L-PBF machine itself, must be fixed while process parameters cannot change. All manufacturing process variables must be understood and, if possible, removed. When qualifying a machine, it is important to visualize the entire manufacturing process that needs to be qualified. This involves powder,

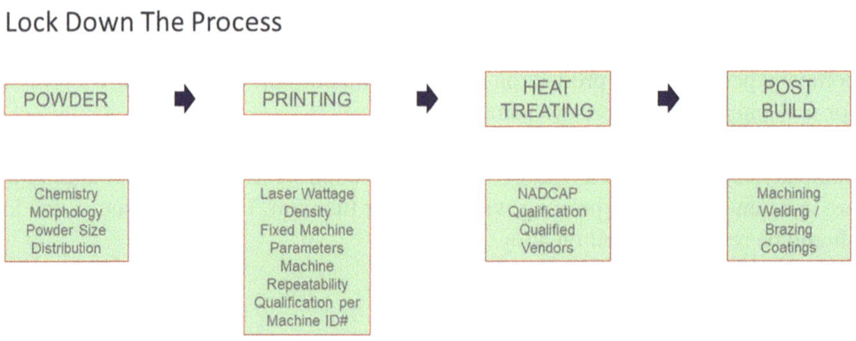

Fig. 8.1 The additive manufacturing process

printing, heat treating, and machining. From the perspective of the flight safety engineer, heat treating and machining are part of the 3D manufacturing process (Fig. 8.1).

8.3 Selecting the Correct Powder Supplier Is an Important First Step

It is essential to select the right powder company to deliver consistent chemistry, morphology, and powder size distribution (PSD). It is especially important to choose a company that employs "air stratification/classification" technology to remove the powder below 8 μm in size. These finer particles of powder inhibit the flowability of the powder, which negatively impacts the machine to properly spread the feedstock across the build plate.

It is important to work with powder companies to ensure that customers receive the powder that works best in each machine.

Powder quality is more than just controlling the fines. One also needs to measure powder size distribution (which can be controlled in large part by sieving) and morphology (powder shape). Powder size (granulometry) and powder chemistry are controlled in the atomization process. The manufacturing process associated with powder atomization has inherent variability. PSD, morphology, powder size, and chemistry can drift over time, so companies must track the quality of powder by inspecting the items mentioned above, so the voice of the process can be captured. Procuring powder from multiple sources makes controlling these variables more difficult.

8.4 Steps Related to Machine Qualification

In the aerospace marketplace, a qualified machine is viewed as an independent entity identified and controlled by a machine serial number. When a machine becomes eligible for production, the engineer cannot move a part to another (similar) machine unless that second machine is also qualified for that alloy. As a result, two machine serial numbers must be qualified for production.

Typically, once a machine and a process have been qualified, a company can place any part on a qualified machine as long as the alloy does not change.

Qualification of a machine begins with producing test specimens so that "performance-based data" can be generated that shows that the machine can produce components that meet design intent. The material properties used as a target must come from the design engineering community. The goal is to create specimens that meet design intent from the perspective of material strengths.

Next, the machine produces a component that is later inspected for validation. The machine can be used to create a component that meets design intent from the perspective of component geometry.

The machine must show the ability to produce the same material property results more than once. Thus, two builds are required for machine qualification.

8.5 Steps Related to Qualifying the L-PBF Printing Process

Qualifying an L-PBF process involves repeatability and currencies. It must be proven via performance-based data that the machine produces the same material properties over repeated builds. That kind of data is only validated through testing, some of which is very expensive.

Printing acceptability begins with the design engineering community establishing material property targets. These targets are a product of the function of the component. Material targets are established, so the component will perform its function when printed and placed into service.

When printing already-designed components, one knows the requirements the machine must produce. If the designer wants to design a new component, they must first understand what material property values are achievable with the new additive manufacturing process.

Designers must have access to material property data involving tensile, LCF, HCF, or creep testing data at various temperature ranges so they can design for success. Through repeated printing and testing, designers will only know what is achievable with printing technology. Designers must have this material property data so components can be designed without fear of failure.

Customers, regulatory agencies, and insurance organizations require performance-based data showing how printed components meet design requirements. As a result, the company implementing 3D printing must work with customers and government agencies to print and test specimens printed across the build plate (X–Y) and the entire height (Z).

Companies should not hesitate to use multiple lasers when printing components. There is now enough data in the marketplace to show with empirical data that material properties do not change when using multiple lasers as long as the machines are calibrated and set up properly. Nonetheless, when one qualifies a machine or a process, the multi-laser technology must *also* be qualified.

The material data requirements for specimens printed with two lasers are the same as those printed with one laser. The machine-process qualification must produce consistent material properties across all seven areas of the build plant.

Simply printing a plate full of specimens is not enough to qualify a machine. There must exist performance-based data to show that the machine and the process can produce the results more than once. Therefore, there must always be at least two builds for machine qualification.

8.6 Qualification Involves Many Experts

Testing machines for acceptability and repeatability is a crucial step of the process qualification.

As stated earlier, the design engineer establishes the material data required. The function and requirements of the component determine the material property requirements. However, it is a team of experts that determines how the process will be qualified and approved. This team can be made up of people working in the fields of structural engineering, quality engineering, manufacturing engineering, and materials engineering.

This team agrees not only on the material property targets but also on how the printed component will be qualified. They determine what technologies and what criteria will be used for component acceptability. Such technologies might include X-rays, computed tomography (CT) scan, fluorescent penetrant inspection (FPI), and roughness (scale).

When setting acceptance limits, some engineers might want to increase the component requirements because they view additive manufacturing as new or novel. This is opening the door for technology rejection. If the current design is a *casting*, then the acceptance requirements should be baselined using *casting requirements*. If the current design is a *forging*, then the requirements should be baselined using *forging requirements*. Try not to require a printed part to meet forging material requirements if the design only requires the strength of a casting. The requirement of the design should determine the acceptance of the printed components.

Each test typically has its own geometry. The ASTM organization is an excellent source for identifying a test specimen shape. Most North American corporations use an ASTM E8 specimen for tensile and LCF testing. This dog bone-shaped specimen can be printed "as built" or as a printed cylinder that is then machined to shape.

8.7 Laser Overlap (Stitching) Area Qualification

Specimen test data must be captured across at least two builds, and a statistically acceptable sample size must come from each build area.

This type of testing and comparison is critical for airworthiness when large components are printed and use more than one laser.

8.8 Building the Material Database

The Battelle Institute is responsible for the Metallic Materials Properties Development Standardization (MMPDS) Handbook, which shows how the A/B basis can be generated. The MMPDS Handbook is recognized throughout the world, and it represents the primary source of statistically based design allowables for metallic materials.

To meet the MMPDS requirement for additive, ten separate builds must be printed. Do not start an MMPDS program without capturing a budget of at least $1,000,000. Some companies realize that their printing of an alloy will continue over a period of years, meaning that this effort can be budgeted across several fiscal-year budgets. Companies printing for internal use may (in the short term) bypass the MMPDS requirement, knowing that the material database will eventually be created.

8.9 Qualification of the Part

Qualification of the machine and of the process is a necessary step. At the end of this process, one still has to qualify the part. This is typically performed by printing test specimens inside or onto a part. Unfortunately, this means that the printed component will have to be destroyed. Typically, one prints specimens next to the printed component, a practice that continues during production. Mechanical properties of these specimens must match the material properties in the process database within an agreed level of tolerance.

8.10 Assistance

Qualifying a machine or a process can be daunting if never attempted before. NASA has two specifications available on the web and accessible to anyone who wants to download them. Simply search for NASA-STD-6030 Additive Manufacturing Requirements for Spaceflight Systems and NASA-STD-6033 Additive Manufacturing Requirements for Equipment and Facility Control. These two standards will help remove some of the confusion and stress engineers and managers might have on this topic.

8.11 Conclusion

When taking an L-PBF process into production, a company will have to qualify the machine, the manufacturing process, and the printed component.

Qualification must be based on a performance-based approach where the printing process, machine, and printed part are shown to produce and to meet design requirements.

Qualification of additive manufacturing machines for airworthiness cannot be separated from the qualification of the process for airworthiness. The process needs to be carefully planned and budgeted. MMPDS creation is expensive.

Chapter 9
Metal AM Applications and Industrialization

9.1 Industry-Specific Requirements

9.1.1 Keeping Perspective on History

Throughout human history, our engineering and science achievements have been tightly paced by the manufacturing technologies at our disposal. Additive manufacturing represents a very powerful tool in the "manufacturing toolbox" to propel our innovative minds and designs. It is however just another tool in the toolbox.

When looking at how metal AM can grow and how it can be implemented, it is always useful to keep perspective and look at previous manufacturing technology emergences throughout history (Fig. 9.1). One important realization is that AM is not new anymore and should be treated as such. Since 1984, with the invention of the stereolithography modality, it unleashed a process of engineering innovation around the basic principle of 3D printing (layer-by-layer material deposition) by machine, materials, software manufacturers, and the academic/research community. This has expanded the AM industry landscape components that give commoditized options to the end users today (Fig. 9.2).

This chapter will aim at providing insights on how users can implement and benefit from metal AM applications in their respective field of use through the establishment of an AM industrial ecosystem and implementation strategy. The reader will likely recognize some patterns that have been the hallmark of previous novel technologies (e.g., composite laminations in the 1970s–1980s) in terms of approach to capture value proposition and industrial elements needed for adoption.

K. Hsu et al., *Additive Manufacturing of Metals*, Springer Series in Advanced Manufacturing, https://doi.org/10.1007/978-3-032-04961-2_9

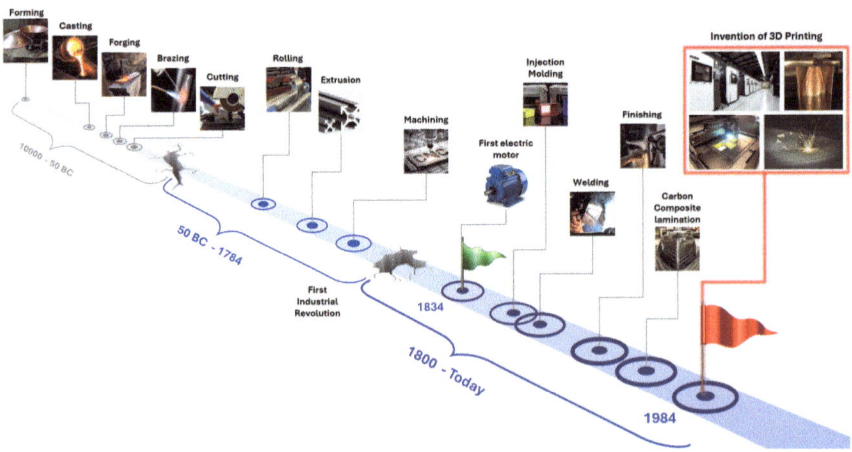

Fig. 9.1 Visual representation of the historical timeline of manufacturing technologies

9.1.2 Industry-Specific Drivers for AM

Metal AM comes in different forms, and the modalities are still expanding and developing [1]. As seen in previous chapters, some metal AM modalities offer various:

- Overall print volume/size
- Feature sizes
- Surface finishes
- Material properties/performance
- Deposition rates
- Equipment architecture
- Complexity of post-processing
- Process throughput capacity
- Cost of manufacturing
- Data connectivity

Depending on the industry, an informed user will use those attributes to determine the value proposition of metal AM. However, for the most part, metal AM remains an expensive manufacturing process and should generally not be considered for prototyping purposes without seriously considering the cost implications and ensuring that the AM benefits are well identified ahead of time. Table 9.1 highlights some of the key drivers of adoptions which can be specific to each industry sector.

Another major influencer when moving from "prototyping" to "production" is the certification/qualification requirements of each industry. Some are lighter than others, and some are more established than others as well. Luckily, some of the metal AM standards are now more common through organizations such as ISO [2], ASTM [3], SAE [4], CEN [5], MIL [6], NASA [7], and NIST [8].

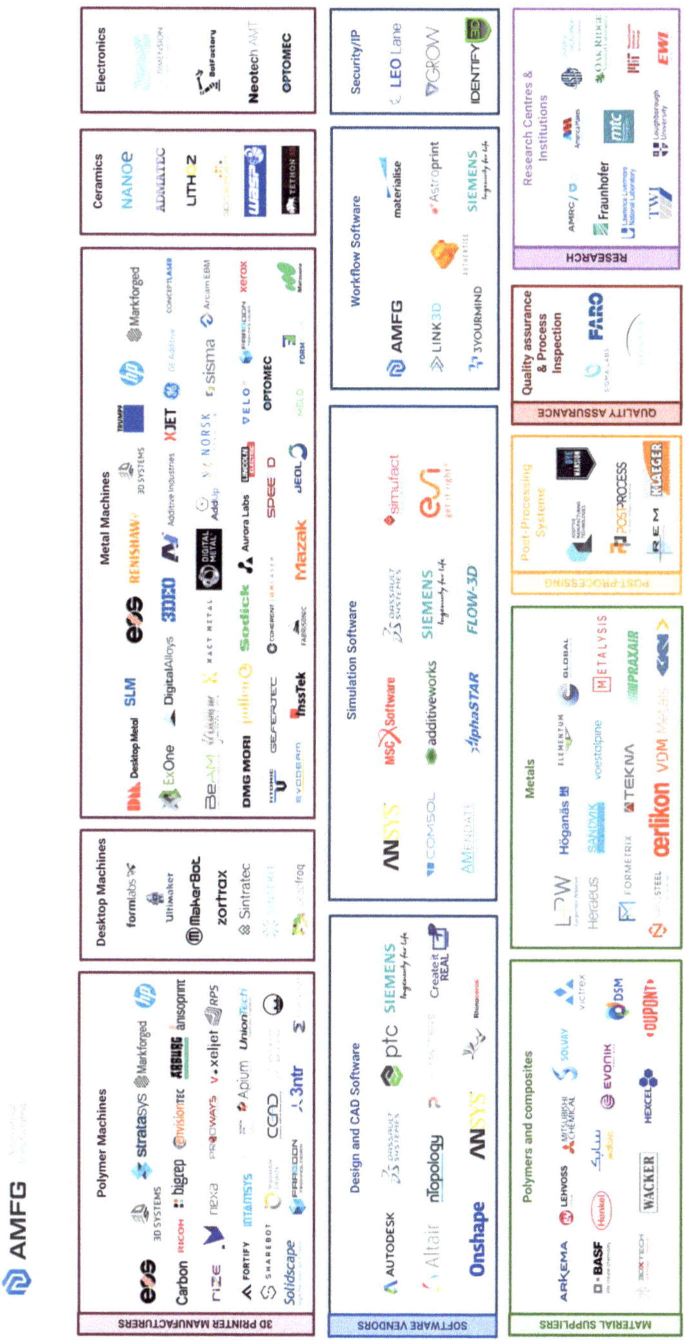

Fig. 9.2 Credits: AMFG. Example of an industry landscape map focused mainly on the manufacturers of AM equipment and related solutions. Such landscape mapping methodology is often used for investment purposes and offers a baseline framework to establish a manufacturing technology-based implementation ecosystem

Table 9.1 Examples of favorable AM adoption factors for key industry segments

Market segment	End-use AM production adoption catalyst factors
 Space	- High cost components offering opportunities to lower cost of manufacturing (Mfg). - Huge opportunity for lightweight solution-related benefits - Uniquely enabled performance improvement through AM such as part consolidation, enhanced functionalities and materials - Low volume production - Part qualification is main practices, enabling faster adoption ecosystem (vs establishing material allowables)
 Aviation	- Huge opportunity for lightweight solution-related benefits - Major business case opportunity to lower cost of Mfg tooling - Uniquely enabled performance improvement through AM such as part consolidation, enhanced functionalities and materials - Massive growth potential due to high volume when AM-only solution are developed. Benefits multiply quickly, and so is the R&D investments to enable that business case.
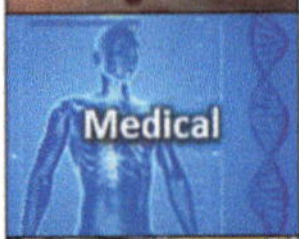 Medical	- High degree of customization product enabled by AM - Enhanced performance of medical component leading to more effective treatments, lower cost overall. - AM offers unique competitive differentiation leading to major growth (e.g. spinal/hip implants, Invisalign®, surgical tools, etc.) - Cost & time savings business case to support hospital operations (e.g. surgeries, prosthetics, visualizations and medical re-dos.)
 Automobile	- Massive opportunity in tooling cost reduction, which is multiplied by the variety of the product portfolio, and the timeline associated with design releases. - Some customizations can offer competitive and differentiated offering, which AM uniquely serve. The sheer volume of demand justify massive and consolidated investment approaches - Performance improvement of some parts will surely ramp, but will face a more challenging ecosystem requirement (see below)
 Oil & Gas	- High cost of tooling repairs offer uniquely-enabled AM solutions - High cost of down time, building AM's capability for fast solution development and distributed spare production - Enhanced functionality industrial tools increasing productivity, which brings significant cost savings / profits, leading to strategic consolidated AM investments
 Consumer products	- Due to the sheer volume of production, AM truly shines in reducing costs, and complementing the current Mfg process workflow (e.g. conformal cooling increase productivity, and lower cost of Mfg tooling) - Cost reduction in Mfg tooling (incl. repairs) - Some growing business case driven by high level of high-end customization and/or where AM offers a unique enabling competitive solution otherwise impossible to achieve (e.g. golfing clubs, car racing, BMW mini custom options, etc.) - The business case for large volume direct AM production is often challenged by the conventional manufacturing technologies such as injection molding and machining. UNLESS added performance and unique high-value functionality is uniquely enable by AM.
 Defense	- Support to the warfighter is paramount, leading to a huge business case potential for faster, cheaper and more distributed Mfg supply chain, uniquely enabled by AM (and other digital Mfg elements). - Support to the logistical defense efforts for spares production and management for legacy equipment position AM uniquely. - The competitive advantage and need for enhanced performance solutions is a great soil for AM to growwinning applications (e.g. lightweight, - Uncontested airspace/environments offer some opportunities to adopt AM in a lower risk environment, which reduce the requirements on the AM ecosystem needed to support this application space (e.g. qualification and materials requrements).

9.2 Industrialization Strategy and AM Ecosystem

9.2.1 *Going from "Prototype" to "Production"*

It is important to mention here that a "prototype" application is defined as a non-functional part/assembly that is part of a product development, a tool, and/or a design iterations for form & fit function. If a part is intended to go into production, it is understood that a "prototype" will be generated along the way, but the end result will be a "production" part. In the polymer AM world, using AM for prototyping a part in early stage of a non-AM production part is very common and well justified due to lower cost of manufacturing and design freedom some of the polymer AM modalities offer. However, when using metal AM (no matter the modality), one should be purposeful to thread the path from prototyping to production stage. Substantial effort is required to adapt Design-for-AM (DfAM) to leverage metal

Table 9.2 Definition of prototype vs production AM metal part

Prototype part	**Definition**: A part that is not intended to be used in the final product, and may or may not meet all the form, fit & functions of the production part. **Requirements**: Typically, the requirements for a prototype are lower (much) than a production part, either from a GD&T or from a material performance standpoint. Qualification requirements are generally not stringent or existent. **Benefits**: The key benefits of using AM for prototypes is time and cost savings (usually using non-metal AM), as well as representation of intent (design reviews, customer demos, form & fits, etc.).
Production part	**Definition**: A part that is intended to be used in the final product using metal AM as manufacturing technology and generally meets all form, fit & functions of the production part, no matter the quantity. **Requirements**: Typically, the requirements are stringent in order to meet intended performances and quality. Quality expectations are generally similar to production. **Benefits**: One-off or multiple early development metal AM parts can lead into larger quantity production. Compatible DfAM from early stage to production will speed the timeline and lower NRE costs.

AM and generally pushes the design outside a "conventional" design. In other words, if the intent is not to use metal AM in production for a given part, don't bother going through the effort to "prototype" that part using metal AM (there are exceptions). A re-design and specific tolerancing challenges will make this effort counterproductive. For this chapter, we therefore will define (Table 9.2):

9.2.2 Industrial Adoption of Metal AM: Starting with the Business Case

The first step in the adoption of metal AM in any given company is to understand your own product and its detailed total manufacturing workflow (a.k.a. AM value stream). With that in mind, it will be possible to identify and formulate the right business cases and see where metal AM can be beneficial. As mentioned above, a one-time metal AM production part can be used as a "bridge solution" to compress schedule, avoid downtimes, and/or grow the production rate up until a transition manufacturing point to a more cost/unit effective conventional modality. AM does not have to produce the final part to show benefits. It can also be impactful to save time and cost in one or more steps of the manufacturing workflow. Tooling is a great example for it. We will review how to maximize these benefits.

Many experienced metal AM users often adopt the following mantra when it comes to using metal AM:

> "Print because you should, not because you can"

From an industrial perspective, the benefits of AM can be summarized in three main categories (Fig. 9.3):

Using this indicator chart (see Fig. 9.3), it is possible to formulate a generic strategy to identify good product/component *families* that would capture as many benefits as possible. As a starting point, the best approach is to target maximum combined benefits and a part family (multiple variances of a similar design) that has the broadest product portfolio penetration and therefore offers the "biggest bang for the buck." Generally, depending on the industry sector, it is wise to start with applications with lower level of criticality (see Fig. 9.4) and/or qualification requirements. Slowly increase the level of application criticality as the AM benefits start "speaking for themselves," and the level of "comfort" among the decision makers is higher from previous successes.

It is also important to position metal AM across the product lifecycle. The value proposition of metal AM may be different for various stages of the lifecycle. For example, for a defense article such as an unmanned aerial system (Fig. 9.5), metal AM can play different roles and will require different levels of stringency in the manufacturing process. However, as mentioned earlier, due to the cost and complexity involved in metal AM (no matter which metal AM modality considered), it is critical to ensure that if metal AM is considered as early as possible in the product lifecycle, ideally at the conceptual stage. Each industry will carry its own opportunities and challenges when it comes to introducing new manufacturing/materials late in the lifecycle, but generally, it is costly and disruptive to use metal AM as an afterthought, unless the value proposition outweighs the burden of requalification (e.g., when a product line is halted, or performance requires re-design, etc.)

9.2.3 AM Ecosystem

Every manufacturing technology has its associated ecosystem as we learned from past new technology introduction. It is common practice in business development and market investment to understand the connectivity between the different elements of a technology, a market, or even a given supply chain. Over the past century, we all got very familiar with conventional manufacturing technologies such as (but not limited to) machining, casting, injection molding, semiconductor manufacturing, thermoforming, composite lamination, and many more. Each presents a specific ecosystem of elements that are synergetic, necessary individually, and that must be connected together to make the industry grow (see Fig. 9.2).

Applying a similar thought process, one can define the metal AM manufacturing technology ecosystem as all elements/components that, when combined and used properly, are required to enable the adoption and effective use of industrial metal AM. In other words, one must connect several key elements if they want to implement metal AM within any given company and move from "Prototype" to "Production." The following framework is proposed for an industrial metal AM ecosystem (Fig. 9.6):

The content of the ecosystem elements needs to be adapted to each specific application space and company. However, the main framework should still apply.

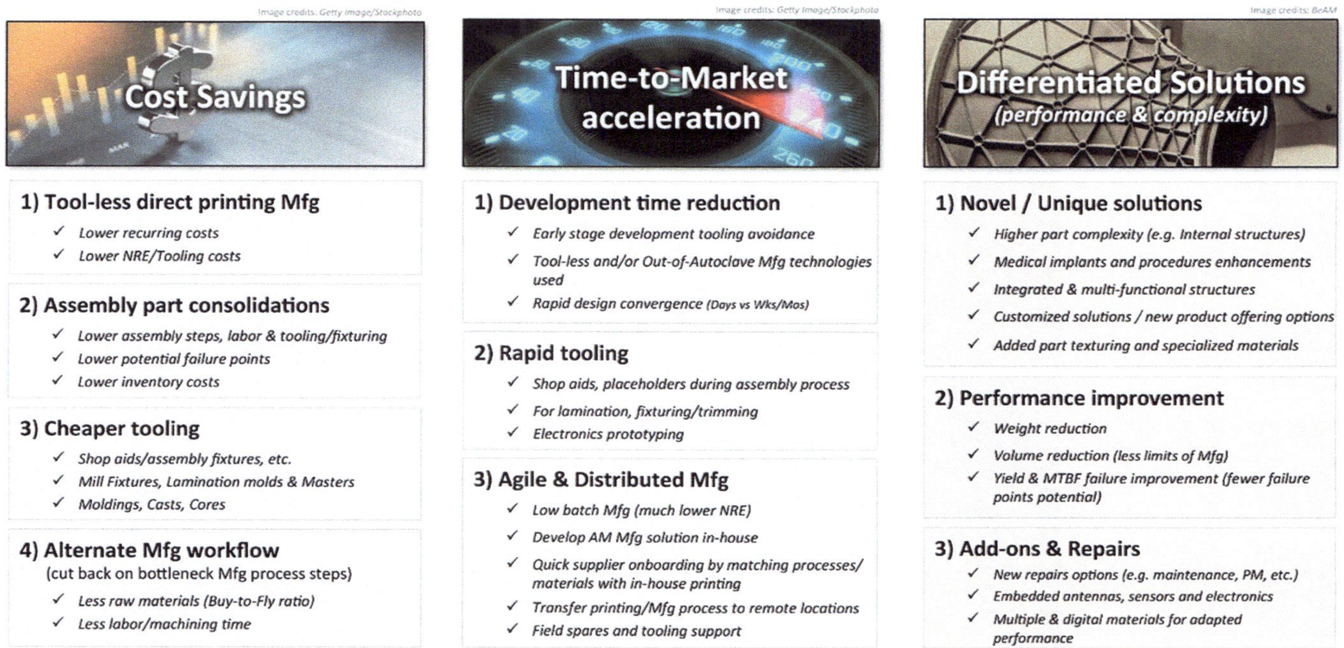

Fig. 9.3 Summary of the benefit categories associated with industrial usage of metal AM

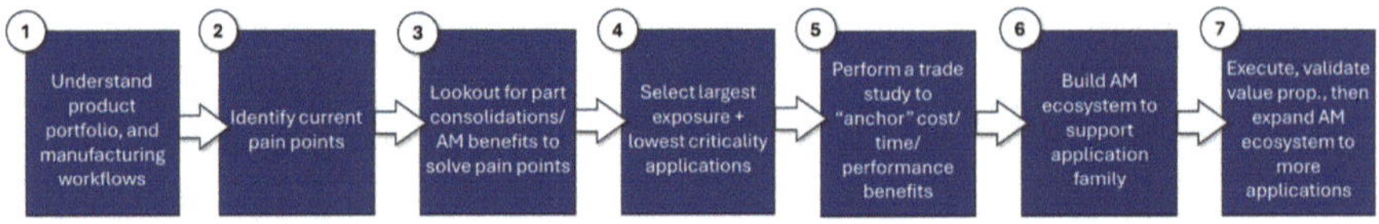

Fig. 9.4 High level implementation strategy for metal AM. Much more detail should be considered and are proposed in the sections below

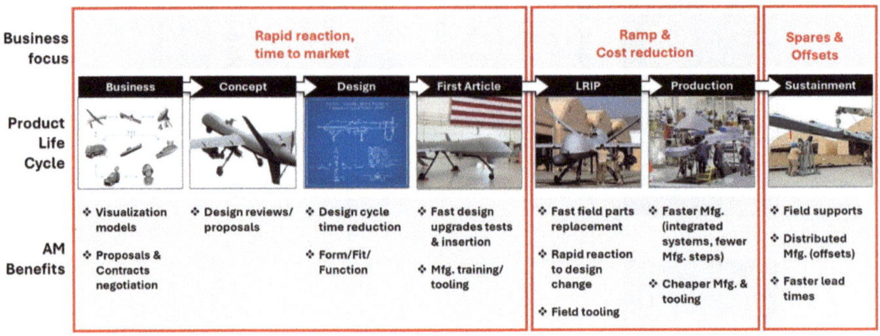

Fig. 9.5 Example of a product lifecycle (unmanned aerial systems for defense) where metal AM (**bold**) offers different levels of value proposition

It was noted earlier that "prototyping" (different from "production") is among the easiest adoption of AM. It is because most, if not all elements of the ecosystem are very "light" or unnecessary. For instance, to print a design form & fit check prototype, product mock-up, or a shop aid/tool using metal AM, datasheet level material is likely sufficient if at all required, with no/limited specialized AM software, and data tracking necessary. In the case of metal AM, a significant level of effort will be required nonetheless for DfAM efforts, build preparation and post-heat treatments, and machining/finishing. It is, however, a very different and much "heavier" burden in the case of fully functional medical, space, machinery, structural, and/or aviation production applications for instance.

Even within a given production application, different AM part criticalities and loading conditions will drive a different depth of the AM ecosystem map (and its associated cost and time investment). For instance, in the medical industry, a part that will stay inside the human body as an implant would require a more extensive set of material data qualification and industrialization testing compared to a medical instrument that leverages the same metal AM material and technology. The same applies in aviation, space, and other industries.

In order to better understand the importance of each ecosystem component, their description is detailed in Table 9.3, how they might differ, and/or adaptation for different AM user market segments.

To make this more concrete, and to understand the connectivity between the different elements of the metal AM ecosystem, let's use a publicly well-known [9] AM metal part from GE Aviation (fuel nozzle [10] for the LEAP engine) and highlight how the AM ecosystem elements notionally* came together to play a role in the making of this stellar assembly-level metal production part (ref. Fig. 9.7).

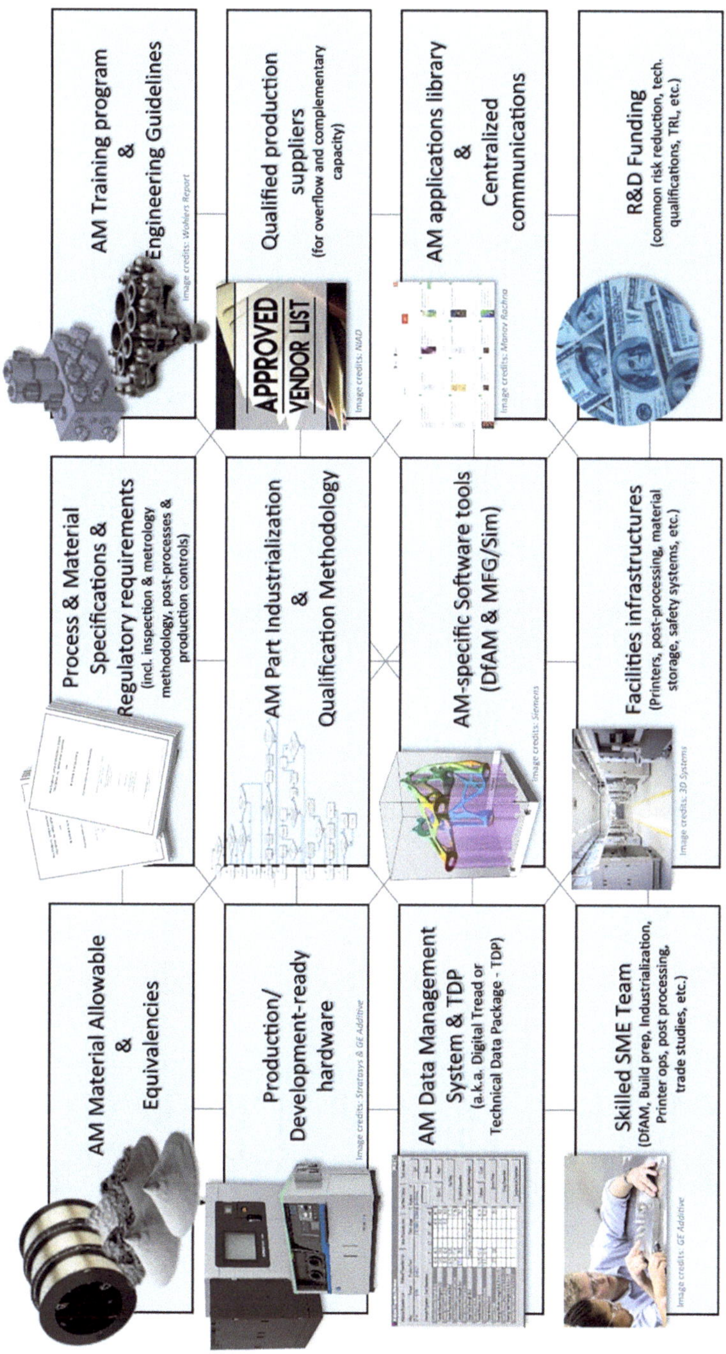

Fig. 9.6 Proposed additive manufacturing ecosystem framework showing the different components required to truly and effectively enable AM within a given company or industry segment

Table 9.3 Ecosystem element descriptions and details, best practices, and variabilities depending on industry segment of the metal AM use

Market segment	Description	Variance with AM adopter market segment
AM Material Allowable & Equivalencies	It is important to recognize that with AM, we do not start with known/certified part material. AM includes not only the process of fabricating the part, but also the process of consolidating the material. Hence, to design and analyze AM parts, it is important to establish a solid material properties foundation. This includes, but not limited to, full material qualification, and/or equivalency testing campaigns to validate that consolidated material properties meet the established baseline, analysis and design knock downs, etc.	- Different market segment will require different level of material data set depth and types. For example, medical market requires different level of cleanliness and biocompatibility than any other market segments. - Within the same market segment, part criticality and requirements will drive different depth of material knowledge. For instance, datasheet-level data are likely sufficient for rapid prototyping for form & fit, while a finished functional part would require a complete material data set and associated statistics.
Process & Material Specifications	Since the part consolidated material is controlled by the combination of the raw material (e.g. filament, powder, etc.) and the AM process specific controls, it is important to establish a process and material specification that will control the yielded part quality consistently. A typical specification framework include **a)** Raw material specs/SCDs **b)** Process control specs and **c)** Consolidated material specification and sometimes **d)** Post-process specifications. A simple way of looking at it is: A + B = C. This equation needs to be controlled to produce quality AM parts for end-use.	- Typically, for any production-level AM part, there will be some level of process documentation. The part criticality, the regulatory agencies involved in the product certification (when applicable) and the customer satisfaction requirements will drive the depth of the controls required for the AM materials and process. - For applications such as shop aid tooling and visual aids, the process controls will likely be very light compared to a serial production application where the printing and its associated post-processing steps shall be controlled to ensure quality. - For FAA or FDA approved products, this is extremely important part of the AM ecosystem, and ultimately the part cost
AM Training program & Engineering Guidelines	Education is paramount to any technology adoption. This component includes education institutions as well as industrial best practice knowledge transferred to the "front line" application developers which are generally the designers of components/products. Establishment of good foundational AM training, as well as specific DfAM (Design for AM) guidelines will be a strong catalyst to the AM application space expansion. This includes a VERY important chapter on how to identify a good AM business case and the establishment of a good AM cost model tool.	- Depending on the market segment's strategy on product design, this component might be more or less of a priority. However, even is design is outsourced, it is essential to establish a basic foundational AM training to help drive the AM benefits through the identification of good business use cases. - There are specialized providers of training, and/or AM consultancy services which can be leverages to jump-start the insourcing of the AM training. - Education institution and research centers (e.g. Penn States, ORNL, MIT, UL, UofT) as well as AM consortiums/organizations (e.g. EWI, America Makes, UL, etc.) can also be great assets.
Production/ Development-ready hardware	This element is typically the most emphasized of the AM ecosystem. However, it should only come together with the remaining components as it can quickly become an expensive "paperweight printing machine" without a solid established ROI (Return on investment). Depending on the production controls needed, different class of printers can serve different types of parts (e.g. prototype vs production vs tooling).	- The level of controllable process parameters determine the level of production readiness/consistency achievable. Its important to understand the requirements before choosing the right AM technology and OEM. - Consistency and repeatability between production assets (when production is insourced) can be an important and expensive factor to consider. For instance, matching production on medical implants require substantial printer qualification of each printers compared to quasi no qualification required for tooling, prototypes or visual aids. - A strategy to align insourced printing assets to the largest application quantity is generally smart, while leveraging qualified vendors for other applications.
AM Part Industrialization & Qualification Methodology	Depending on the AM process used, there is typically an "industrialization" process which consist of identifying all risks associated to the AM production of a given part, and the establishment of a iterative plan to reduce those risks through either simulations, development/prototyping builds or both. This process is usually associated to part development cost or NRE (Non Recurring Expense).	- Some AM process have lower NRE than others. Generally, the more thermally challenged is the process, the more NRE will be required. - The industrialization process is also proportional to the criticality and regulation associated to the production part. - Low criticality or prototyping parts usually do not require industrialization. - Generally, the larger the production volume, the easier it is to justify the industrialization NRE cost amortization. This is why in low volume applications, it is important to have a strong business case and/or have a very effective industrialization NRE phase. - When dealing with service bureaus, the industrialization process is often offered as a "black box" cost included in the recurring part cost.
Qualified production suppliers	Depending on the sourcing strategy, it will be important to consider AM <u>production</u> supplier. It is generally unrealistic, especially at the inception of an AM strategy, to serve all slowly growing applications by insourcing all printing assets required: multiplied by the variety of materials for each printing technologies involved. A given qualified supplier shall be able to meet all AM process and material controls in place within your company/industry and able to assist through the industrialization process.	- Depending on the part criticality and controls required, different class of AM printing service provider can be leveraged. For instance, regular Service bureau may be great at providing "quick" prototyping units, while not able to meet process & material controls required for serial production. Also, the same AM printing service provider may be qualified for an AM technology, and not another. Same is true for different materials. - Sourcing strategy offering complementary and matching AM platform and materials to the one used in-house might save cost of qualification in the early stage of an AM strategy. However, over time, it is wise to spend time and money to expand the variety of suppliers and associated AM platforms to enable a more competitive sourcing plan.
AM Data Management System & TDP	A digital database solution that is capable to store, track and report on all relevant AM process step parameters in order to provide quality SPC controls, traceability, as well as a framework for the establishment of a Technical Data Package (TDP). Over time, this solution will provide the basis to establish and augment a material properties database used for design and analysis.	- Although there are numerous existing production and database tracking solutions, it is important to understand the extent of the functionalities and data content required to support the specific AM market segment user. For instance, a user focused on rapid prototyping applications will likely need more printer loading and work order management functionalities compared to a Space market part production control requiring raw materials, process, and post processing operations metrics to ensure quality.
AM-specific Software tools	Existing CAD SW (Software)package solutions now include modules specifically aimed at covering SOME aspects of the AM design and Mfg workflow. It is sometimes needed to use additional SW plug ins or external SW niche solutions to: **a)** support DfAM to reach design freedom offered by AM Mfg (e.g. 2.5D/3D lattice generation, topology optimization, generative designs, and **b)** AM Mfg operation support solutions such as: build orientation optimization, file fixing, nesting, slicing, support generation, and build simulation and pre-compensation..	- Depending on the functionality, the maturity level of the SW solution provider is variable. - There is no good substitute to a knowledgeable DfAM engineer specialized in the specific AM process in question. However, some SW can enhance significantly the Mfg file preparation efficiency, and provide a good "intuition calibration" to the CAD designer or the Mfg AM engineer while optimizing the part design or AM build. - Build simulation SW, although not quite there yet, can provide opportunities to reduce the industrialization NRE steps by avoiding build failures or giving indications on how the AM parts will "behave" throughout the build.
AM applications library & Centralized communications	Depending on the size of the AM user's company, it may be challenging to learn efficiently from each other without a centralized structure to support the growth of AM. At the very least, establishing a centralized location for AM information, AM SME contacts, best practices and application library to be shared within the organization will speed up tremendously the knowledge transfer. Furthermore, AM "mistakes" due to lack of knowledge, although a good learning opportunity, can be damaging for an emerging AM strategy within a company, especially when customers and executives are exposed to the negative impacts it may have and reflect on AM.	- Most AM industry users, after spending a number of years working with AM in group clusters, have found much increased synergies across divisions/groups within an organization when a dedicated centralized group of AM SME (Subject Matter Experts) is established. - It is most effective when high visibility and high benefits application FAMILIES are well documented to serve as training/PR materials - Most training consultants and OEMs extensively use application use cases as a source of best practices and teaching. However, due to the IP sensitivity of most industry users, it is most important to enhance external training material with internal, meaningful/relevant documented use cases in order to best reach the internal application development community.
Skilled SME Team	No matter if an organization insources designs, manufacturing and processing of AM parts, it is paramount to plan for dedicated AM SME to help keep an holistic strategy on the AM ecosystem establishment and growth within the organization. This will help to ensure that there is a good resource internally to drive application development, funding, CAPEX, business cases, design efforts, process controls, supplier qualification, training, and also to promote AM at all levels of the organization.	- Depending on how vertically integrated the production is within an organization, it is most effective to merge design/engineering development AM teams with the Mfg AM teams in order to ensure compatibility of processes and design approaches and avoid duplication of NRE costs and time delays during the industrialization phase of a part production. - Useful forums that can be used by the AM SMEs to grow AM awareness within an organization are: online internal training, lunch & learn, show & tell sessions with leadership, use case publication and their associated business benefits, , press releases, and other internal communications.
Facilities Infra-structures	If an organization wants to benefit from AM, at the least from development time and cost savings, there will be a need to invest in printing equipment and therefore in a facility suitable to sustain the level of AM activities desired. To be most effective, such facility should consider printing preparation, printing and post-processing steps of the AM workflow specific to the AM process insourced.	- It is most effective planning for a new AM facility to support early engineering development phases as well as production operations under nearby facilities instalations in order to favor process match, information exchanges between teams, as well as optimize the printer utilization. - Production often requires stricter process controls than engineering development equipment. It is important to plan accordingly. - Metal printing of reactive materials present specific safety and funding challenges that should not be underestimated. Best practices and AM-knowledgeable safety consultants should be used to ensure code compliance without going overboard.
R&D Funding	Depending on how an organization structure project development and funding, one general reality is the need for any production project to be cost and risk sensitive, while meeting schedule constraints. Sometimes this leads to an adverse reaction to AM when the TRL (Technology Readiness Level) is too low and/or the main ecosystem elements are not in place. Therefore, it is hugely useful and effective to invest R&D funds to lower the AM technology risks associated to the application space of interest. Over time, funding will succeed to implement AM within production program by bringing the risks level to an affordable level.	- For a specific application where the business case benefits, combined with the production volume, lead to an NRE level representing a small proportion of the expected benefits, AM-related risk reductions are typically covered by the part industrialization process. - In settings when diversity is high, and volume of production are low, it is much more taxing to the specific AM part production business case when large NRE is required to industrialize the part. This is why it is important to focus the R&D funds towards reducing risks associated to part <u>families</u> with the most benefits and most part variances so that part qualification "by similarity" can be achieved and NRE cost lowered.

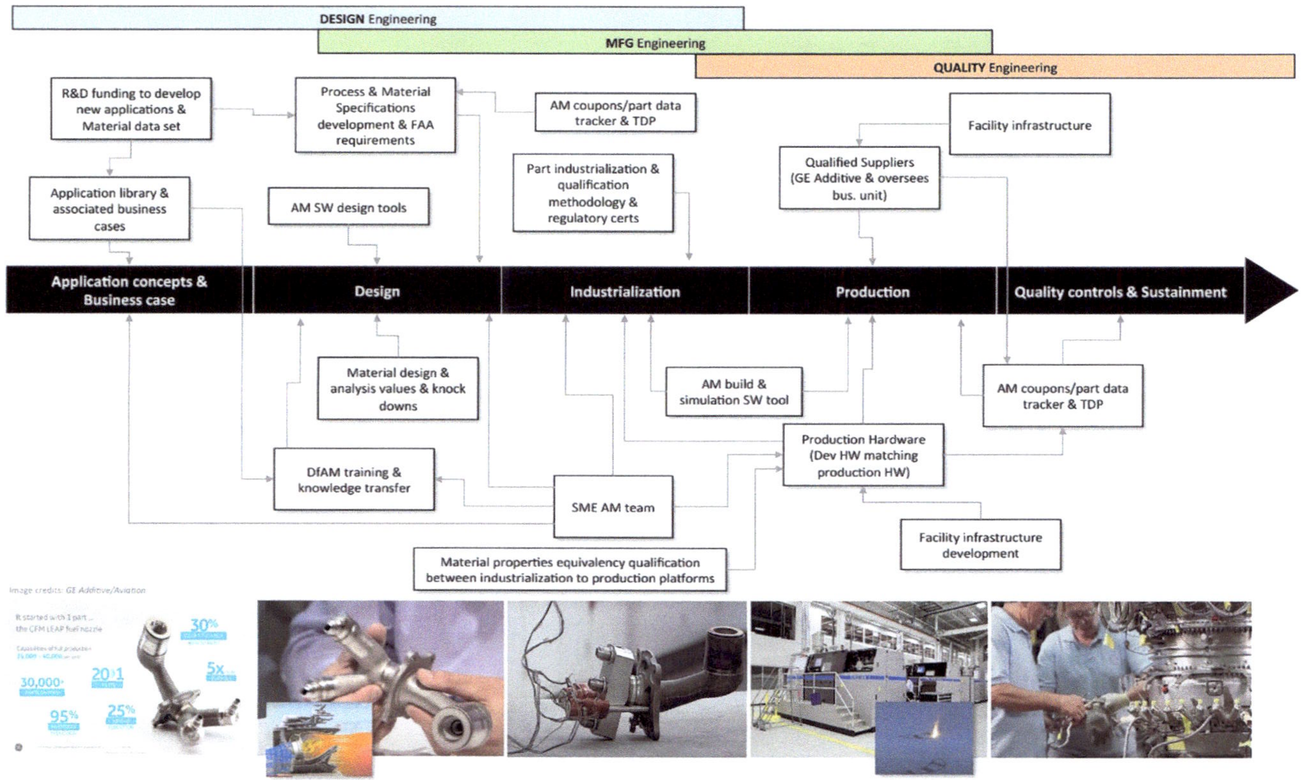

Fig. 9.7 Simplified illustration of how the different elements of the industrial AM ecosystem came together to enable GE Aviation fuel nozzle production application. Image credits: GE Aviation/Additive

*Disclaimer: This illustration was not reviewed by GE for this book and only uses publicly obtained information through press releases, articles, videos, and other published data. Insider and/or confidential knowledge was not used for this discussion. This use is only intended to be educational example and is not intended to represent GE's opinion or strategy

9.3 Application Space Mapping

9.3.1 Identifying the Right Applications

Before launching into the establishment of a metal AM ecosystem, it is useful to classify the levels of AM applications for your specific industry. This will influence how to approach the business case assessment, as well as the level of complexity of the qualification process associated with these different types of applications.

Consider the classification structure proposed in Fig. 9.8 below. Coupled with considerations of where those applications are with respect to the product lifecycle (Fig. 9.5), it becomes useful to establish a general approach to metal AM application onboarding. Agnostic to the industry, it is possible to summarize it as:

When considering the application space of any given industry sector, it is worth noting the following:

- Foundation AM application exists in almost all industry sectors and typically should not be using metal AM processes. Applications within that level can/should however help metal AM production applications. Average cost and time savings are on the order of ~85% in that class of non-metal applications.
- Depending on when the metal AM application is applied during the product lifecycle (ref. Figs. 9.5 and 9.9), there will be more or fewer options for design

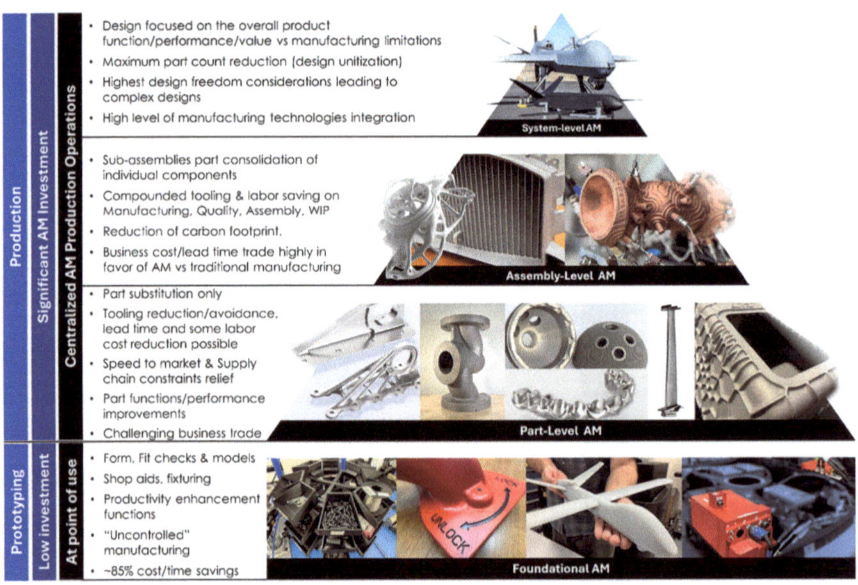

Fig. 9.8 Representation of classes of metal AM application and their associated attributes of (**a**) key benefits and examples, (**b**) best infrastructure to help onboard the application, (**c**) design and qualification implications. Images courtesy of: nTopology, LEAP 71, GE Additive, Airbus, EOS, Stryker, 3D Systems, and General Atomics Aeronautical Systems Inc

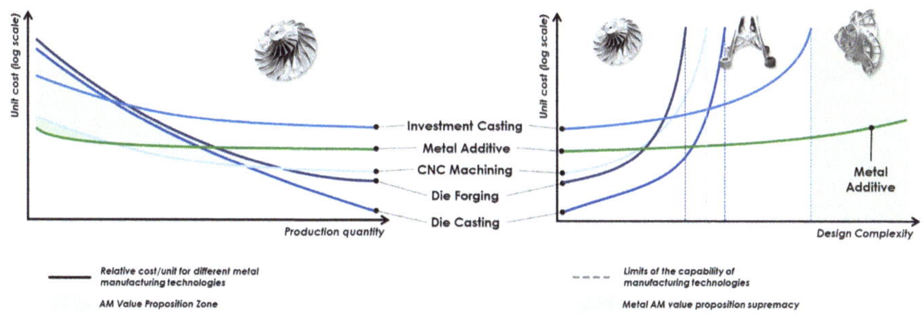

Fig. 9.9 Highlight mapping of difference of metal AM application attributes depending on its position in the product lifecycle. Image credits: General Atomics Aeronautical Systems, Xometry

freedom and complexity. Therefore, the manufacturing options and AM value proposition of AM might vary accordingly. The higher the complexity, the best chance of achieving "AM supremacy" in terms of manufacturing option (ref. Fig. 9.10).

- AM value proposition is directly linked to part count reduction and design complexity.
- When considering larger system-level applications, not only is the fabrication of the complex structures important, the total fabrication value chain must also be considered, including printing, post-processing, inspection, assembly, and integration. Multiple manufacturing modalities might be required/integrated to provide best-in-class cost and cycle time efficiencies.
- To achieve AM value proposition for most lifecycle use cases, part count must be reduced. To achieve part count reduction, it is imperative to intersect the design cycle as early as possible to avoid re-designs, and overall barriers to entry. One must design for AM.
- For some industries, when considering system-level applications, AM print volume will become a driving cost and overall manufacturing complexity factor.
- For anything above foundational AM level, it is imperative to put emphasis on design for AM, material selection, process controls, reliability, and efficiency in order to ensure best Return-On-Investment (ROI) on the AM effort.

9.3.2 AM Application Criticality Dependencies

As described in Fig. 9.4, it might be strategically (acceptance by stakeholders) and practically (faster and lower cost) important to consider the criticality level of the AM application considered. Most especially when dealing with metal AM (no matter the modality), it usually relates to an application that has a certain level of functional loading requirements. Figure 9.11 offers a simplified way of looking at criticality levels, while considering the risks associated with the use of metal AM

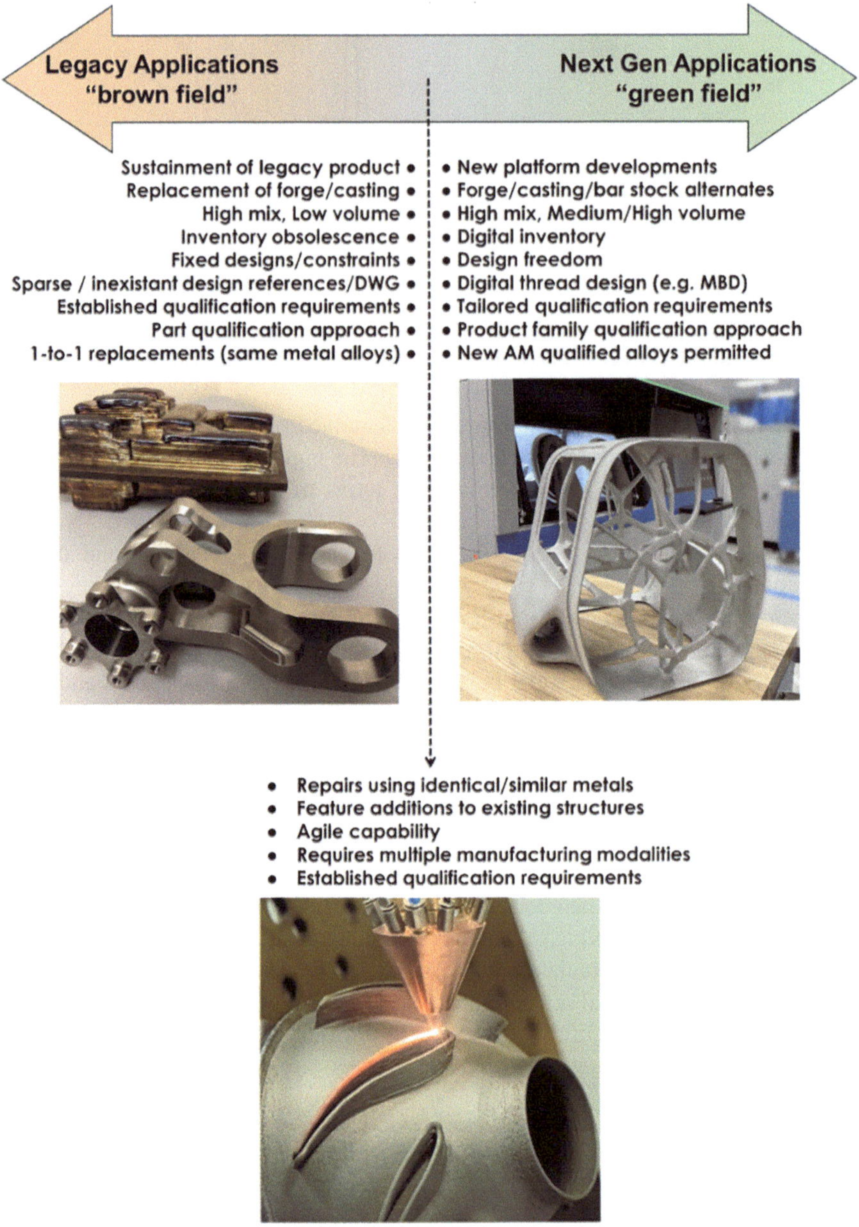

Fig. 9.10 (**LEFT**) Relative metal part manufacturing cost/unit as a function of production volume for different manufacturing technologies, including "generic metal AM," including the influence of upfront non-recurring expenses (NRE) such as tooling and programming. (**RIGHT**) Impact of design complexity, driven by part consolidation, on unit cost per component, clearly showing limitations of conventional manufacturing technology and where AM value proposition dominates most

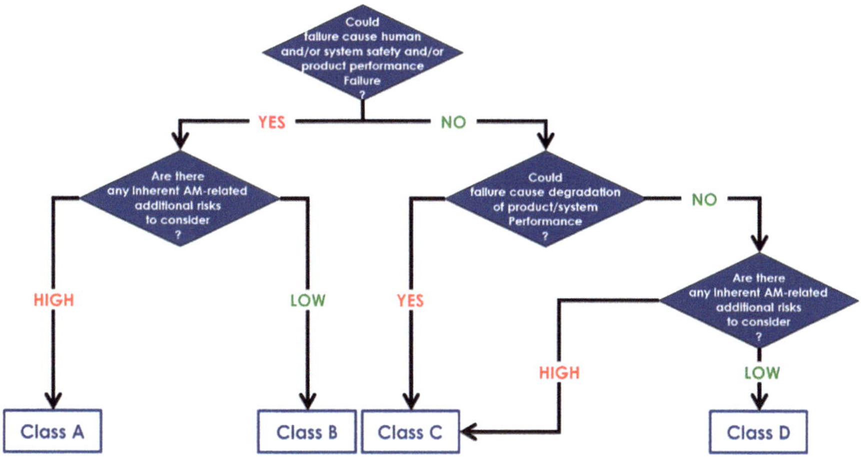

Fig. 9.11 Simplified AM application criticality assessment which affects the depth of AM ecosystem requirements

itself as a factor. The risk of metal AM in this assessment will depend on the maturity level of the metal AM manufacturing processes used within your organization.

Although a part criticality assessment can vary depending on the industry sector, generally the impact of a failure will drive the criticality level of a given AM application/part. In turn, the criticality level of the application will drive:

- Qualification requirements of the part testing, the process, and the material. In case of large-size fracture critical applications (too big/complex to test), it may require the development of a damage tolerance analysis model which can be a substantial investment.
- Time and funding required to fulfill all qualification requirements, leading to time-to-market challenges.
- Ability for key deciding and funding stakeholders to accept the introduction of a newer manufacturing technology. The higher the bar, the more nervous people generally get. Introducing metal AM for end-product applications will generally go smoother when its criticality is low, and its value proposition (Fig. 9.3) is highest.
- Post-processing requirements (e.g., finishing, heat treatment, machining, etc.) might be higher when fatigue loading is a concern.
- The amount of documentation required to support the establishment of the AM ecosystem represented in Fig. 9.6. For instance, specifications are generally imposed in production, depending on the industry. Nowadays, some industry standards for metal AM are available that facilitate the process. However, it remains burdensome to establish supplier qualification, inspection plans, work instructions, test, and engineering documentation, which typically scale in complexity and approval review cycles with the criticality level of an application.

In the case of a legacy product replacement and/or maintenance using metal AM, it is important to realize that the legacy qualification requirements and design constraints will be associated with how it was originally manufactured, qualified, and brought to production. It is useful to approach such a use case by looking at how the criticality and risk factors were evaluated for the baseline manufacturing approach and build the AM version as a "buy-in" considering the AM considerations only, if applicable.

9.3.3 *Total Cost Considerations*

As shown in Fig. 9.4, when identifying good AM applications, it is important to perform a trade study to validate, at least on paper, that AM is the best manufacturing option based on established experience with previous/similar applications and other manufacturing metrics. Specifically, when considering the cost aspect of metal AM, it is important to understand the different components of the "total cost" of manufacturing, which goes well beyond the printing operation (see Fig. 9.12).

In production manufacturing, there are typically two main types of cost: (a) nonrecurring and (b) recurring. Table 9.4 summarizes the types of cost elements to consider and compare conventional/baseline manufacturing when performing a trade study:

It is also important to note that there are other required metal AM production costs that are not mentioned in Table 9.4 and are not directly associated with any given AM applications. Items such as machine-to-machine, material, process qualifications, facility-related costs, safety, consumables, engineering support, etc. are

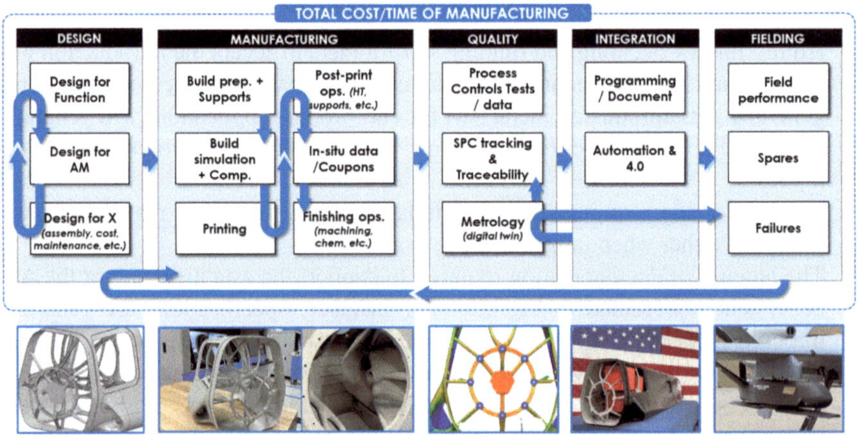

Fig. 9.12 Illustration of the total cost consideration for a metal AM part/application when including the full value stream of manufacturing and its associated cost. Image credits: General Atomics Aeronautical Systems

Table 9.4 Types of cost elements to consider during a metal AM trade study "total cost" analysis associated to any given production application/part. Some costs translate as material expenses, labor, or outside order procurement costs

Non-recurring cost	• **Tooling**: Although one of the appealing aspects of AM is to avoid tooling, it is still often required to execute post-printing operations such as machining, assembly, welding, drilling, etc. • **Build iterations**: Given that most metal AM modalities are challenged when dealing with in situ/post-print thermal stresses, as well as powder evacuations, support removals, and post-heat treatment GD&T, it is often necessary to go through several trials and prototyping builds before stabilizing the production technical data package (TDP). These builds generally fall under the "industrialization" section of the AM ecosystem (ref. Fig. 9.8). It is imperative to consider simulation and possibly pre-compensation software when dealing with metal AM to reduce as much as possible the iteration count. • **Engineering, manufacturing, and quality files/TDP**: In production, generally, process controls and associated production data package are required in order to ensure repeatability and quality. This usually is the outcome of the industrialization process. • **Fielding/testing**: Generally, prior to going to production with metal AM, a significant amount of testing is performed to ensure all risks are reduced to an acceptable level. • **Part qualification**: For part qual. Approach, there will be a major NRE component associated to material qualification and process optimization.
Recurring costs	• **Manufacturing execution system (MES)-related cost**: Depending on the level of digital connectivity and automation of the MES process, there may be recurring work-order specific labor involved in processing any production order of an AM part, including outside operations management (e.g., operations not supported internally). • **Printing-related cost**: Which include pre-build material handling, printing, and post-print operations such as depowdering, support removal, part labeling, consumables, raw material consumption and waste, electrical power, etc. • **Post-processing**: Generally, metal AM requires significant post-print operations such as heat treatment, machining, welding, marking, surfacing, etc. • **Quality (QA)**: Depending on the criticality level of the production part, there will be more or less requirements for metrology and inspection tasks for each metal AM part, which can accumulate to a significant amount of the total cost. Tasks such as GD&T metrology, material coupon testing, non-destructive metrology (NDT), destructive tests, certifications, and process control validations contribute to these costs. A smart way to lower quality costs sometimes involves the establishment of a quality plan which tracks the quality metrics of a given metal AM application over time, and reduces progressively the frequency of QA-related tasks as the statistical process control data justify the lower risk of failure.

some examples. Depending on the business model and associated financial allocation practices, there will be different ways to charge all "other" related costs to a specific metal AM application which will vary the total cost/part produced.

When referring to Fig. 9.12, Table 9.4 mainly covers the "Manufacturing" and the "Quality" portion of the total cost considerations. However, when comparing an AM application vs a traditional manufacturing workflow, it is often critical to broaden the perspective of the full associated value chain. For instance, as mentioned previously, if considering a metal AM application that would improve performance and/or reduce the part count of an assembly, there will be cost savings on the "design" end of the chain (e.g., # of drawing release and validation reviews cycles, etc.), as well as on the "integration" end (e.g., warehousing of parts, # of parts to handle and integrate together, assembly documentation, etc.). These supply chain cost reductions can often be significant. If there are less failures due to performance improvement, there could also be significant cascading benefits which are often neglected in the trade analysis. However, the latter benefits are often hard to quantify and track due to time elapsed between design and fielding stages of the product value chain.

9.3.4 Business Model Considerations

A few business model-related learnings should be considered when thinking of using metal AM to support the business objectives. Several companies have adopted metal AM in many different industries, and each has specific ways to adopt.

For metal AM production operations, the "cost-to-entry" can be quite significant depending on the onboarding approach. In many cases, AM gets implemented over time through critical mass growth following where the applications are most potent. It is also usually associated with people that have an interest in metal AM technologies pushing its implementation.

In large organizations, it sometimes leads to clusters of AM operations in different departments leveraging separate funding sources, assets, facilities, and people. Over time, large enterprises will foster inter-departmental communications and knowledge transfers which can enhance the efficiencies of the metal AM adoption overall and accelerate the enterprise-level business benefits. For smaller enterprises, it is generally more centralized and can be a powerful tool to accelerate its metal AM adoption. Table 9.5 summarizes new metal AM adopters of business model recommendations.

When it comes to prototyping AM (generally not focused on metal AM), a decentralized model is most definitely more effective at spreading the business value proposition (~85% cost and time savings) across the enterprise, while a centralized operation will favor faster implementation of metal AM due to the amount of skills and infrastructure required to support metal AM. As the enterprise matures in their AM implementation, one generally sees diversification and multiplication of metal AM operation centers, following a more uniform approach, which leverages common resources such as specifications, qualified supply chain for material and contract manufacturing, as well as AM-specific design software, guidance, and skills.

Table 9.5 Recommendations of possible business models for the adoption of AM (metal and non-metal) across an enterprise, as well as associated lessons-learned and attributes based on industry knowledge and experience. This model may be helpful to new adopters to accelerate their journey into metal AM

Model	Main Functions	Attributes
De-centralized AM	▪ Non-production AM (prototyping) ▪ Form/Fit/some functions ▪ Mainly polymers ▪ Lower product criticality	▪ Low cost of equipment (desktop/office class assets) ▪ Rapid reaction ▪ Lower process controls / total costs ▪ At point of use, application- specific to area ▪ Local operators, low training requirements, low processing complexity/costs ▪ Low/no facilities costs ▪ Fewer safety considerations
Centralized AM	▪ Focused on production AM and/or high requirements applications ▪ Fully functional capabilities ▪ Incl. metal AM ▪ Incl. full part processing value chain operations ▪ Higher product criticality	▪ High of equipment cost (especially for metal AM, industrial class assets) ▪ Longer lead times for metal AM ▪ Tight process controls / higher total costs ▪ Consolidated specialized skills & assets for best manufacturing efficiencies serving the enterprise ▪ Enterprise processing workflow must be enabled to request work ▪ High/specific facilities costs and engineering controls required ▪ Higher safety considerations ▪ Accelerated coordination and strategy alignment around AM, knowledge transfer and implementation

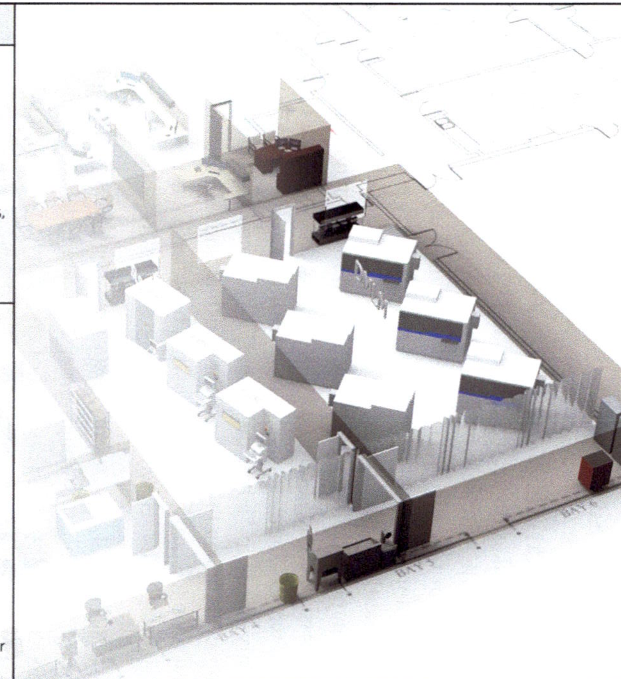

9.3.5 Pragmatic Considerations to AM Implementation

9.3.5.1 Setting up for Success

Establishing a metal AM ecosystem (ref. Fig. 9.6) might appear overwhelming, and the reality is that it takes dedication, time, and “a village” to do it. For example, it took GE Aviation approximately 6–7 years to establish their first ecosystem for the fuel nozzle AM part (see Fig. 9.7). However, with today’s level of maturation and the industrial commodity level of some metal AM, it is possible to accelerate this timeline significantly. It is generally through the catalyst effect of a winning business case that an organization realizes the need for the establishment of an ecosystem and progressively starts building it up. The good news is that there is no need to “re-invent the wheel” for every company wanting to get into industrial use of metal AM. One only needs to “adapt the wheel.” With the right strategy, partners, and business approach, it can bring significant business advantages. It does not have to happen all at once to start yielding industrial benefits. The key to success is to identify a good starting point. Here are a few ingredients for success:

1. As shown in Fig. 9.4, it is useful to start by identifying a strong business case, even if it is based only on a rough-order-of-magnitude (ROM) or notional analysis, using a part with the lowest level of criticality and highest potential benefits. There are plenty of good examples published, and use cases exist in every market segment to build that first business case. This will establish a good “bang for the buck” effect and fuel momentum to attack the remaining gaps in the ecosystem to enable other part families within your application space.
2. Establish the ecosystem components required for that part (re. item 1 above). The lower the part criticality and requirements, the easier the point of entry will be, the lower the R&D funding, work force, and industrialization cost needed.
3. While focusing on the few starting winning/promising business case part families, map out a progressive strategy to assess applications with more demanding requirements. As a result, you will fill out the gaps left in the ecosystem. It may take years to expand materials, processes, suppliers, printing hardware and facilities, etc. But, with a progressive roadmap, and a few strategic “AM wins” along the way, leadership support and associated funding will follow most of the time (with the use of an “ounce of visionary ingredient”).
4. Establish “champions” across your organization, at ALL levels. The best way to do that is to bring benefits to THEM. For instance, a design engineer will quickly become a champion if you can prove that there are better ways to build a part, and faster ways to design it through providing sample AM parts. Engineers typically appreciate simplification of the complex, the “look & feel” of optimized parts and having another tool at their disposal to meet requirements. Providing specialized and enabling software tools and training for DfAM will also gain their support. An executive leader on the other hand will focus more on cost savings, time-to-market, competitive advantages, and customer satisfaction demonstrations.

5. For many early industrial metal AM adopters, although sometimes it took them several years to organically come to this realization; one emerging common "note" is that, at some point, they eventually consolidated AM activities with dedicated team members in a Center of Excellence-like structure (e.g., BMW, GE, Honeywell, SpaceX, Stryker, etc.). This is one excellent way to build momentum and synergy to accelerate the pace of the ecosystem establishment and adoption within an organization and avoid mistakes.
6. Whenever possible, consider elements of the ecosystem where you can get a "jump start" or an acceleration boost. For instance, depending on the market segment of the user, some of the expensive and lengthy areas of the ecosystem (Fig. 9.6) are material properties, process specifications, and DfAM methodologies. Fortunately, there are consulting groups, standards organizations, and consortiums/research institutions that can be useful to "acquire" knowledge. Worthy of mention also are industry partnerships that can help tremendously when it is acknowledged and understood that the ecosystem elements shown in Fig. 9.6 are not the intellectual property (IP) components of the metal AM work (for the most part), but rather, the IP resides in what we do with it (a.k.a. the applications). Strategic Original Equipment Manufacturers (OEMs) and printing service providers (a.k.a. additive contractors) can also be a great source of "win-win" knowledge transfer, acceleration, and business opportunities.

9.3.5.2 Key Takeaways

There are a lot more nuances to the metal AM ecosystem that highly depend on your market sector, your business model(s), your specific applications/products, your customers, and other factors. This chapter aimed at providing a useful framework to start from when standing up a strategy within an organization toward industrial uses of metal AM. There are different perspectives on the road to AM success. Here are some possible key practical metal AM implementation strategy takeaways to consider:

- Metal AM is generally too expensive to use for simple prototyping not aimed at production-level use. The NRE associated with a "prototype-only" metal AM is generally too high for the value proposition it provides for forward design applications. However, for legacy applications, it may offer a supply chain constraint resolution (parts not available through traditional sourcing paths) and/or performance issue resolution opportunities that make the NRE worth it.
- Metal AM can sometimes be a replacement to current conventional manufacturing technologies. But the TOTAL COST business trades are often difficult to win at the part level (Fig. 9.8). Generally, the more complex the design, the better chances the metal AM application has to win the trade analysis. Metal AM most often offers a disruptive way to think differently about part design and manufacturing and comes as a powerful complement to existing manufacturing methods. Without design for AM (DfAM) considerations, the AM benefits are hampered.

- As shown in Fig. 9.8, the key to metal AM value proposition for new designs is through part consolidation. Reducing part counts has a significant cascading effect on business value proposition that often makes metal AM and its qualification worth the investment.
- Some of the metal AM processes should no longer be considered as "immature" manufacturing technologies (e.g., DED repairs, metal LPBF, etc.) and have become commoditized. However, its level of integration in our respective organization's business practices and processes still requires some work. It will most likely take another decade before it becomes mainstream and standardized across all markets and educational systems. The good news is that, today, organizations do not have to go through the decade-long process of maturation before benefiting from AM. There are best practices and standards that can be leveraged. Like other manufacturing technologies, one way to accelerate the process is to establish a metal AM ecosystem and consider AM as holistically as possible. In general, not one application will justify the investment of metal AM qualification, but with the right application targets, it can be done without too much pain.
- The establishment of the AM ecosystem does not have to happen all at once, and gaps within each ecosystem element can be filled over time, while growing the application space it serves.
- In order to move into industrial metal AM the "right way" and ensure AM part quality, it is paramount to avoid hasty uses of AM without proper controls and understanding of the key variables that will lead to part quality and its performance. It is very counterproductive when public failures of AM parts occur (toward customers and/or leadership). It's harder to gain support to establish an AM ecosystem than to lose it.
- Start with low criticality parts that present the highest business benefits. By focusing on part families vs unique "one-off" applications, it will soon become obvious how metal AM can positively move the "business benefit needle" for your company's leadership.
- Plan for a dedicated team/person acting as metal AM Subject Matter Expert (SME) internally to help build an adoption strategy and execute it, together with the rest of your organization. Plan on building champions and knowledge with all relevant stakeholder groups of your organization (e.g., design engineering, manufacturing engineering, materials & process, system engineers, program management, executives, procurement, communications, and finances) and potentially with your customer as well. This will help accelerate the ecosystem build-up and its acceptance.
- If your business model or size does not support the establishment of a metal AM ecosystem internally; through the help of your internal AM SME, search for suppliers/partners that have it in place and can get you started at a much lower cost-to-entry. Invest in an internal AM SME first in order to understand your business need and how to identify the right external resources to select. Otherwise, you might be driven by your suppliers/partners/consultants. AM consultancy services can be of benefit, but spend time ahead of any engagement to understand how the information and data received will fit into your metal AM ecosystem.

This way, you will spend wisely and effectively your R&D funding and resources of your ecosystem.

- Encourage your leadership to invest first in the establishment and understanding of a metal AM ecosystem for your organization rather than rushing into buying printing hardware. Printers without an ecosystem will either limp along for a long time (from a business benefit and growth perspective), or may end up in failure.
- To create momentum and to grow the number of metal AM applications within your organization, it may be useful to consider an "80%–90% solution." Tangibly, this means that it may be wise to choose materials and printers that may not be perfectly tuned to your applications but can allow for faster qualification and deployment while you refine the ecosystem over time. For example, having lower mass or strength margins in your design using an alternate already qualified alloy may be just fine for certain lower criticality applications and will help build momentum.
- Education is key to success within an organization, especially within the designer's community. Understanding key DfAM rules and details such as appropriate GD&T drawing notes and associated requirements will greatly impact the value proposition of a metal AM application and shorten the metal AM application onboarding.

References

1. https://ampower.eu/download/5098/
2. https://www.iso.org/committee/629086.html
3. https://www.astm.org/products-services/standards-and-publications/standards/additive-manufacturing-standards.html
4. http://profiles.sae.org/teaamsam-m/
5. https://www.cencenelec.eu/about-cen/
6. https://www.cto.mil/wp-content/uploads/2021/01/dod-additive-manufacturing-strategy.pdf, https://www.esd.whs.mil/Portals/54/Documents/DD/issuances/dodi/500093p.PDF
7. https://standards.nasa.gov/standard/NASA/NASA-STD-6030
8. https://www.nist.gov/additive-manufacturing/products-resources-0/standards-benchmarks
9. https://www.ge.com/news/reports/epiphany-disruption-ge-additive-chief-explains-3d-printing-will-upend-manufacturing
10. https://youtu.be/rMzVSbNebCg

MIX
Papier aus verantwortungsvollen Quellen
Paper from responsible sources
FSC® C105338

If you have any concerns about our products,
you can contact us on
ProductSafety@springernature.com

In case Publisher is established outside the EU,
the EU authorized representative is:
Springer Nature Customer Service Center GmbH
Europaplatz 3, 69115 Heidelberg, Germany

Printed by Libri Plureos GmbH
in Hamburg, Germany